International Perspectives on Aging

Volume 16

Series Editors
Jason L. Powell
Sheying Chen

More information about this series at http://www.springer.com/series/8818

Kate O'Loughlin • Colette Browning
Hal Kendig
Editors

Ageing in Australia

Challenges and Opportunities

Springer

Editors
Kate O'Loughlin
Ageing, Work and Health Research Unit
Faculty of Health Sciences
University of Sydney
Lidcombe, NSW, Australia

ARC Centre of Excellence in Population Ageing Research (CEPAR)
University of New South Wales
Kensington, NSW, Australia

Hal Kendig
ARC Centre of Excellence in Population Ageing Research (CEPAR)
University of New South Wales
Kensington, NSW, Australia

Centre for Research on Ageing, Health and Wellbeing (CRAHW)
Research School of Population Health
Australian National University (ANU)
Canberra, ACT, Australia

Colette Browning
ARC Centre of Excellence in Population Ageing Research (CEPAR)
University of New South Wales
Kensington, NSW, Australia

Royal District Nursing Service Institute
VIC, Australia

International Institute for Primary Health Care Research Shenzhen, China

Peking University
Beijing, China

ISSN 2197-5841 ISSN 2197-585X (electronic)
International Perspectives on Aging
ISBN 978-1-4939-6464-2 ISBN 978-1-4939-6466-6 (eBook)
DOI 10.1007/978-1-4939-6466-6

Library of Congress Control Number: 2016954328

Printed on acid-free paper

This Springer imprint is published by Springer Nature
The registered company is Springer Science+Business Media LLC
The registered company address is: 233 Spring Street, New York, NY 10013, U.S.A.

Foreword

As a subject of robust political and policy debate, population ageing has spent much of the past couple of decades flying under the radar. In large part, that is because the last decade of the twentieth century and the first decade of the new century saw Australia cash in on the demographic dividend flowing from the Baby Boomer generation and their children (Generation X) both occupying the working age cohort—pushing that cohort to its largest ever size at two thirds of the overall population, and helping to underpin 20 years of strong economic growth. This decade, by contrast, has seen the large Baby Boomer generation start to qualify for the Age Pension, ushering in a period of substantial growth in the absolute and relative size of Australia's older age cohort.

The orthodox view of many public commentators and much of Australia's media is that this trend will cripple the economy, thin out the labour market, bust the Federal Budget and usher in a gerontocracy, whereby older voters use their weight of numbers to entrench their own privileges at the expense of everyone else. *Ageing in Australia* calmly and persuasively addresses each of those assertions, and many others as well. The truth is that Australia is better positioned than almost any other nation to manage this substantial shift in our population profile, thanks to strong economic foundations, robust demographic projections and sound public policy stretching back decades, particularly around retirement incomes.

But Australia is not as well positioned to help ensure that the additional years that have been added to our lives are good years—that we are able to lead lives in older age that are secure, healthy, active and connected to our communities. This important book focuses as much attention on that human or social dimension as it does on the macro-economic and fiscal implications of ageing. In doing so, the authors remind us of the incredible cultural, ethnic and socio-economic diversity of older Australians and of the importance of valuing Australians across the life course, instead of assessing their contribution simply at a particular age.

The ageing of our population is substantially a product of longer life expectancy—the longevity revolution of the twentieth century that the World Health Organisation aptly describes as 'one of humanity's greatest triumphs'. A nation as secure and prosperous as Australia should be approaching this transition alive to the challenges it presents, but celebrating the societal and individual opportunities that flow from that revolution as well.

For much of the early part of this century, Australia saw less academic work published on ageing than many other comparable nations. That thankfully has started to change with more published academic work now complementing and analysing the regular publications from government agencies like Treasury and the Productivity Commission. *Ageing in Australia* marks a high point in academic work on ageing in recent years. This volume brings together Australia's leading academics in the area, canvassing the broader issues of ageing as well as particular policy dimensions that range across income security, health, transport, housing and more. For students and others keen to know even more, the authors provide extensive sources for additional reading and research.

Population ageing will be an enduring feature of Australian public policy and politics for many years to come. As we grapple with its complex challenges and opportunities, it is essential that policy makers and the general community have access to high quality, independent academic work in the area. *Ageing in Australia* is an outstanding example of such work.

Mark Butler

Acknowledgments

We would like to express our thanks and gratitude to the many people who contributed to this book, as without their continued interest and support it would not have been written.

Thanks to the chapter authors for their expertise, tolerance and patience, and especially to those who stepped in at the last to produce a chapter in record time.

Our publisher Springer stayed with us through what turned out to be quite a marathon and we thank them for their patience and guidance along the way.

Thanks to our Universities for their continued support and to our colleagues for their advice and good humour around all matters related to 'the book'! Also thanks to our students whose input and lively discussion have challenged us over the years.

Support from the ARC Centre of Excellence in Population Ageing Research (CEPAR) is gratefully acknowledged and a special thank you to our CEPAR collaborators who contributed chapters.

This book would never have reached publication stage if not for the contribution made by Jayne Groves. Jayne's incredible eye for detail and continued editorial assistance was invaluable to chapter authors and to us as editors, and we thank her most sincerely for her contribution. Lisa Cannon assisted ably in completing key chapters.

We also wish to thank the many older people who have inspired and informed us about ageing and contributed to the messages presented in this book.

Kate O'Loughlin, Hal Kendig, Colette Browning

Contents

Contributors

Hazel Bateman ARC Centre of Excellence in Population Ageing Research, School of Risk and Actuarial Studies, University of New South Wales, Kensington, NSW, Australia

Philip Bohle Ageing, Work and Health Research Unit, Faculty of Health Sciences, The University of Sydney, Lidcombe, NSW, Australia

Colette Browning ARC Centre of Excellence in Population Ageing Research (CEPAR), University of New South Wales, Kensington, NSW, Australia

Royal District Nursing Service Institute, VIC, Australia

International Institute for Primary Health Care Research, Shenzhen, China Peking University

Peking University, Beijing, China

Julie Byles Research Centre for Generational Health and Ageing, The University of Newcastle, Newcastle, NSW, Australia

Hunter Medical Research Institute, Newcastle, NSW, Australia

Rafal Chomik ARC Centre of Excellence in Population Ageing Research, Sydney, NSW, Australia

Kathleen Clapham, B.A. (Hons.), Ph.D. Australian Health Services Research Institute, University of Wollongong, Wollongong, NSW, Australia

Graham Currie Public Transport Research Group, Institute of Transport Studies, Monash University, Clayton, VIC, Australia

Jenny Davis College of Healthcare Sciences, James Cook University QLD, Australia

Alexa Delbosc Public Transport Research Group, Institute of Transport Studies, Monash University, Clayton, VIC, Australia

Stephen Duckett, Ph.D., D.Sc., F.A.S.S.A., F.A.H.M.S. Health Program, GRATTAN Institute, Carlton, VIC, Australia

Cathy Duncan, Grad. Cert. H.S.R. Australian Health Services Research Institute, University of Wollongong, Wollongong, NSW, Australia

Debbie Faulkner Centre for Housing, Urban and Regional Planning, Faculty of Arts, University of Adelaide, Adelaide, SA, Australia

Susan Feldman Monash University, Melbourne, Victoria, Australia

Yun-Hee Jeon Faculty of Nursing and Midwifery, The University of Sydney, NSW, Australia

Hal Kendig ARC Centre of Excellence in Population Ageing Research (CEPAR), University of New South Wales, Kensington, NSW, Australia

Centre for Research on Ageing, Health and Wellbeing (CRAHW), Research School of Population Health, Australian National University (ANU), Canberra, ACT, Australia

Peter McDonald Crawford School of Public Policy, Australian National University, Canberra, ACT, Australia

Jack Noone Sydney, NSW, Australia

Ageing, Work and Health Research Unit, Faculty of Health Sciences, Lidcombe, NSW, Australia

Kate O'Loughlin Ageing, Work and Health Research Unit, Faculty of Health Sciences, University of Sydney, Lidcombe, NSW, Australia

ARC Centre of Excellence in Population Ageing Research (CEPAR), University of New South Wales, Kensington, NSW, Australia

John Piggott ARC Centre of Excellence in Population Ageing Research, School of Economics, UNSW Business School, University of New South Wales, Kensington, NSW, Australia

Harriet Radermacher Monash University, Melbourne, Victoria, Australia

Shane Thomas International Institute for Primary Health Care Research, Shenzhen, China

Primary Health Care Research, Monash University, Clayton, VIC, Australia

Peking University, Beijing, China

Chapter 1
Challenges and Opportunities for an Ageing Australia

Kate O'Loughlin, Hal Kendig, and Colette Browning

1.1 Introduction

Population ageing is a global phenomenon whose impact is only now being fully recognised and understood. Increased longevity, aspirations for improved quality of life, advances in health and better welfare provision are generally accepted as parts of a variable success story that will demand ongoing action with the expectation that populations worldwide will live to increasingly older ages (OECD 2015; WHO 2015). Governments around the world are now grappling with the significant social and economic challenges raised by population ageing: some of the notable policy responses in Australia to date include encouraging people to stay in the labour force beyond the

K. O'Loughlin (✉)
Ageing, Work and Health Research Unit, Faculty of Health Sciences, University of Sydney, Lidcombe, NSW 2141, Australia

ARC Centre of Excellence in Population Ageing Research (CEPAR), University of New South Wales, Kensington, NSW 2033, Australia
e-mail: kate.oloughlin@sydney.edu.au

H. Kendig
ARC Centre of Excellence in Population Ageing Research (CEPAR), University of New South Wales, Kensington, NSW 2033, Australia

Centre for Research on Ageing, Health and Wellbeing (CRAHW), Research School of Population Health, Australian National University (ANU), Canberra, ACT 0200, Australia

C. Browning
ARC Centre of Excellence in Population Ageing Research (CEPAR), University of New South Wales, Kensington, NSW 2033, Australia

Royal District Nursing Service Institute, 31 Alma Road, St. Kilda, VIC 3182, Australia

International Institute for Primary Health Care Research, Shenzhen, China

Peking University, Beijing, China

K. O'Loughlin et al. (eds.), *Ageing in Australia*, International Perspectives on Aging 16, DOI 10.1007/978-1-4939-6466-6_1

traditional retirement age of 65 years, increasing the pension eligibility age, promoting healthy and active ageing, supporting older people to 'age in place' and the introduction of consumer-directed care and user-pays into aged care services.

1.2 Rationale for the Book

The purpose of this book is to present older people as part of the possible solution rather than simply the problem in population ageing. Who are we referring to when we speak of 'older people'? Entering older age is usually linked to retirement age, which in most developed countries has been between 60 and 65 years for men and 55 and 60 years for women. With population ageing and increased longevity, we have become more age conscious and have introduced new demographic categories to define and distinguish different life stages. We now refer to mature age, young–old, old, old–old and Laslett's (1989) conception of a Third Age and Fourth Age or to generational categories such as the post-World War II baby boomers, Generation X and Generation Y. For the purpose of providing a chronological definition of 'older people' in the Australian context, we will use 65 years as a somewhat arbitrary entry point, as this is still considered by many to be the expected retirement age and for now denotes eligibility for the age pension.

While there are many challenges associated with population ageing, there also are opportunities that are often overlooked. Rather than seeing all older people as a 'burden' or 'unproductive' once they leave the paid workforce, we need to acknowledge the social and economic contributions they continue to make as consumers, carers and volunteers and draw on and put to good use the knowledge, skills and experiences they have acquired across their life course. However, we do have to also acknowledge that as people age they are more likely to live with a number of chronic health and often disabling conditions that will affect their lifestyle and bring them into increased contact with the health and social care systems. In order to better understand these changes, we want to provide a new perspective on ageing in the Australian context by focussing on multidisciplinary perspectives (psychology, sociology, epidemiology, gerontology, economics, demography, medicine, allied health) and the challenges and opportunities related to ageing and aged care. This approach will show the developmental and interdependent nature of ageing at the individual, societal and structural level. In doing so, it addresses both structural dimensions and individual experiences that are integral to an understanding of ageing for future healthcare practitioners and policymakers.

Both the challenges and opportunities associated with population ageing require innovative perspectives to meet the needs of the increasing numbers of older people and those who will interact with them. This book aims to provide an up to date overview, commentary and analysis on ageing to inform both policies and constructive action by individuals, service providers and policymakers to respond more positively to ageing and places Australian developments in ageing in an international context. The book sets out to provide a strong evidence base and thought-provoking discussion points for you as students in health sciences and related disciplines, as you undertake education and training to take up practitioner roles in clinical and community health settings or policymaking roles in health and aged care services.

1.3 Background and Contextual Factors

Written by Australia's leading researchers and policy analysts, the chapters included here are intended to provide a framework for addressing the interconnecting themes associated with understanding and approaching ageing as a developmental, lifespan and social process and not a fixed chronological point. That is, we are emphasising the importance of understanding individuals in diverse social contexts and placing Australia in an international and historical context. This approach is central to gaining an understanding of what is required to address the needs and recognise the potential of the increasing numbers of older people and those who interact with them; to identifying the social and individual dimensions of ageing, health and well-being and the transitions that occur in later life taking into account psycho-social and physical aspects; to examining the policy and practice implications of an ageing society and the role of public and private providers in care and service provision to older people and to developing a critical understanding of the issues related to ageing and the life course and policy debates around population ageing and quality of life for older people, their families and carers.

Previous publications on ageing in the Australian context, including a series of edited books since the 1990s, have synthesised our knowledge on ageing. The publication of *Longevity and Social Change in Australia* (Borowski et al. 2007) provided a comprehensive account of the diverse impacts of ageing including income, health and care policies as well as assessment of the dimensions of ageing related to gender and ethnicity. The scope of thinking about ageing was widened with the inclusion of chapters on life-long learning, ageing and the law, the politics of ageing and older indigenous Australians. The published compendium *Contemporary issues in gerontology: Promoting positive ageing* (Coulsen and Minichiello 2005) provided a multidisciplinary focus on ageing and aged care with a particular emphasis on professional practice issues for those working in health and aged care. More recently, research evidence and policy options were reviewed in *Population Ageing and Australia's Future* being released by the Academy of Social Sciences in Australia (ASSA) and the ARC Centre of Excellence in Population Ageing Research (Kendig et al. 2016).

1.4 Australia: Historical and Population Profile

Australia is considered a developed country but one with several features that distinguish it from other countries. We are growing older with a predominantly Western heritage and an increasingly Asian future (Kendig and Lucas 2014). While Australia has a large landmass, it has a small population of approximately 24 million and a quite low overall population density of 2.9 people per square kilometre; the major population concentrations are in the large coastal cities of Sydney, Melbourne, Brisbane, Adelaide and Perth with few people living in the sparsely populated interior. A relatively small Indigenous population (2–3 %) experiences severe disadvantage and has very low life expectancy. Successive waves of immigration, initially from Britain during the nineteenth century, then from Central and Southern Europe following WWII and more recently from Asia, have driven population growth. From the late 1940s to the mid-1960s, Australia had a major, sustained baby boom.

Population ageing has been modest in Australia relative to some other countries notwithstanding decades of low fertility rates. Demographic ageing has accelerated in recent years as the large baby boom cohort of the 1950s and 1960s has begun to enter later life and with the increased life expectancy at older ages. Life expectancy at age 60 years in Australia is now more than 25 years, and it is projected to rise further. As of 2015, 20 % of the Australian population was aged over 60, and this is projected to increase to 29 % by 2050 (United Nations 2015). The proportion aged over 80 years has been increasing and it will again rise with ageing of the baby boom cohort (United Nations 2015).

Australian data indicate that a male child born in 2012 has a life expectancy of 79.9 years and a female child 84.3 years (Australian Institute of Health and Welfare (AIHW) 2014). For those who had survived to the age of 65 years in 2012, the expectations were on average for an additional 19.1 years for men and 22 years for women (AIHW 2014).

Approximately, 90 % of older Australians live in private households while only 6 % live in healthcare establishments such as aged care facilities (AIHW 2015). Even among older people with a substantial disability, only a minority reside in long-term care accommodation such as hospitals or nursing homes. Widows and never-married men are the mostly likely groups to be in residential care, reflecting the importance of family care networks for older Australians. Overall, most people continue to live in the community through a long period of later life. Only a third of older women and a quarter of older men are projected to ever live in residential care before death.

Sustained economic growth has underpinned substantial rises in real incomes, and Australia has been affected less than other developed countries by global financial crises. The unemployment rate has been relatively stable at between 4 and 6 % since 2004. Personal wealth has increased along with high home ownership rates and rising property values, to the benefit of the sizeable majority of older Australians, and a substantial number enjoy a high standard of living in single-family housing although this may be more difficult to achieve for younger cohorts now aspiring to buy homes.

While Australia is still seen by some as the 'Lucky Country', we have to acknowledge that there are considerable and increasing socio-economic inequalities including significant pockets of disadvantage. Long-term renters are likely to have insecure housing and high housing costs relative to income. There is increasing recognition that young and middle-aged people—notably those who are unemployed and/or single parents and their children—are among the groups who are most likely to be impoverished. With an ageing Australia, many of the vulnerabilities of older Australians—seen most starkly among older Aboriginal people—arise earlier in life while others arise with health transitions, widowhood and other adversity in later life.

1.5 Australia in a Global Context

To understand Australia's position in a global context, and as an indication of the effectiveness of our national responses, we can make comparisons across other relatively wealthy nations that are at the forefront of population ageing. For this purpose, we consider New Zealand, the United Kingdom (UK), the USA and Germany in terms of comparable information available from the Global Age Watch Index (2015). As shown in Table 1.1 and using 2015 as a benchmark, the five countries have had comparable population ageing although Germany stands out for its significantly older population. The countries also have comparable levels of Gross National Income (GNI) per capita, although by this measure the USA was doing significantly better at that time. By mid-century, population ageing is expected to have advanced even further in Germany, to 40 % aged 60 years or over, while the other countries are projected to have ageing populations that by then will be comparable to that of Germany at present.

The Global Age Watch Index also provides indications of the well-being of older people in four broad areas: income security; health status; capability and enabling environment. Overall, Australia ranks near the top of the 96 countries listed: 17th as compared to tenth for the UK, ninth for the USA, fourth for Germany and 12th for New Zealand. While these ratings have many limitations, they do provide some indications of our strengths and weaknesses as a country in which to grow older:

- Australia ranked very highly at fifth in the world for health status. The four benchmark countries had similar levels of life expectancy at age 60 years (23–25 years); similar levels of healthy life expectancy (17–19 years) and high levels of self-assessed psychological/mental well-being for people aged 50 years or over.
- Australia had a surprisingly low rank of 62 for income security, due to comparatively low pension rates, and relatively low income/consumption of older people relative to younger people. These findings are explained largely by Australia not having a universal, contributory social security system (as found in the other countries), but the consequences are ameliorated by high rates of home ownership and hence low outlays on housing.

Table 1.1 Australia in a Global Context

	Year	Australia	New Zealand	UK	USA	Germany
Number of (million) people over 60	2015	4.9	0.9	14.9	66.5	22.3
Percentage (%) of the population over 60	2015	20.4	20.3	23.0	20.7	27.6
	2050	28.3	29.4	30.7	27.9	39.3
GNI per capita ($US)	2015	$41,241.9	$30,885.9	$37,052.8	$51,484.3	$44,400.7

Source: Global Age Watch (2015)

- Australia rates eighth in the world in terms of the 'capability' of the older population. Among the four benchmark countries, Australia rates mid-range in terms of the employment rate for 55–64 year olds, while the UK rates much lower than the others primarily on the basis of low secondary education attainment.
- Australia has a comparatively low rank (26th) for 'enabling environment', lowest along with New Zealand (30th) of the benchmark countries, as indicated by measures of social connectedness, perceptions of safety when walking alone, freedom in one's life and satisfaction with public transport.

In considering our strengths and weaknesses relative to comparable developed countries, we can conclude that while Australia remains a wealthy country, there is increasing concern for inequalities between advantaged and disadvantaged groups, for the fiscal sustainability of governments and for intergenerational equity between birth cohorts.

1.6 Structure of the Book

The structure of the book groups the chapters under four key headings: dimensions of ageing; living and lifestyle; policy and practice in health and care; and actions and future directions. Each chapter provides an overview of the topic and draws out what are the challenges and opportunities for older people, their families, governments and the broader community. Particular emphasis will be placed on the policy and practice implications of an ageing society and the role of various providers of services and care to older people including governments, healthcare practitioners, family, employers and not-for-profit organisations. Common and recurring themes of diversity addressed in each chapter include gender, socio-economic resources and geographical location.

- ***Gender*** is central to understanding differential exposures and experiences at all stages of the life course including older age where women are much more likely than men to reach advanced later life and to live without a partner, in poor health and in poverty.
- ***Socio-economic diversity*** is seen in the advantages and disadvantages relating to income including superannuation, wealth including home ownership and health and financial literacy based mainly on earlier life experiences.
- In Australia, we need to account for the ***spatial patterns and geographic diversity*** of the population notably the shifting population patterns such as ageing in middle suburbs, retirement migration and the relatively poor access to services in rural and coastal areas.

The substantive chapters will include a case study based on the author(s) knowledge and expertise; this will allow students to understand and critically evaluate a 'factual' case and propose solutions. Key websites are included for students to access the most up to date resources related to each chapter topic. Structuring the book in this way and including evidence-based material on specific topic areas allows us to include resources and material that facilitate the development of a critical understanding of issues related to ageing and the life course and the initiatives and policy debates relevant to population ageing.

1.7 Chapter Overview

The seven chapters included under *Dimensions of Ageing* outline the interconnectedness of all aspects of ageing and provide a conceptual and historical framework for understanding the population profile of older Australians from an age, gender and health perspective and the influence of age-based attitudes and stereotypes on the lived experience of older Australians. Ageing in the context of cultural diversity, indigeneity and employment is also addressed in this section.

Chapter 2 provides a 'big picture' of developments in ageing over the post-World War II era for individuals and Australian society and establishes core concepts and information on lifespan development, social change and historical legacies in order to set a context for later substantive chapters on specific topic areas. **Chapter 3** provides an overview of what are age-based attitudes and stereotypes and how they originate and are perpetuated in society. The chapter examines the available evidence on how these attitudes and stereotypes are manifested in the workplace, media, public institutions, the healthcare system and policy discourses and strategies. Legislative and policy initiatives and older people's experiences will be drawn on in identifying the challenges and opportunities in addressing attitudes and age-based stereotypes.

Chapter 4 examines population ageing and demographic change in the context of national and global trends and considers the challenges and opportunities these present in Australia. A particular focus is on the spatial patterns and geographic diversity of Australia's ageing population. **Chapter 5** provides an overview of the health profile of older people in Australia including analysis and discussion of epidemiological evidence of changes in the pattern of health-related issues and the challenges and opportunities these present. Multiple comorbid chronic conditions are common among older people in Australia; the chapter shows how promotion of healthy ageing across the life course is needed and relies on a prevention focus, comprehensive health care and supportive environments to maintain quality of life and independence into older age. **Chapter 6** provides an overview of the history of migration to Australia and examines the available research evidence around the influences of ethnicity and the migration experience on health and ageing. It identifies the challenges and opportunities that population ageing presents for individuals, families and policymakers in a culturally diverse nation such as Australia.

Chapter 7 addresses some of the assumptions made about old age within an Indigenous context and identifies the challenges and opportunities for making improvements in the lives of Indigenous Australians as they age. The available evidence indicates that many in the Aboriginal and Torres Strait Islander population experience significant health, income and social disadvantages that persist from early and midlife and intensify in later life. Current policy initiatives and strategies will be drawn on in considering the challenges and opportunities in addressing the needs of older Indigenous Australians. **Chapter 8** looks at the position of older workers and identifies the challenges and opportunities associated with a mature and older aged labour force in Australia. With government policy encouraging older workers to remain in paid work for longer, we need to look at the available evidence to understand current patterns and types of employment undertaken by older workers including participation rates, employment sector, health status and also the legislative and policy initiatives that specifically address the employment of older workers.

The next three chapters on *Living and Lifestyle* examine a range of issues associated with the social and economic circumstances of people as they age and how these may impact on their living conditions and lifestyle, particularly if they are living with chronic or disabling health conditions and some distance from health and care services. **Chapter 9** provides an overview of Australia's retirement income system including pensions, superannuation and taxation and examines the available evidence on the policy and planning activities needed to provide equitable and sustainable standards of living for older people. The provision of retirement income at a time of significant demographic change has become one of the major challenges in public policy, and many countries are now trying to balance equity, adequacy and sustainability in their retirement income provision models by reforming their retirement income systems.

Chapter 10 examines housing and the residential environments important to older people and the challenges and opportunities they present in making available a range of supportive environments that meet both the changing needs of older people and the government's policy direction of ageing in place. The chapter provides an insight into the current housing circumstances of the older population, the influences on the direction of future housing and locational choices and the need to move towards, and plan for, age friendly communities. Planning for this will challenge policymakers across a number of policy arenas and at all levels of government. **Chapter 11** provides an overview of the transport options available to older people including driver ability and access to public and private transport taking into account the geographical diversity of Australia. It discusses the challenges and opportunities in providing a range of options for drivers and non-drivers to maintain access to affordable mobility options.

The following three chapters on *Policies and Practice in Health and Care* consider the health implications of population ageing from organisational, policy and practice perspectives. **Chapter 12** provides an overview of national responses to ageing in health policies with a particular focus on the implications of ageing for hospital care. While older people use hospitals more, the chapter shows that change in hospital admission patterns is occurring slowly and that small incremental changes to policies are the appropriate response, rather than crisis-response type changes to the health system. Policy responses to the implications of ageing for hospital care include a focus on reducing preventable admissions and diversion strategies to increase use of alternate sites of care.

Chapter 13 identifies and describes the health services available and utilised by older people in community settings and critically reviews the focus on primary health care initiatives in terms of access, affordability and efficacy. The chapter explains how primary health care operates in Australia; types of services available and their use by older people; workforce challenges associated with it and recent policy reforms. A key challenge concerns co-ordination between the health and aged care system in the delivery of primary health care services to older people and the need for integrated care around issues of multi-morbidity and multi-professional input to the care of older people. Addressing these issues is vital and will assist in the prevention of entry of older people into the more expensive secondary and tertiary health care and residential aged care systems.

Chapter 14 considers what we mean by 'aged care' from an individual (self-help, family) and organisational (care accommodation and services) perspective and critically reviews policy and practice developments from consumer, industry and government perspectives. Following on from the Productivity Commission Report *Caring for Older Australians* (2011) and subsequent reforms, the chapter assesses issues concerning the introduction of programs for community and residential care including consumer-directed care; user-pays strategies; quality assurance and rebalancing provision across care settings. Implementation and practice issues associated with new models of care are identified and discussed and supported by the inclusion of case studies looking at, for example, accommodation and home support, dementia care, re-enablement, quality improvement and aged care management.

The concluding chapter examines *Actions and Directions* and returns to the issues raised in the book's introductory two chapters and builds on the substantive chapters that followed to examine some contrasting views on likely futures from several perspectives. This includes social framing and related prescriptions for 'solving the ageing problem' as a fiscal problem as viewed by government, experts and think tanks. The discussion focuses on the trends and politics underlying intergenerational equity, given its centrality to ageing issues at this critical turning point in Australia's history, and provides a review of promising international approaches to population ageing in addressing challenges and opportunities of individual and societal ageing. The chapter concludes by critically assessing recent experiences and near term prospects for national leadership on ageing—both within and beyond government—including actions by individuals, employers and the broader community as well as governments.

As you read through the chapters in the book, you will see that Australia faces challenges ahead with population ageing; however, it is also important to appreciate that, by comparison to relatively wealthy European countries, Australia's population is ageing at a more modest rate and our economic and fiscal outlook is encouraging. We know that there will be challenges and adjustments, but arguably there are more opportunities than crises ahead. Australia has several decades or more before it reaches the proportions of older people already present in Europe or Japan, so investing in human capital at all points in the life course (education, health, employment, age-friendly communities) is an effective and equitable way to manage demographic pressures in the future. By taking this approach, we can work towards an Australian society that realises and promotes the benefits of an older society.

References

Australian Institute of Health and Welfare. (2014). *Australia's health 2014*. Australia's Health Series No. 14 (Cat. No. AUS 178). Canberra: AIHW.

Australian Institute of Health and Welfare. (2015). *Australia's welfare 2015*. Australia's Welfare No. 12 (Cat. No. AUS 189). Canberra: AIHW.

Borowski, A., Encel, S., & Ozanne, E. (2007). *Longevity and social change in Australia*. Sydney: UNSW Press.

Coulsen, I., & Minichiello, V. (Eds.). (2005). *Contemporary issues in gerontology: Promoting positive ageing*. Sydney: Allen & Unwin.

Global Age Watch. (2015). *Global age watch index 2015*. Help Age International. Retrieved from http://www.helpage.org/global-agewatch/.

Kendig, H., & Lucas, N. (2014). Individuals, families and the state: Changing responsibilities in an ageing Australia. In A. Torres & L. Samson (Eds.), *Aging in Asia-Pacific: Balancing the state and the family* (pp. 211–224). Quezon City: Philippine Social Science Council.

Kendig, H., McDonald, P., & Piggott, J. (Eds.). (2016). *Population ageing and Australia's future*. Canberra: ANU Press.

Laslett, P. (1989). *A fresh map of life: The emergence of the third age*. London: Weidenfeld & Nicolson.

OECD. (2015). *Health at a glance 2015: OECD indicators*. Paris: OECD.

Productivity Commission. (2011). *Caring for older Australians: Overview* (Rep. No. 53). Canberra: Productivity Commission.

United Nations. (2015). *World population prospects: The 2015 revision, key findings and advance tables* (Working Paper No. ESA/P/WP.241). New York: Department of Economic and Social Affairs, Population Division, United Nations.

WHO. (2015). *World report on ageing and health*. Retrieved from http://www.who.int/ageing/publications/world-report-2015/en/.

Part I
Dimensions of Ageing

Chapter 2
Australian Developments in Ageing: Issues and History

Hal Kendig

2.1 Introduction

The experiences and social constructions of ageing and older people have changed significantly over the lifetimes of people now entering later life. The profound truth is that our images of ageing and the life chances of ageing people are inseparable from the periods of history in which they have lived their lives. These realities are obscured as we continue to misinterpret the 'facts of ageing' as intrinsically a matter of biology (and hence largely inevitable), discounting what is recognised increasingly by researchers as the malleability of ageing experiences. The optimistic implication here is that constructive actions improving socio-economic circumstances and healthy ways of living have good potential for improving experiences of ageing over the entire life course.

The most remarkable post-World War II (WWII) development on ageing arguably has been the substantial extension of later life. For millennia, the maximum lifespan had been considered to be the biblical 'three score and ten'. A century ago, the average life expectancy in Australia was only in the 50s with death rates highest for babies and young mothers: the relatively few older people at the time were in some senses 'fortunate survivors'. Over the post-war years, however, life expectancy at birth has increased steadily. In an unexpected historical turn since the 1980s, this increase in life expectancy has been greatest for those from 60 years of age (Australian Bureau of Statistics (ABS) 2011; Australian Institute of Health and

H. Kendig (✉)
ARC Centre of Excellence in Population Ageing Research (CEPAR), University of New South Wales, Kensington, NSW 2033, Australia

Centre for Research on Ageing, Health and Wellbeing (CRAHW), Research School of Population Health, Australian National University (ANU), Canberra, ACT 0200, Australia
e-mail: hal.kendig@anu.edu.au

K. O'Loughlin et al. (eds.), *Ageing in Australia*, International Perspectives on Aging 16, DOI 10.1007/978-1-4939-6466-6_2

Welfare (AIHW) 2014). These gains have been much greater for higher socio-economic groups and much less for Indigenous Australians, thus heightening social inequalities. We have yet to fully recognise, let alone respond to or plan very effectively, for the radical consequences of living longer for reshaping later life.

This chapter considers individual ageing, including the 'new' life course, quality of life and life chances as *goals* for an ageing Australia, as well as looking at the increasing diversity, social inequalities and disadvantaged populations that are important foci for potentially improvable aspects of ageing. It also considers societal ageing and social change, including processes of cohort succession with an emphasis on the large baby boom cohort. It then turns to issues of intergenerational equity and outcomes for disadvantaged groups. The chapter concludes with an overview of post-war policy developments on ageing and recent political aspects of ageing.

2.2 Change over the Life Course

2.2.1 The New Life Course

The 'new' life course for most people now includes several decades of life after the age of 60 years. In 2012, men aged 65 years could expect another 9 years without disability and another 15 years without severe core activity limitations; comparable expectations for older women were 10 and 16 years, respectively (AIHW 2014). Women had a longer expectation for life with a severe disability, 6 years versus 4 for men, with high levels of dependency likely only for a limited period near the end of life. From the late 1990s, the increases have been greater for disability-free years than for years with disability for both men and women during later life.

The encouraging news is that there also have been expectations for more years without perceived poor health, without activity limitations and also without limiting health conditions. Nonetheless, there also have been increases in the expected length of time with long-term health conditions, disabilities and with core activity limitations in daily living (AIHW 2014). For people aged 60–64 years, the improved expectancies without activity limitations are notable, as they have amounted to more than 2 years for both men and women from 1998 to 2012 (AIHW 2014). The mixed findings indicate some increased capacities to work longer and to participate in social life on entry to later life, but also more years with needs for assistance in later life.

In 2003, the Prime Minister's influential *Towards Healthy Ageing Working Group* (Prime Minister's Science, Engineering and Innovation Council (PMSEIC) Independent Working Group), comprised of health and policy experts, set a national vision for 'an additional 10 years of healthy and productive life expectancy by 2050' (PMSEIC 2003, p. 2). The Working Group outlined a research agenda which guided development of the National Health and Medical Research Council (NHMRC)/Australian Research Council (ARC) Ageing Well, Ageing Productively

Research Program to inform strategies for improving ageing well and other positive outcomes. To achieve this ambitious national vision in an equitable way, it will be particularly important to achieve better outcomes for Aboriginal people (see Chap. 7) and those in lower socio-economic groups as they are lagging behind the health improvements being seen for more advantaged groups.

There has been a notable re-shaping of the life course along with the lengthening lifespan. There have been increasingly extended periods of education, later entry to the workforce and later marriage and childbearing (Gong and Kendig 2016). In midlife, there has been increasing variability in forming and disbanding relationships and households, more complex patterns of workforce participation and a notable rise of chronic disease from middle age onwards. Fundamental to the massive post-war changes has been the reduction in the fertility rate from 3.5 to 1.9 births per woman, along with women's increasing educational attainment, workforce participation and conceptions of their place in society. Across virtually all spheres of the life course, patterns have become more complex over recent decades in terms of gender, ethnic and socio-economic variations.

Changes in labour force participation in mid- and later life have been one of the most significant developments for an ageing Australia. Labour force participation from 1979 to 2007 for men aged 55–64 years fluctuated but overall remained relatively steady (68 %). However, for women at these ages there was remarkable change, with the rise from 20 % to 49 % as more women stayed in the workforce or returned after childbearing (Taylor 2010). The economic downturn after the 2007 Global Financial Crisis had serious impacts on older people on the verge of retirement from paid work, inducing adaptive responses such as working longer; those who had recently retired with suddenly reduced value in superannuation and other assets found their standards of living seriously challenged (Kendig et al. 2013; O'Loughlin et al. 2010).

The Australian Institute of Health and Welfare (AIHW 2015) has provided a comprehensive report on developments in ageing and related population and social topics over recent years. Facts and implications are examined for welfare expenditure, labour force participation and caregiving among other topics important for ageing. The report documents the increasing age of the workforce, life expectancies, aspects of aged care, ageing in the 'welfare system', mental health and palliative care.

Life history research on the individual lives of ageing baby boomers has examined the consequences of diversity earlier in life—notably in terms of education and health and then in terms of occupations, childbearing and caregiving through middle age (Kendig et al. 2016). These early and mid-life experiences have had continuing influences on workforce participation and its income and social participation benefits on entry to later life (Majeed et al. 2015). With labour force participation increasing for women in middle age, the stresses of work and caregiving—and choices between them—are particularly acute for women in middle and older age (O'Loughlin et al. in press). Limitations on family support for frail older parents arise mainly from women (and men's) high labour force participation in mid-life, often occasioned by financial necessity, notably the costs of buying housing and self-funding retirement.

Leading overseas work complements Australian research on challenges and opportunities of ageing. Phillipson's book *Ageing* (2013) reviews international developments in the social constructions and inequalities arising from the new 'third age' and the lengthening of the late 'fourth age'. He makes a case for 'rebuilding our institutions' in order to enable new and better pathways for later life. Paid work and family care policies have been compared for Australia, Canada and other countries highlighting the stresses on caregivers and the value of flexible working arrangements, carer leave provisions and other supportive responses (Kröger and Yeandle 2013). *The Sage Handbook on Social Gerontology* (Dannefer and Phillipson 2010) provides a comprehensive international review of ageing topics and policies including contributions from Australia and New Zealand. A Special Issue of the Journal of Cross-cultural Gerontology (in press) on 'Older Workers and Caregiving in a Global Context' (O'Loughlin and Phillips in press) brings together researchers in ageing from Africa, Australia, Canada, China, New Zealand, Singapore and UK to examine the impact on those taking on extended caring roles within the family as a consequence of population ageing.

2.2.2 Quality of Life and Life Chances

Public conceptions of ageing continue to revolve largely around vulnerabilities to poor health, low income and losses of family and other social support. These limitations and responses to them are important and will be examined in the challenges of ageing in following chapters in this book. Yet, as we will explore further in Chap. 3 on attitudes, the negative aspects of ageing that predominate notably in the health field serve to reinforce 'problem-focused' public policy. It is important to appreciate that the negatives of ageing are only part of the story: understanding of positive, aspirational dimensions of ageing are necessary to better balance knowledge and outcomes on ageing.

A central message on ageing is the fact that, in Australia, older people consistently emerge as happier and report higher levels of subjective well-being than other age groups (Fig. 2.1). The happiest group is comprised of those over age 76, living with a partner and in good health with average or higher living standards. Poor health, low income and limited social support do reduce levels of subjective well-being for older people, but this proves to be the case for people at other ages too (Cummins et al. 2013). Overall, people in a wide variety of circumstances, including those in late life, have capacities to maintain their subjective quality of life, making adjustments during times of adversity (Cummins 2014).

The normally high levels of subjective well-being among older people relate to the processes of subjective well-being homeostasis (Cummins 2014), modest expectations as well as stability, manageable stress and mature coping capacities in daily life. It can also reflect 'successful ageing' in so far as processes of life review, and comparative assessments can enable older people to positively assess their lives *in their own terms*. There is increasing Australian evidence that well-being and life

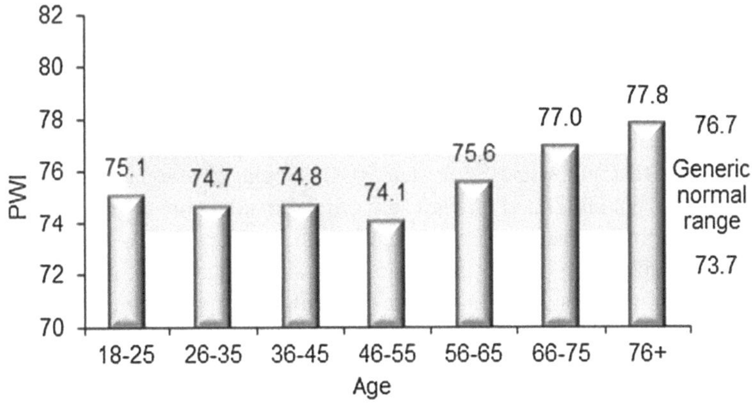

Fig. 2.1 Subjective well-being by Age Groups (N=58,493). *Source*: Cummins et al. (2013: 156) (*Note*: The Personal Wellbeing Index [PWI] comprises seven domains rated on satisfaction. All results from the Index are standardised into a scale from 0 to 100)

satisfaction in later life (as well as earlier in life) are enhanced by a sense of control and choice in decision making in a variety of contexts including retirement from paid work (de Vaus et al. 2007; Quine et al. 2007); moving into supportive housing (Gardner et al. 2005); living at home with community care support (Brooks and Kendig 2004) and availability of health and care programs (see Chap. 14; Kendig and Browning 2016).

Our qualitative and quantitative research demonstrates the high value older people place on what has been termed 'ageing well', conceptualised in terms of ongoing psychological well-being, good self-rated health, independence in daily life and remaining in their own homes (Kendig et al. 2014). While health declines and eventual death are inevitable, our longitudinal research and related studies demonstrate that ageing well can persist through most of later life. Moreover, personal resources and healthy life styles can make a critical difference in not only ageing well but also enabling the independence, participation and contributions that are so highly valued by older people themselves. A 'sense of purpose' is an important psychological resource for ageing well as demonstrated by a major Adelaide study (Windsor et al. 2015). At the core of adaptations in ageing is the striving to maintain lifelong identities and continuity notwithstanding the transitions and personal challenges of ageing.

Life history research on Australia's baby boomer cohort, now entering later life, has examined determinants of quality of life and life satisfaction inclusive of higher order outcomes such as perceived self-realisation, control and personal autonomy (Kendig et al. 2016). The health and socio-economic resources that people bring to later life emerge as the most significant factors for well-being, but behind these more immediate influences are social factors and life events tracing back through midlife to childhood. The findings underscore the value of a lifespan approach in understanding ways in which personal and social resources can be built throughout life and recovered after life setbacks. These capacities and resilience yield social as well as individual benefits at each life stage including eventual ageing well and productive ageing.

Outcomes in later life thus reflect the systematic accumulation of economic and social advantage or disadvantage as people progress through social structures over the life course (Dannefer and Settersten 2010). Social class-related opportunities and resources from childhood onwards are crucial to educational and occupational attainment and their consequences for income and wealth; these resources can have compounding effects on health, family stability and almost all dimensions of life. Home ownership and public housing are particularly important resources as both provide security and low housing outlays in later life; however, there have been few studies of socially disadvantaged groups from their own perspectives, such as older homeless men in Sydney (Russell et al. 2001). It is important to recognise the ways in which age and ageing differences interlock with other dimensions of social structures—including gender, area of residence, ethnicity and migration—as they influence multiple vulnerabilities of disadvantaged groups in later life.

Gender is fundamental to understanding variations in ageing experiences. For example, older men are more likely to view health as central to continuing physical capacities, while older women are more likely to view health as a capacity for continuing social connectedness (Kendig et al. 2014). The health promoting actions important for ageing well and threats to ageing well vary between older men and women and also between singles and couples: it is important to appreciate 'his', 'hers' and 'their' experiences and views of ageing. The social processes that generate difference and inequalities extend beyond individuals and households to embrace multiple generations of privilege or deprivation in family lineages. These intergenerational processes extend from the transmission of social class opportunities early in life through to family inheritances typically received in late middle age.

Bengtson and Kuypers' (1971) now classic term 'generational stakes' recognises that older generations value highly their continuity with the next generations and that they can care more about the futures of their children and grandchildren than themselves. By most forms of social accountancy, older Australians (as with their counterparts in other countries) in fact provide more support *down* the family generations than they receive *upwards* from younger relatives (Kendig and Lucas 2014). The long-term trend towards separate households and mobility among the generations reflects rising real incomes of older people and patterns of 'intimacy at a distance' rather than the caricature of 'family abandonment'. Some families and lineages, however, have little capacity for this intergenerational support as a result of multi-generational deprivation that can be very difficult to address (Gong and Kendig 2016).

A national study examined social factors in productive activities across age groups (Loh and Kendig 2013). From midlife onwards, people were found to have high rates of productive activities with the balance shifting from paid work towards caregiving and volunteering in later life. Consistent with human capital theory, productive activities were higher among those having more resources in terms of good education and good health. Analyses of the baby boomer cohort, when they were in middle age, found that productive activities were also associated with having higher incomes and higher status occupations. Women were particularly likely to be making contributions across several spheres of life. The findings underscore the contributions made by ageing people beyond the paid workforce—volunteering, caregiving, child care and domestic work—and the value of investing in human capital across the life course.

In assessing our knowledge on ageing, it is important to appreciate the paucity of 'critical' research in Australian gerontology. Aberdeen and Bye (2013) argue that our research has generally accepted and worked within a research context that has not been critical of the policy framework of positive ageing in which economic and medical perspectives predominate. Further, there is a dominance of survey research funded by governments which perhaps inevitably have a conservative, 'problem' focus (Kendig and Browning 2016). We have a long way to go in the multidisciplinary understanding of how positive individual and social action at a societal level can improve outcomes in later life.

2.3 Social Change and Outcomes for Social Groups

2.3.1 Cohort Life Chances

Australia's massive socio-economic development over the post-war era has yielded enormous benefits, as well as some social costs, and we should recognise that the consequences of change have fallen differentially on age groups and other social groups. The concept of cohort succession is central to understanding ageing and social change because younger groups generally take up emergent social opportunities, while older people are more likely to retain social positions established when they themselves entered adulthood. Over the post-war years as a whole, successive cohorts generally benefited economically from rising real incomes, more dual income households and increases in the value of owner occupied housing.

The continuing influence of the past is further revealed in how our images and assumptions regarding ageing and older people, as well as their resources through life, have been shaped by their social experiences over the post-war era. We can begin with those who are currently in advanced old age (say aged 80 years and older), recognising that these people, born before WWII, have largely paved the way in our current understandings of ageing. As a broad group generalisation, they are notable for:

- Lifelong attitudes and life directions set in childhoods during the Depression of the 1930s and WWII war years—hence their characteristic stoicism which reflects earlier life experiences as well as age-related vulnerabilities and perspectives in their now being very old
- Entry into adulthood before the educational and other opportunities that emerged for young adults in the 1960s and 1970s (broadly disadvantaged by their birth cohort)

- Generally, steady employment and rising real incomes through the post-war economic boom, albeit with great variation between social groups, with entry into home ownership early enough to benefit from housing asset increase from the 1980s (advantaged by their birth cohort)
- Late-middle age during the early 1990s economic recession, with subsequent precarious employment, particularly for older workers, and limited accumulation of the superannuation benefits that did not become widely available until the 1990s

The large baby boomer cohort that is now beginning to enter later life has had a dominant influence on Australian society and public policies throughout their lives. Their large numbers mean that they place pressure on aspects of Australian society as they enter each age group, be it on suburban housing and schools during their childhoods from the 1950s, through to emerging workforce shortages and pressure on income support, as they are now reaching what has traditionally been 'retirement age'. Through the mid-twenty first century, the baby boom cohort will dominate the landscape of ageing, as Australia becomes one of the oldest countries (see international context in Chap. 1).

In addition to the impact of their sheer numbers, the baby boomers are already beginning to transform fundamental realities and assumptions concerning later life. Their cohort as a whole has been at the forefront of social progress since the 1960s as indicated by: increasing educational and employment opportunities; the rise of the feminist movement; reduced (but continuing) gender discrimination in employment; choice over fertility and accumulation of superannuation. Their lifelong orientations, opportunities and resources have increased—with widening variability and inequalities—the personal and socio-economic resources of older individuals. They are challenging traditional assumptions of passive and dependent ageing, which has resulted in concepts such as active ageing and productive ageing being developed and promulgated as baby boomers approached later life.

The younger generations of today, variously termed Generations X and Y, are facing an apparent reversal of what had previously been a steady and seemingly inexorable rise of economic fortunes. There is increasing recognition that more recent birth cohorts are facing much of the brunt of downturns of employment markets, expensive entry into home purchasing and the prospects of decreasing public benefits and services together with likely increases in taxation. While governments continue to forecast rising real incomes over the longer term, which would advantage these younger cohorts, at present older people retain a dominant share of Australia's wealth, primarily in home ownership.

A report by the Grattan Institute (Daley and Wood 2014) demonstrates that the boom in house prices and changes in government spending patterns are the main reasons why older households have had substantial wealth increases from 2003–2004 to 2011–2012, while younger households had reductions of wealth over this same period. These historic changes in economic fortunes caution against assumptions that older people are necessarily disadvantaged relative to younger people in terms of their birth cohort as well as their increasing age.

Baby boomers and the cohorts that follow them reflect the changing patterns of ageing and intergenerational relationships. With the life time prospects of younger generations becoming more uncertain, there has been a notable rise in the popular press against what have been termed 'selfish baby boomers' who are said to have had privileged lives (Hamilton and Hamilton 2006). The Global Financial Crisis of 2007 arguably accentuated this key turning point in perceptions of intergenerational relations given its impacts on opportunities for younger people.

2.3.2 *Disadvantaged Groups and Life Chances*

The Productivity Commission has provided an account of social disadvantage in Australia as a multidimensional concept, including poverty, deprivation, socio-economic capabilities and social exclusion (McLachlan et al. 2013). It is important to recognise that these disadvantages represent intense levels of hardship that are at the extreme lower levels of living standards.

- A small proportion of Australians—between 5 % and 10 % of all adults—were reported to experience 'deep and persistent' disadvantage, with major factors being long-term health conditions or disability and low education.
- Indigenous people and public housing residents were heavily overrepresented among the disadvantaged population.
- People aged 65 years and older were included among the vulnerable groups, but compared to the others they had lower rates of 'relative income poverty' (13 %), 'multiple deprivation' (8 %), 'deep social exclusion' (8 %) and 'persistent social exclusion' (6 %).
- Single older Australians were noted for their high rates of poverty and low income, while many older people were noted as having wealth that they could draw on in retirement.

The authors comment on the impact of education as a key factor in social disadvantage and its importance as an early life investment in productivity and well-being over the life course.

2.4 Age and Intergenerational Equity

Intergenerational equity is arguably the major issue ahead for an ageing Australia. A definition developed in the UK (Piachaud et al. 2009) distinguishes four types of intergenerational equity: equity between different living generations (e.g. government benefits age groups); equity between living generations and those not yet born (e.g. environmental issues and burdens of public debt); private transfers between generations (e.g. financial support and caregiving) and public transfers between generations (e.g. government debt, taxes and pension costs inherited by younger generations).

Authoritative evidence on the overall resource flows between age groups, which clarifies this ideologically laden issue, is available from National Transfer Accounts based on labour incomes and consumption (Rice et al. 2014, p. 11).

The findings can be summarised as follows:

- Younger people, up to age 21 years, as well as older people over 60 years, on average consume more than they produce while those in the middle years are net contributors.
- Levels of consumption rise through the childhood years and then remain relatively stable through adulthood before rising again in the late 70s as people approach the end of life.
- Productivity, as indicated by labour income, rises quickly from age 21 years, as people enter the workforce. It then rises steadily further to the early 50s, before declining steadily from the late 50s, to very low levels by the time they reach age 70 years.
- From 2003–2004 to 2009–2010, the most notable changes were significant increases in productivity for the 50s and older age group and also increases in their consumption; children up to 15 years had comparable rises in consumption.

To understand the economic and policy implications of demographic dependency ratios, it is important to take account of the income and consumption patterns of the age groups (Rice et al. 2014). In 2010, private incomes and transfers within households were the main funding sources for consumption by those in middle age and also for those aged 65–75 years. While reliance on public funding increases in the older age group, it is not until the 75 years and over age group (and also for those aged 0–15 years) that consumption is funded mainly by public taxes and household transfers. Private savings were significant in funding consumption at ages 45–64 years, but they were very small for the two older age groups.

2.5 Social Change and Public Policy Developments in Ageing

An historical perspective enables us to appreciate that many societal views and social structures retain outdated images of ageing grounded in earlier historical circumstances. The eminent American gerontologist Matilda White Riley drew on the concept of 'cultural lag' in her interpretation that outdated and inaccurate views on ageing are major impediments in constructively responding to population ageing (Riley et al. 1994). In Australia, the evolving social and policy treatment of older people over the post-war era has created ongoing legacies of interests and cultural vestiges that are still apparent today. Views of 'needy and deserving' old people were underpinned by beliefs that age-related vulnerabilities were not the 'fault' of people themselves and that support had been 'earned' through earlier contributions in war, taxpaying and nation building.

Contemporary accounts of Australian developments in ageing, as summarised below, are available in earlier compendiums on ageing including Howe (1981), Kendig and McCallum (1986, 1990) and Borowski et al. (1997, 2007).

- In the 1940s and 1950s, older people were seldom represented in public affairs except in relation to some residual health and welfare matters, motivated by (inaccurate) perceptions of family abandonment of 'the aged'. Responses in the Menzies period of government (1949–1966) were free public hospitals and modest pensions as well as age-specific accommodation augmenting efforts by church and other charitable bodies.
- The 1960s saw greater recognition of systemic old age poverty, along with the notable growth of subsidies for the provider-driven nursing home industry that has led public conceptions of not only aged care but also more widely of ageing and older people generally.
- The short, comparatively radical Whitlam Labor government in the early 1970s further recognised older people as 'needy and deserving' while building a foundation of universal health, income and care programs.
- The Fraser Coalition government of the late 1970s and early 1980s actually expanded public expenditure with few policy redirections on ageing, notwithstanding concern for government deficits and some policy reviews that became influential in later years.
- The still influential Hawke era of Labor government in the 1980s set a basic foundation for the mainstream health, income, care and superannuation policies as reviewed in later chapters of this book. Notable initiatives were aged care reforms (aimed partly at restricting expenditure growth), improved/restored universal health funding and historic employer and public-funded superannuation schemes that aimed to improve retirement incomes and 'age proof' Australia's public expenditure.
- During the Recession of the early 1990s, the Keating Labor government set new directions with income support and care programs combined with tighter means testing. Targeting public resources to the most needy, along with modest tax reform, set economic rationalist approaches to maintaining social equity as well as restraining public expenditure.
- The Howard Coalition government (1996–2007) shifted policy most conspicuously towards extending tax subsidies for superannuation (benefiting most those with more private resources), while the Rudd Labor government (2007–2010) increased pensions and relaxed means testing on the pension during the political aftermath of the 2007 Global Financial Crisis.
- The Abbott Coalition government's (2013–2015) avowed priorities of (eventually) reducing public debt, through restraining expenditure growth while continuing existing taxation arrangements, have set the current policy context (and contests) that have wide-reaching significance for individual and population ageing. Ageing has now emerged centre stage as a public issue as discussed in the chapters in this book.

Throughout these post-war developments, there was relatively little explicit leadership within government on behalf of older people, while pensions and nursing homes dominated in government thinking. Although earlier portfolios included responsibilities for 'the Aged' in various combinations with housing and health, it was not until 1988 that a Minister of Aged Care was established, and then a Minister on Ageing was designated from 2001 to 2010. Labor then established a Minister on Mental Health and Ageing, elevated to Cabinet, while a returned Coalition government subsequently established a junior Ministry for Disability and Ageing within the Social Services portfolio from 2013. As of late 2015, the Minster for Health was also designated as Minister for Aged Care.

In the 1980s, a Commonwealth Office on Ageing played a leading part in bringing consumer interests into Labor's aged care reforms. Under the Coalition, the Office was responsible for the ambitious *National Strategy on Ageing* 2001, although the Strategy had little subsequent impact on policy development. The Office continued with variable responsibilities until it was disbanded by the Coalition government in 2013. Various offices in State governments, notably in Victoria, have been more significant in bringing grassroots involvement into active service development and coordination (Encel and Ozanne 2007).

There has been increasing involvement of peak advocacy groups around ageing and aged care including: 'consumer' organisations, notably the Council on the Ageing (Australia) (COTA) and National Seniors Australia (NSA); 'provider' lobbies (Aged and Community Services Australia and Leading Age Services Australia); professional bodies (nursing and other professions) and special purpose advocates such as Alzheimer's Australia and Carers Australia (Encel and Ozanne 2007). The National Aged Care Alliance currently has a key role in bringing these diverse interests into a valuable coalition that relates closely to government. Mark Butler (2015), as a former Minister of Mental Health and Ageing, provides an astute overview of the political activity and contributions to policy development of COTA, NSA and other advocacy groups. The Australian Association of Gerontology (AAG) is Australia's peak national body working to expand knowledge on ageing and linking professionals working across the multidisciplinary fields of ageing.

2.6 Political Directions

Ageing has had relatively little political prominence in Australia so far over the post-war era in contrast to the 'age and generational politics' that are so prominent in Europe and the USA. A useful review of developments through the 1990s concluded that the politics of ageing and mobilisation of older Australians had largely followed, rather than led policy developments (Encel and Ozanne 2007). Butler (2015) made a case that current and recent generations of older voters have had strong party identifications that have not changed much over time. He also noted that Australia's compulsory voting system limits the power of older Australians relative to the voluntary USA voting system, where older people have a high turnout relative to younger people.

In the 2013 election of the Abbott government, voters aged 65 years and over were strong supporters of the Liberal-National parties (56 %) in contrast to voters aged under 25 years (35 %), 25–44 years (42 %) and 45–64 years (45 %) (Bean and McAllister 2014). McAllister (2011) noted the stability of voting patterns over time and the lack of much evidence supporting either ageing effects (in which people would become more conservative with age) or major cohort effects (continuing political identities formed when people came of age). However, over recent elections both baby boomer and younger voters have evinced less identification with major parties, relative to their counterparts from earlier cohorts. Butler (2015), drawing on separate evidence, noted that the baby boom cohort had come of age mainly during the Labor governments of the Whitlam/Hawke/Keating era. He further noted that this cohort could present challenges for conservative parties, although boomers' votes could become more contestable in the future.

2.7 Conclusion

The longevity revolution and cohort succession have raised new challenges and opportunities for ageing individuals and an ageing Australia. This chapter has aimed to provide an overview of the social, economic and political developments of an ageing Australia over the post-WWII era. It has examined ways in which the lives of diverse groups of ageing people have unfolded during periods of considerable social change. One of the most important insights is that ageing is a social process which, over the life span, can be improved by positive action. Another is that ageing is highly variable with an accumulation of advantage or disadvantage for different groups over their lives. Uncertainty over Australia's social and economic future adds to the complexity and stakes in actions on ageing.

References

Aberdeen, L., & Bye, L. (2013). Challenge for Australian sociology: Critical ageing research—Ageing well? *Journal of Sociology, 49*(1), 3–21.

Australian Bureau of Statistics. (2011). *Life expectancy trends-Australia. Australian social trends* (Cat. No. 4102.0). Canberra, Australia: ABS. Retrieved from www.abs.gov.au/socialtrends.

Australian Institute of Health and Welfare. (2014). *Healthy life expectancy in Australia: Patterns and trends 1998 to 2012.* AIHW Bulletin No. 126 (Cat. No. AUS 187). Canberra, Australia: AIHW.

Australian Institute of Health and Welfare. (2015). *Australia's welfare 2015.* Australia's Welfare Series No. 12 (AIHW Cat. No. AUS 189). Canberra, Australia: AIHW.

Bean, C., & McAllister, I. (2014). Documenting the inevitable: Voting behaviour at the 2013 Australian election. In C. Johnson & J. Wanna (Eds.), *Abbott's gamble: The 2013 federal election* (pp. 411–424). Canberra, Australia: ANU Press.

Bengtson, V., & Kuypers, J. (1971). Generational differences and the developmental stake. *The International Journal of Aging and Human Development, 2*, 249–260. doi:10.2190/AG.2.4.b.

Borowski, A., Encel, S., & Ozanne, E. (1997). *Ageing and social policy in Australia.* Cambridge, England: Cambridge University Press.

Borowski, A., Encel, S., & Ozanne, E. (2007). *Longevity and social change in Australia.* Sydney, Australia: University of New South Wales Press.

Brooks, L., & Kendig, H. (2004). Understanding community nursing for older individuals and carers. In R. Nay & S. Garrett (Eds.), *Nursing older people: Issues and innovations* (pp. 135–150). Melbourne, Australia: Maclennan and Petty.
Butler, M. (2015). *Advancing Australia: The politics of ageing*. Melbourne, Australia: Melbourne University Press.
Cummins, R. A. (2014). Homeostasis. In A. C. Michalos (Ed.), *Encyclopedia of quality of life and well-being research* (pp. 2901–2904). Dordrecht, Netherlands: Springer. Retrieved from http://springerreference.com/docs/html/chapterdbid/412582.html&gt.
Cummins, R. A., Woerner, J., Weinberg, M., Collard, J., Hartley-Clark, L., & Horfiniak, K. (2013). *Australian unity wellbeing index: Report 30.0—The wellbeing of Australians: Social media, personal achievement, and work*. Melbourne, Australia: Australian Centre on Quality of Life, School of Psychology, Deakin University. ISBN 978-1-74156-186-9. Retrieved from http://www.deakin.edu.au/research/acqol/reports/auwbi.php.
Daley, J., & Wood, D. (2014). *The wealth of generations*. Carlton, VIC, Australia: GRATTAN Institute. Retrieved from grattan.edu.au/report/the-wealth-of-generations/.
Dannefer, D., & Phillipson, C. (Eds.). (2010). *The SAGE handbook of social gerontology*. London: Sage.
Dannefer, D., & Settersten, R. (2010). The study of the life course: Implications for social gerontology. In D. Dannefer & C. Phillipson (Eds.), *The SAGE handbook of social gerontology* (pp. 4–21). London: SAGE.
de Vaus, D., Wells, Y., Kendig, H., & Quine, S. (2007). Does gradual retirement have better outcomes than abrupt retirement? Results from an Australian panel study. *Ageing and Society, 27*(5), 667–683.
Encel, S., & Ozanne, E. (2007). The politics of ageing. In A. Borowski, S. Encel, & E. Ozanne (Eds.), *Longevity and social change in Australia* (pp. 296–315). Sydney, Australia: University of New South Wales Press.
Gardner, I., Browning, C., & Kendig, H. (2005). Accommodation options in later life: Retirement village or community living? *Australasian Journal on Ageing, 24*(4), 188–195.
Gong, C., & Kendig, H. (2016). Ageing and social change in Australia. In H. Kendig, P. McDonald, & J. Piggott (Eds.), *Population ageing and Australia's future*. Canberra, Australia: ANU Press.
Hamilton, M., & Hamilton, C. (2006). *Rich boomer, poor boomer: Retirement prospects for the not-so-lucky generation*. Bruce, Australia: The Australia Institute.
Howe, A. (Ed.). (1981). *Towards an older Australia: Readings in social gerontology*. St Lucia, Australia: University of Queensland Press.
Kendig, H., & Browning, C. (2016). Social research and actions in ageing well. In H. Kendig, P. McDonald, & J. Piggott (Eds.), *Population ageing and Australia's future*. Canberra, Australia: ANU Press.
Kendig, H., Browning, C., Thomas, S., & Wells, Y. (2014). Health, lifestyle, and gender influences on aging well: An Australian longitudinal analysis to guide health promotion. *Frontiers in Public Health, 2*, 2–70. doi:10.3389/fpubh.2014.00070.
Kendig, H., Loh, V., O'Loughlin, K., Byles, J., & Nazroo, J. (2016). Pathways to well-being in later life: Socioeconomic and health determinants across the life course of Australian baby boomers. *Journal of Population Ageing, 9*(1), 49–67. doi:10.1007/s12062-015-9132-0.
Kendig, H., & Lucas, N. (2014). Individuals, families and the state: Changing responsibilities in an ageing Australia. In A. Torres & L. Samson (Eds.), *Aging in Asia-pacific: Balancing the state and the family* (pp. 211–224). Diliman, Philippines: Philippine Social Science Council (PSSC).
Kendig, H., & McCallum, J. (1986). *Greying Australia: Future impacts of population ageing*. Canberra, Australia: Australian Government Publishing Service. Retrieved January 1, 1987, from www.popline.org/node/349069#sthash.fufCTojC.dpuf.
Kendig, H., & McCallum, J. (1990). *Grey policy: Australian policies for an ageing society*. Sydney, Australia: Allen & Unwin.
Kendig, H., Wells, Y., O'Loughlin, K., & Heese, K. (2013). Australian baby boomers face retirement during the global financial crisis. *Journal of Aging & Social Policy, 25*, 264–280.

Kröger, T., & Yeandle, S. (Eds.). (2013). *Combining paid work and family care: Policies and experiences in international perspective*. Bristol: Policy Press.

Loh, V., & Kendig, H. (2013). Productive engagement across the life course: Paid work and beyond. *Australian Journal of Social Issues, 48*(1), 111–137.

Majeed, T., Forder, P., Mishra, G., Kendig, H., & Byles, J. (2015). A gendered approach to workplace participation patterns over the life course for an Australian baby boom cohort. *Journal of Vocational Behavior, 87*, 108–122.

McAllister, I. (2011). *The Australian voter: 50 years of change*. Sydney, Australia: UNSW Press.

McLachlan, R., Gilfillan, G., & Gordon, J. (2013). *Deep and persistent disadvantage in Australia*. Canberra, Australia: Productivity Commission Staff Working Paper.

O'Loughlin, K., Humpel, N., & Kendig, H. (2010). Impact of the global financial crisis on employed Australian baby boomers: A national survey. *Australasian Journal on Ageing, 29*(2), 88–91.

O'Loughlin, K., Loh, V., & Kendig, H. (in press). The interrelations between caregiving, paid work, and health status for Australia's baby boomers. *Journal of Cross-Cultural Gerontology*, in press.

O'Loughlin, K., & Phillips, J. (Eds). (in press). Older workers and caregiving in a global context. *Journal of Cross-Cultural Gerontology*.

Phillipson, C. (2013). *Ageing*. Cambridge, England: Polity Press. ISBN 978-0-7456-3084-7.

Piachaud, D., Macnicol, J., & Lewis, J. (2009). *A think piece on intergenerational equity*. Manchester, England: Equality and Human Rights Commission.

Prime Minister's Science, Engineering and Innovation Council (PMSEIC) Independent Working Group. (2003). *Promoting healthy ageing in Australia*. Canberra, Australia: Department of Education, Science and Training.

Quine, S., Wells, Y., de Vaus, D., & Kendig, H. (2007). When choice in retirement decisions is missing: Qualitative and quantitative findings of impact on well-being. *Australasian Journal on Ageing, 26*(4), 173–179.

Rice, J., Temple, J., & McDonald, P. (2014). *National transfer accounts for Australia: 2003–04 and 2009–10 detailed results*. Canberra: ARC Centre of Excellence in Population Ageing Research (CEPAR). Retrieved from www.cepar.edu.au/media/134354/nta_report_2014.pdf.

Riley, M., Kahn, R., & Foner, A. (1994). *Age and structural lag: Society's failure to provide meaningful opportunities in work, family, and leisure*. New York: Wiley.

Russell, C., Touchard, D., Kendig, H., & Quine, S. (2001). Foodways of disadvantaged men growing old in the inner city: Policy issues from ethnographic research. In D. Weisstub, D. Thomasma, S. Gauthier, & G. Tomossy (Eds.), *Aging: Culture, health, and social change* (pp. 191–215). Dordrecht, The Netherlands: Kluwer Academic.

Taylor, P. (2010). Cross-national trends in work and retirement. In D. Dannerfer & C. Phillipson (Eds.), *The SAGE handbook of social gerontology* (pp. 540–562). London: SAGE.

Windsor, T., Curtis, R., & Luszcz, M. (2015). Sense of purpose as a psychological resource for aging well. *Developmental Psychology, 51*(7), 975–986.

Chapter 3
Attitudes to Ageing

Kate O'Loughlin and Hal Kendig

3.1 Introduction

The rights, involvement in society and self-esteem of older people depend largely on their social treatment in everyday life. Notwithstanding close relationships within families, negative public attitudes and discrimination can limit older people's social and economic contributions and may impact on their health and well-being. In Australia, older people can be depicted as a 'social problem' as a consequence of public concern about the costs to government of demographic change and an ageing population (Australian Treasury 2010, 2015). While Australia has legislation in place that is intended to address age discrimination at both a national level (Age Discrimination Act 2004) and within States and Territories, negative attitudes and portrayals of older people, particularly as represented in the media, are considered to be a major factor impacting on mature age and older people, their experiences in workplaces and the quality of care they receive in the health care system.

The impetus for improving expectations and attitudes concerning ageing and older people arguably rests in a sense of social justice, but it is also underpinned by profound social changes underway in Australia (Gong and Kendig 2016). First, the remarkable extensions of life expectancy at older ages are redefining later life with

---

K. O'Loughlin (✉)
Ageing, Work and Health Research Unit, Faculty of Health Sciences, University of Sydney, Lidcombe, NSW 2141, Australia

ARC Centre of Excellence in Population Ageing Research (CEPAR), University of New South Wales, Kensington, NSW 2033, Australia
e-mail: kate.oloughlin@sydney.edu.au

H. Kendig
ARC Centre of Excellence in Population Ageing Research (CEPAR), University of New South Wales, Kensington, NSW 2033, Australia

Centre for Research on Ageing, Health and Wellbeing, Research School of Population Health, The Australian National University, Canberra, ACT 2601, Australia

K. O'Loughlin et al. (eds.), *Ageing in Australia*, International Perspectives on Aging 16, DOI 10.1007/978-1-4939-6466-6_3

several decades of active ageing becoming recognised as the new norm with new opportunities. Second, the baby boom cohort with its higher expectations is succeeding the Depression cohort of people notable for their stoicism. Third, with labour force shortages and government austerity looming for decades ahead, there are strong structural and political forces driving demands for more contributions during later life. Improving attitudes is central to addressing outdated practices and beliefs, which still present formidable barriers in adjusting to an ageing Australia.

3.2 Formation of Attitudes

In social psychological terms, an attitude refers to a range of emotions, beliefs and behaviours that we have about or towards people, objects or events (Vaughan and Hogg 2013). Attitudes are generally formed as a result of our socialisation and experiences growing up within particular social groups such as the family, as well as broader social and cultural influences. A major model applied in understanding attitude formation is the ABC, which is centred on the three core components of *affect* (feelings, emotions), *behaviour* (experiences, past behaviours) and *cognition* (beliefs, thoughts, attributes). For most people, the core of age-related attitudes is formed early in childhood through relationships with grandparents and great grandparents, engendering positive feelings along with awareness of age vulnerabilities and social connections to the past. Attitudes including prejudices and discrimination continue to be shaped by experiences through life. They are an inextricable part of human thought, reflecting a basic tendency to categorise people, objects and situations as the fundamental way of making them meaningful; we construct our worldviews around social entities and experiences (Vaughan and Hogg 2013).

A report produced by the Australian Human Rights Commission (2013) found that most Australians (71 %) felt that age discrimination was common, and over a third of those aged 55+ years reported having experienced age-related discrimination. Almost half of all Australians felt that discrimination was present within the health care system, within government policy or in access to services. The most common types of age-related discrimination reported by older people included being turned down for a position, being treated with disrespect (64 % aged 65+) and being subjected to jokes about ageing. Stereotyping and associated discrimination can be more damaging to an individual than the actual experience of ageing and older age itself (Minichiello et al. 2000) with recognised consequences for age equality and social inclusion.

3.3 Stereotypes

In a similar way to attitude formation, stereotypes are socially constructed beliefs and expectations that we build up through the process of socialisation and input from our social and cultural environments. This includes what we hear and experience within our family and other social groups, from personal exposure to the stereotyped individual or group, and from the information and images we see in various forms of media (Posthuma and Campion 2009). Stereotyping is something that we all do, whether consciously or not, and has been referred to as a form of shorthand in the way we 'sort' and evaluate people. An extensive literature identifies how we evaluate people based on their behaviours, competencies and physical appearance and attractiveness (Kite et al. 2005). This is particularly the case in developed (Western) societies that are socially organised around individualist rather than collectivist norms and values (Markus and Kitayama 1991, 2003) and are also very much youth centred (Macnicol 2006; Calasanti 2005; Twigg 2008).

Age-based stereotypes may originate as early as childhood, and these stereotypes are reinforced throughout the life course (e.g. Levy 2003). They can create a fear of a *future self* (Nelson 2005) which, in turn, reinforces a fear of ageing at both the personal and societal level. A term used to describe these conscious or unconscious fears is *benign ageism* (Nelson 2005). Levy (2003) adds that *ageing self-stereotypes* carried from childhood can become the expected reality for us as we enter older age. They can operate below conscious awareness and can influence social interactions even when we do not engage in overt discriminatory behaviour.

Negative stereotypes and systematic discrimination can be directed towards any age group (Taylor et al. 2013) as has been recognised in age discrimination legislation. The focus is usually not on those in 'middle age'—the norm—but rather on those we deem to be 'too young' or 'too old' (Iversen et al. 2009). For example, in the workplace being too young is often associated with a lack of knowledge and experience, while being too old is associated with being less capable of taking up new skills and more inflexible (Posthuma and Guerrero 2013).

Age-based stereotypes are also evident in social and health policy discourses and strategies which refer to ageing as a 'social problem', with references regularly made to the *burden* of an ageing population, the *dependency* of older people, particularly on the finances of others, and to the *frailty and ill health* associated with ageing. In presenting older people in this way, we are in effect engaging in stereotypical and potentially discriminatory behaviour that can have an exclusionary or marginalising effect on older people.

3.4 Ageism

In a social and legislative sense, Australia, and other developed societies, have addressed issues around discrimination based on race and ethnicity, gender, religion, sexuality and living with a disability. However, age and ageism have not received the same attention to date (Iversen et al. 2009; Abrams et al. 2011). Until fairly recently most developed countries had a mandated retirement age that was linked to eligibility for the Age Pension. While age discrimination legislation no longer permits an enforced retirement age, in Australia we still tend to associate entry to older age with this socially constructed retirement age of 65 and over.

The concept of ageism was introduced by the American gerontologist Robert Butler as "a process of systematic stereotyping and discrimination against people because they are old" (1969:243) and likened it to racism and sexism. Since that time Butler and others have continued to address ageism and its impact on individuals and communities (Butler 2005; Minichiello et al. 2000; Iversen et al. 2009).

Advocacy groups, and older people themselves, also have been speaking out about ageist attitudes and discriminatory behaviour. For example, a survey on ageism and its impact on travel insurance was carried out in 2012 by National Seniors Australia and COTA Australia (National Seniors Australia and COTA 2012a, b). This survey reported that people aged 70 years and over were more likely to experience difficulties in obtaining travel insurance, with one-third of respondents reporting that they had to pay higher travel premiums because of their age. A survey respondent commented:

> Less than 12 months ago I sought a quote on line and when I gave my age at 68 they point blank refused any type of travel insurance offer.

3.5 Employment

> The right to work, free from discrimination on any basis, is a fundamental human right. Individuals who are denied the right to work are denied the independence, dignity and sense of purpose that work brings. (Australian Human Rights Commission 2015a. Issues Paper: Employment discrimination against older Australians)

Individual and government expectations that mature and older workers will remain in paid work face a number of barriers, including perceptions and experiences of ageism and age discrimination in workplaces (Australian Human Rights Commission 2016; Adair and Temple 2012). Notwithstanding age discrimination legislation in Australia and other developed countries (Filinson 2008) there is still ample evidence to suggest that age discrimination remains common across all types of industries and professions and a majority of countries (see Loretto and White 2006; Buyens et al. 2009; Shah and Kleiner 2005; Syed 2006; Maurer et al. 2008; Taylor et al. 2013; Iweins et al. 2013; Marcus and Fritzsche 2015). What this tells us is that while governments are encouraging mature and older workers to remain in paid work, the evidence points to reluctance on the part of employers to change their attitudes and employment practices. In an address to the Australian Human Resources Institute in 2012, the Age Discrimination Commissioner The Hon Susan Ryan commented: "It appears we have a serious disjunction between the policy of raising the pension age, the needs of the economy and the persistence of age discrimination in the workforce".

3.5.1 Evidence of Age Discrimination in the Workplace

Australia's first national prevalence survey of age discrimination in the workplace (Australian Human Rights Commission 2015b) clearly demonstrated the extent and nature of age discrimination being experienced by older workers. Key findings are shown in Box 3.1.

Box 3.1 *National Prevalence Survey of Age Discrimination in the Workplace: Key Findings* Age discrimination is ongoing and a common occurrence in the Australian workforce.

- Current levels of age discrimination experienced by people aged 50 years and older are similar to the levels experienced in the past.
- Anticipation of discrimination can discourage some older people from looking to enter the workforce.
- Age discrimination is more likely to be experienced by those aged between 55 and 64 years old and those looking for employment.
- Men and women are equally likely to be subjected to age discrimination.
- The groups most vulnerable to experiencing discrimination are typically people who are in a lower income bracket or in a single parent household.
- Managers aged 50 years or older reported that they took an employee's age into consideration on a regular basis when making decisions.

Source: Australian Human Rights Commission (2015b)

While there is clear evidence of age discrimination, it appears that very few people who experience age discrimination in their employment take action through the legislative mechanisms available to them. .In 2014-2015, the Australian Human Rights Commission received only 149 complaints under the Age Discrimination Act (AHRC 2015). The majority (63 per cent) of these complaints concerned employment, with 73 per cent of these coming from people aged 45 years and over, and the highest percentage (38 per cent) being from those aged 55-64 years. The NSW Anti-Discrimination Report for 2014-2015 also shows very few complaints on the basis of age ($N=79$); of these, 66% related to employment (Anti-Discrimination Board of NSW 2015).

Within the workplace, age discrimination can be present in one of two ways: direct or indirect age discrimination (Australian Human Rights Commission Good Practice Good Business Factsheets). Direct discrimination occurs when an individual or group is treated less favourably because of their age. An example of this relating to mature or older age would be targeting an employee(s) over the age of 50 years for redundancy rather than applying it across the entire workforce. With indirect discrimination, an employer may introduce certain conditions or requirements that, on the surface, apply to all employees but in effect may disadvantage a person(s) if they cannot comply with it. An example here could be requiring an older worker to undergo a physical fitness test even when a standard of physical fitness is not required or has not been specified for the particular job. In this situation younger people may be more likely to meet the fitness requirement thereby potentially disadvantaging older workers doing the same job.

Much of the published research on older workers broadly finds evidence for one of two stereotypical images: the positive one, depicting the older worker as reliable, loyal, punctual, highly experienced and possessing a good work ethic; the negative one, depicting the older worker as being resistant to change, inflexible, unable to learn new skills and, in the view of younger workers, often in the way of their own progression up the career ladder (Patrickson and Ranzijn 2005; Posthuma and Guerrero 2013). Research on workplace age discrimination has addressed a number of issues associated with ageist attitudes and stereotypes held by employers, co-workers and society. These include evidence of younger job applicants receiving more positive responses, with older women being the most discriminated against (Gringart and Helmes 2001). A qualitative study revealed managers' bias and age discrimination when asked about older workers (Loretto and White 2006).

A survey of decision-makers and their hiring practices across Australian industries showed systematic negative stereotyping of older workers (Gringart et al. 2005) with older workers being seen as inferior when it came to trainability, adaptability, creativity and showing an interest in new technology. Another study found a mismatch between what older job seekers believe employers want and what employers are actually seeking (Ranzijn et al. 2004).

3.6 Intergenerational Attitudes: Is There Conflict Between Young and Old?

As outlined in Chaps. 2 and 8, governments in Australia and elsewhere are aiming to mitigate the economic consequences of population ageing by encouraging people to remain in paid work for longer. However for these policy directions to succeed, the negative stereotypes and discriminatory behaviours directed at older people will need to be addressed, especially among younger generations (Posthuma and Campion 2009; Iweins et al. 2013; Australian Human Rights Commission 2013; Kite et al. 2005).

Age-related policy directions arguably are influenced by underlying, longer term public attitudes towards social equity between generations and cohorts. So what do we know about intergenerational attitudes towards the differing opportunities afforded to younger and older generations in Australia? A national study by Kendig et al. (2015) investigated the age and generational differences in attitudes towards Australia's baby boomers held by generations both younger and older than the boomers. The data came from the 2009 Australian Survey of Social Attitudes, a biennial national mail out survey of Australians aged 18 years and over ($N=1525$). The purpose of the study was to provide evidence of the social treatment of older people, cohort equity and the views of different generations towards Australia's baby boomers who might be viewed as a 'privileged' cohort (see Chap. 2).

Figure 3.1a, b shows views on the overall lifetime social and economic opportunities of the baby boom cohort, as compared to the preceding and following cohorts. Nearly 60 % of the respondents felt that the boomers had better lifetime prospects

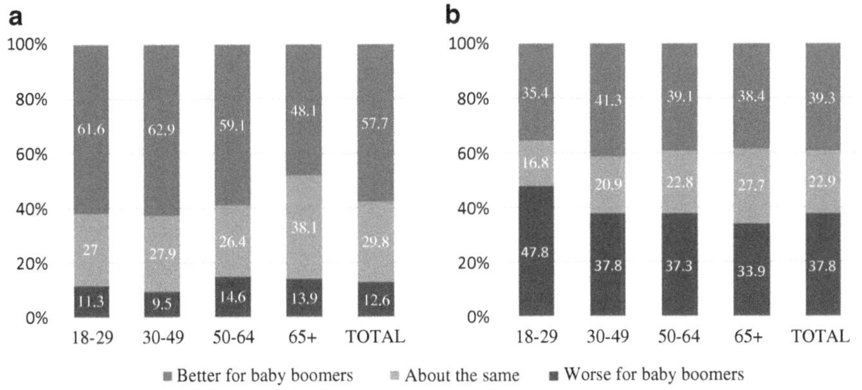

Fig. 3.1 (**a**) Life-long opportunities for baby boomers compared to older people who have retired. (**b**) Life-long opportunities for baby boomers compared to younger people today

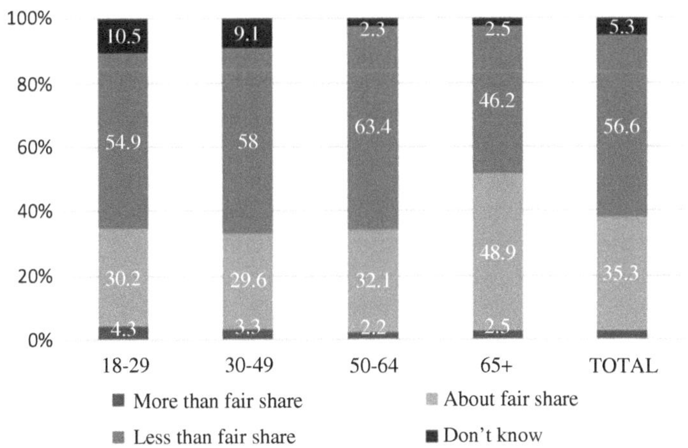

Fig. 3.2 Are older people getting their fair share of government benefits?

than those who were already retired; very few thought that the boomers had fared worse. Interestingly, more of those retired thought that the prospects were about the same for these two cohorts, while younger people thought the boomers were doing better. Comparing boomers to now younger people, there were more divided views on the generations' lifelong prospects, with younger adults being relatively more likely to report better lifelong prospects for their own cohort than for the boomers.

Figure 3.2 shows the population as a whole has divided views on the fairness of age distributions of government benefits. More than half felt that older people were receiving less than their fair share, reflecting the traditional attitude towards older people as being 'needy and deserving'. Those who were already aged 65+ years were the group with the highest proportions, at nearly half, reporting that the age shares of older people were about right, while baby boomers in late middle age felt

Table 3.1 Conflict between young and old

	Very strong/strong	Not very strong	No conflict	Can't choose
18–29	29.6	56.6	12.6	1.2
30–49	21.1	58.4	17.9	2.6
50–64	28.0	53.1	16.5	2.4
65+	26.1	55.3	14.9	3.7
Total	25.7	55.6	16.1	2.6

that older people were more likely to report receiving less than their fair share. The controversial issue at the time—policy to progressively raise pension age eligibility—attracted widespread opposition although it was supported by 40 % of those aged 65 and over. These findings indicate the political challenges for governments in their efforts around reducing public provisions for older people in the context of population ageing.

Finally, we directly asked about perceptions of intergenerational conflict. As shown in Table 3.1, 26 % perceived 'strong or very strong' levels of intergenerational conflict, 56 % 'not very strong' conflict and 16 % 'no conflict'. Perceived intergenerational conflict had little association with respondents' gender or housing tenure (an indication of wealth). Perceptions of intergenerational conflict tended to be higher among people on relatively lower incomes and younger people aged in their twenties. It is important to note, however, that this Australian survey was conducted before the Abbott government heightened efforts to reduce national debt.[1]

Based on this evidence, we have to consider how attitudes towards older people might improve or deteriorate if there was increasing intergenerational competition, for example, in the labour force. This raises questions around what political, social and economic actions could improve attitudes towards age and ageing and reduce age discrimination. As a start, the Australian Human Rights Commission is working with employers on case studies that demonstrate the value of older workers. In the end the attitudes of employers and the broader public may follow their direct economic interests in recognising the value of older workers.

3.7 Ageism in Health Care

The evidence suggests that those in the caring professions view ageing as a process of inevitable decline and this, in turn, shapes the negative attitudes they hold towards older people and their health and well-being. In fact, it is argued that health professionals may be more vulnerable to ageist attitudes because of their increased

[1]These survey items on intergenerational attitudes have been replicated in the 2015–2016 Australian Survey of Social Attitudes to provide comparative data on any change in attitudes over the period, especially in the context of policy changes related to population ageing.

exposure to ill and infirm older people (Giles et al. 2002). Older people themselves report that health professionals are the group most likely to make them 'feel old' rather than 'normal ageing' (Minichiello et al. 2000).

Ageism in health care also can be viewed at the system level in terms of the priority and treatment accorded to older people in hospitals and primary care (see Chaps. 12 and 13). There is some evidence from the US that older people are often the target of rationing; however, there is also evidence that older people can receive forms of positive discrimination when it comes to accessing and paying for services that are not always available to the general population (Kane and Kane 2005). We need to think about the different sites of practice (primary care, acute care, rehabilitation, residential care, community care) and type of care being provided (e.g. general or specialised geriatric care) and how this might shape the attitudes and behaviours of health professionals in their interactions with older people. Further, new approaches such as reablement in community care (Chap. 14) and health promotion in primary care (Chap. 13) are fundamentally based on attitudes recognising that ageing does not necessarily involve decline and that people can recover even after health and social setbacks.

We might be surprised to hear concerns about negative attitudes to ageing among health professionals because we have an expectation that those who choose to enter the health professions do so because they want to work with people with illness or disability. Further, we might expect that, through the process of professional socialisation, they will acquire a better understanding of the significance of poor health on a person's life and what is needed to provide quality care for the best possible patient outcomes.

From the available evidence, health professionals express ageism in several ways. Older people are seen as 'bed blockers' (see Gething et al. 2002; Tadd et al. 2011), a response generally associated with the view that the often complex and comorbid conditions of older age cannot be 'cured' (Le Couteur et al. 1997; Leung et al. 2009). Care for older people is seen as a 'Cinderella service', a low status option for specialisation (Le Couteur et al. 1997; Arino-Blasco et al. 2005) and, for nurses, is associated with low professional self-esteem (Wells et al. 2004). Evidence suggests that most health professionals have little or no general or specialised knowledge of ageing and older people (Courtney et al. 2000; Gething et al. 2002; Hobbs et al. 2006; Stewart et al. 2005; Deasey et al. 2014). While these attitudes may vary in some subtle ways between health professions, the important point to make is that practitioners in medicine, nursing and allied health share them (Kearney et al. 2000) and that these attitudes have been found in most developed countries.

An ethnographic study by Tadd et al. (2011) looked at the experience of older people in acute care settings in four National Health Service (NHS) facilities in the UK. Their findings showed that the attitude of acute care staff was that such settings are not the 'right place' for older people, especially those with a cognitive impairment, and that the staff response to this was to focus on tasks rather than the person. The authors concluded that staff attitudes in these settings resulted in the potential for older people to not receive 'dignified care' and that NHS facilities needed to address the quality of care provided to older people who were their major client base.

Arino-Blasco et al.'s (2005) study on the importance of dignity, and how this can be maintained while providing appropriate health care services for older people, explored the attitudes and behaviours of health professionals (N=424) in six European countries. The key points to emerge from participants identified both negative and positive views. Most held negative views of the lives of older people; a clear distinction was drawn between older people who were deemed to be fit and not frail. Old age was seen to signify ill health, dependency, vulnerability, frailty and loss of competence, which impacted on the older person's dignity. Many older people were seen as not being familiar with technological advances and information technology. However, many health professionals reported that, while working with older people was given low status, they found it enjoyable and satisfying, and it offered variety and an intellectual challenge. In the views of these health professionals there was a need for more professional education around person-centred care, with an emphasis on supporting older people to live with dignity and maintain their independence and quality of life for as long as possible (Arino-Blasco et al. 2005).

Australian research has also investigated the attitudes and behaviour of health practitioners across professions and care settings. Gething et al. (2002), in their study of registered nurses in Australia and the UK, applied three instruments intended to measure knowledge of, and attitudes towards the ageing process and older people: Reactions to Ageing Questionnaire (Gething 1994), Facts about Aging Quiz (Palmore 1977, 1988) and Aging Semantic Differential (Rosencranz and McNevin 1969). This study found that registered nurses in both countries had misconceptions and used negative stereotypes about the ageing process and older people, particularly in the way they underestimated and devalued the capacities and capabilities of older people. It identified a need for education on ageing and older people, in both the pre-employment and post-employment periods.

A literature review by Deasey et al. (2014) examined the knowledge and understanding of the ageing process and attitudes to older people of nurses in emergency care departments (EDs). Older people coming into EDs may present with an acute episode and, at the same time, with a range of complex and comorbid conditions. The research shows that ED nurses are not always prepared to consider or accept that a 'quick fix' may not be possible and that a more holistic approach might be required to meet the needs of older people in this setting. Another key finding is that managerial style, past experiences and the centrality of the curative approach of the medical model in health care service delivery were factors that also influenced nurses' negative attitudes towards older people.

Research involving medical doctors and allied health professionals in Australian health care settings reports similar findings: generally negative attitudes, formed around ageist stereotypes and knowledge deficits, that significantly influence practice, and potentially the quality of care, provided to older people. Research has shown that both medical students (Le Couteur et al. 1997; Tam et al. 2014) and practising doctors (Leung et al. 2009) generally have neutral to negative attitudes towards older people, and do not consider geriatric medicine as a career choice because of its 'low tech' nature, with practice based around multiple, chronic conditions rather than the curative focus of acute care. On a more positive note, participants in these and other studies who had been exposed to geriatric medicine through a clinical posting (see Lui and Wong 2009), or a geriatric medicine course during training (Tam et al. 2014), had a better understanding and more positive attitude.

Research among allied health professionals indicates that physiotherapists (Morris and Minichiello 1992; Hobbs et al. 2006) and occupational therapists (Giles et al. 2002; Stewart et al. 2005) hold neutral to slightly more positive attitudes to older people than nurses or doctors, although both professional groups lack knowledge about the ageing process and older people. Of concern were the potentially ageist attitudes shown by psychologists (Gething et al. 2003; Helmes and Gee 2003; Koder and Helmes 2006). While not having the same level of involvement with older people as other health professionals, psychologists' attitudes towards treating older clients were less favourable than treating younger clients, with older clients being seen as 'less appropriate' for therapy because of the likelihood of a poorer prognosis. The value of early exposure in the professional education and clinical settings to improve knowledge and attitudes was shown in Hobbs et al.'s study (2006) of first year and fourth year physiotherapists. There was a considerable improvement in knowledge and attitudes over the 4 years; however, the authors report that it was not as much as expected.

What does this mean for an older person? Older people report that their sense of dignity and identity can be challenged in negative interactions they have with health professionals. This is very often expressed as relating to their sense of a lack of privacy, poor communication and a general insensitivity to their needs (Woolhead et al. 2004).

3.8 Media Representations of Older People

Media, in all its forms, plays an important role in people's lives, irrespective of their age, gender, ethnicity or geographical location, and this is particularly so for those with access to television and the internet. The Australian Human Rights Commission Report (2013) on stereotypes of older Australians found that all forms of media (television, radio, magazines, digital) influence negative perceptions of older Australians and that these media portrayals of ageing and older people need to be challenged. The key findings of the report include:

- Older Australians are underrepresented and often poorly portrayed in the media; this contributes to a sense of invisibility through limited, homogeneous portrayals of older people.
- The media influences negative perceptions of older Australians.
- Social media portrays older people as vulnerable and as victims (crime, physically vulnerable, risk of illness).
- The portrayal of older people in advertising is 'unfair'; it should depict older people as living 'normal' lives, as contributors to society rather as victims.

The perceptions reported are ageist in that older people feel that they are not valued for their lived experience and contributions to the social, economic and cultural life of their community. Media representations often operate on the same stereotypes of older people that are found in the workplace; that is, they are seen in either a very positive way (affluent, healthy, active, happy) or a more negative way (burden, sick, frail, poor, dependent), rather than depicting the heterogeneity and diversity that would be found in any other group in the population.

3.8.1 *Baby Boomers and the Media*

The baby boomer cohort has come in for particular attention in the way that they are portrayed in the media, particularly around attitudes and lifestyle (Tavener et al. 2014). As Australia's baby boomers are in the 'young-old' age group (born 1946–1965), they are often represented as being very different to previous generations (old-old age groups) as they have lived through extended periods of economic prosperity and extensive social change (e.g. birth control, second wave feminism). Because of this they are depicted as being more affluent, productive, healthy and innovative (Tavener et al. 2014; Healy 2004) and are considered the demographic dream for commercial and marketing interests. However, there is also evidence that they are not as healthy as expected (Australian Human Rights Commission 2016; Humpel et al. 2010) and many are not as financially prepared for retirement as they and others, especially governments, expect of them (Hamilton and Hamilton 2006; Noone et al. 2013; O'Loughlin et al. 2010). Further, characterisations of 'the baby boomers' are yet another stereotype that belies the great diversity among individuals and social groups depending on their earlier and current life experiences.

3.9 Conclusion

As shown throughout this chapter the experience of ageing is heavily structured by prevailing attitudes and expectations that in turn depend on social position and relations. Early life experiences within families and communities generally engender positive orientations towards ageing and older people, but these attitudes can be overlaid later by perceptions of dependency and decline. The existence and perpetuation of age-based stereotypes result in ageist attitudes and behaviour such as age discrimination in many areas of social and economic life including the workplace, health care system and media. The consequences of this provide challenges to individuals, governments and society as a whole in terms of addressing the barriers and lack of fair treatment to people based solely on their chronological age.

There are indications that attitudes towards ageing are improving in some respects in line with the rising expectations, incomes, education and other social resources of many individuals in the baby boom cohort. There also are opportunities to proactively address some of the fundamental challenges at the socio-cultural and structural levels. There are complex two-way relationships in the changing attitudes and social positions of ageing individuals, and public policies are important influences in the social construction of ageing. On the one hand, governments concerned with the fiscal costs of demographic change can come very close to scapegoating older people for rising public expenditure. On the other hand, there are proactive efforts to address ageism and age discrimination by advocacy groups such as COTA and the Australian Human Rights Commission's leadership through the work of the Age Discrimination Commissioner. An understanding of prevailing attitudes and ideas in the way older people are depicted and treated and the efforts to address this are central to all of the aspects of ageing considered throughout this book.

3.10 Case Study

Catherine is 58 years old and has been employed for the past 18 years in a full-time position with a major international financial institution. In that time she has progressed through the company ranks and has undertaken both internal and external education and training to gain general and specific knowledge and skills related to her work. She supervises a team of five younger employees (three women, two men) who have entered the organisation with tertiary qualifications and Catherine feels that she has a good working relationship with them, and also with her immediate manager and others at senior executive level. Catherine is married with two adult children, two grandchildren and provides care and support for her elderly parents.

In the past 3 months, Catherine has taken more time off than usual to assist her father who has cancer and required surgery and follow-up treatment, including periods of daily visits to the hospital for treatment. She tried to minimise this by asking other family members to assist and also by scheduling her father's appointments early in the morning so she could get to her office no later than 10 a.m. She discussed the need for this flexibility with her work team and her immediate manager and all agreed that it was workable and that she had their full support. As the weeks passed Catherine noticed that the team was scheduling meetings as early as 8 a.m. or 9 a.m., that is, before she arrived, and that decisions were being taken with her immediate manager rather than with her. Also, one of her team members, Peter, was assuming responsibility for many of the tasks associated with her position; he explained that he was doing this as a way of making it less stressful for her during this difficult time with her father's illness.

Catherine sought a meeting with her manager to discuss what was happening and was told that while the company valued her length of service and contribution, perhaps it was time for her to think about reducing her workload and moving away from a leadership role given the additional family responsibilities she now had and the time that these were clearly taking from her paid work. He suggested she have a chat to the HR manager about her options for transitioning to retirement and reassured her that there would be part-time work available in various sections of the company should she want to continue working.

3.11 Questions

1. What are the underlying messages being conveyed by the manager to Catherine?
2. What are the underlying messages being conveyed to Catherine's work team?
3. Provide three points to support the argument that the manager is simply showing concern for Catherine's situation; provide three points to support the argument that the manager is displaying discriminatory behaviour.
4. What options does Catherine have to address her situation in the workplace? Which would you take and why?

References

Abrams, D., Sophieke Russell, P., Vauclair, C.-M., & Swift, H. (2011). *Ageism in Europe: Findings from the European social survey*. London, England: European Research Group on Attitudes to Age (EURAGE).

Adair, T., & Temple, J. (2012). *Barriers to mature age labour force engagement in Australia: Report on the 2011–12 national survey on the barriers to employment for mature age people*. Canberra: National Seniors Productive Ageing Centre.

Age Discrimination Act 2004. (Australia) https://www.legislation.gov.au/Details/C2016C00746.

Anti-Discrimination Board of NSW (2015). Annual Report 2014-2015. Retrieved from www.antidiscrimination.justice.nsw.gov.au.

Arino-Blasco, S., Tadd, W., & Boix-Ferrer, J. (2005). Dignity and older people: The voice of professionals. *Quality in Ageing and Older Adults, 6*(1), 30–36.

Australian Human Rights Commission (AHRC). (2013). *Fact or fiction? Stereotypes of older Australians*. Research Report. Sydney: AHRC.

Australian Human Rights Commission (AHRC). (2015a). Issues paper: Employment discrimination against older Australians. In *Willing to work: National inquiry into employment discrimination against older Australians and Australians with disability*. Sydney: AHRC.

Australian Human Rights Commission (AHRC). (2015b). *National prevalence survey of age discrimination in the workplace: The prevalence, nature and impact of workplace age discrimination amongst the Australian population aged 50 years and older*. Sydney: AHRC.

Australian Human Rights Commission (AHRC). (2015c). Annual report 2014–2015. Sydney: AHRC.

Australian Human Rights Commission [AHRC] (2016). National Inquiry into Employment Discrimination Against Older Australians and Australians with Disability 2016. Sydney, AHRC.

Australian Human Rights Commission. Good practice good business factsheets. https://www.humanrights.gov.au/sites/default/files/GPGB_age_discrimination.pdf.

Australian Treasury. (2010). *Intergenerational report, Australia to 2050: Future challenges*. Canberra: Australian Government.

Australian Treasury. (2015). *2015 intergenerational report, Australia in 2055*. Canberra: Australia Government.

Butler, R. N. (1969). Age-ism: Another form of bigotry. *Gerontologist, 9*(4), 243–246.
Butler, R. N. (2005). Ageism: Looking back over my shoulder. *Generations, 29*(3), 84–86.
Buyens, D., Van Dijk, H., Dewilde, T., & De Vos, A. (2009). The aging workforce: Perceptions of career ending. *Journal of Managerial Psychology, 24*(2), 102–117.
Calasanti, T. (2005). Ageism, gravity, and gender: experiences of aging bodies. *Generations, 29*(3), 8–12.
Courtney, M., Tong, S., & Walsh, A. (2000). Acute-care nurses' attitudes towards older patients: A literature review. *International Journal of Nursing Practice, 6*(2), 62–69.
Deasey, D., Kable, A., & Jeong, S. (2014). Influence of nurses' knowledge of ageing and attitudes towards older people on therapeutic interactions in emergency care: A literature review. *Australasian Journal on Ageing, 33*(4), 229–236.
Filinson, R. (2008). Age discrimination legislation in the U.K.: A comparative and gerontological analysis. *Journal of Cross-Cultural Gerontology, 23*(3), 225–237.
Gething, L. (1994). Health professional attitudes towards ageing and older people. Preliminary report of the reactions to ageing questionnaire. *Australasian Journal on Ageing, 13*(2), 77–81.
Gething, L., Fethney, L., McKee, K., Goff, M., Churchward, M., & Matthews, S. (2002). Knowledge, stereotyping and attitudes towards self-ageing. *Australasian Journal on Ageing, 21*(2), 74–79.
Gething, L., Gridley, H., Browning, C., Helmes, E., Luszcz, M., Turner, J., et al. (2003). The role of psychologists in fostering the wellbeing of older Australians. *Australian Psychologist, 38*(1), 1–10.
Giles, L. C., Patson, J. E., Butler, S. G., & Stewart, J. J. (2002). Ageism among health professionals: A comparison of clinical educators and students in physical and occupational therapy. *Physical & Occupational Therapy In Geriatrics, 21*(2), 15–27.
Gong, C., & Kendig, H. (2016). Ageing and social change in Australia. In H. Kendig, P. McDonald, & J. Piggott (Eds.), *Population ageing and Australia's future*. Canberra: ANU Press.
Gringart, E., & Helmes, E. (2001). Age discrimination in hiring practices against older adults in Western Australia: The case of accounting assistants. *Australasian Journal on Ageing, 20*(1), 23–28.
Gringart, E., Helmes, E., & Speelman, C. (2005). Development of a measure of stereotypical attitudes towards older workers. *Australian Psychologist, 48*(2), 110–118.
Hamilton, M., & Hamilton, C. (2006). *Rich boomer, poor boomer. Retirement prospects for the not-so-lucky generation*. Canberra: The Australia Institute.
Healy, J. (2004, March). *The benefits of an ageing population* (Discussion Paper No. 63). Canberra: The Australia Institute.
Helmes, E., & Gee, S. (2003). Australian therapists' attitudes toward older clients: Educational and training imperatives. *Educational Gerontology, 29*(8), 657–670.
Hobbs, C., Dean, C. M., Higgs, J., & Adamson, B. (2006). Physiotherapy students' attitudes towards and knowledge of older people. *The Australian Journal of Physiotherapy, 52*(2), 115–119.
Humpel, N., O'Loughlin, K., Wells, Y., & Kendig, H. (2010). The health of Australian baby boomers. *Australasian Journal on Ageing, 29*(1), 8–13.
Iversen, T. N., Larsen, L., & Solem, P. E. (2009). A conceptual analysis of ageism. *Nordic Psychology, 61*(3), 4–22.
Iweins, C., Desmette, D., Yzerbyt, V., & Stinglhamber, F. (2013). Ageism at work: The impact of intergenerational contact and organizational multi-age perspective. *European Journal of Work and Organizational Psychology, 22*(3), 331–346.
Kane, R. I., & Kane, R. A. (2005). Ageism in healthcare and long-term care. *Generations, 29*(3), 49–54.
Kearney, N., Miller, M., Paul, J., & Smith, K. (2000). Oncology healthcare professionals' attitudes towards elderly people. *Annals of Oncology, 11*(5), 599–601.
Kendig, H., O'Loughlin, K., Hussain, R., Heese, K., & Cannon, L. (2015). Attitudes to intergenerational equity: Baseline findings from the Attitudes to Ageing in Australia (AAA) study.

CEPAR Working Paper 2015/33. Sydney, Australia. ARC Centre of Excellence in Population Ageing Research.

Kite, M., Stockdale, G., Whitley, B., & Johnson, B. (2005). Attitudes toward younger and older adults: An updated meta-analytic review. *Journal of Social Issues, 61*(2), 241–266.

Koder, D. A., & Helmes, E. (2006). Clinical psychologists in aged care in Australia: A question of attitude or training? *Australian Psychologist, 41*(3), 179–185.

Le Couteur, D., Bansal, A., & Price, D. (1997). The attitudes of medical students towards careers in geriatric medicine. *Australasian Journal on Ageing, 16*(4), 225–228.

Leung, S., LoGiudice, D., Schwarz, J., & Brand, C. (2009). Hospital doctors attitudes towards older people. *Internal Medicine Journal, 41*(4), 308–314.

Levy, B. (2003). Mind matters: Cognitive and physical effects of aging self-stereotypes. *Journal of Gerontology: Psychological Sciences, 58*(4), 203–211.

Loretto, W., & White, P. (2006). Employers' attitudes, practices and policies towards older workers. *Human Resource Management Journal, 16*(3), 313–330.

Lui, N. L., & Wong, C. H. (2009). Junior doctors' attitudes towards older adults and its correlates in a tertiary-care public hospital. *Annals of the Academy of Medicine, 38*(2), 125–129.

Macnicol, J. (2006). *Age discrimination: An historical and contemporary analysis*. Cambridge: Cambridge University Press.

Marcus, J., & Fritzsche, B. (2015). One size doesn't fit all: Toward a theory on the intersectional salience of ageism at work. *Organizational Psychology Review, 5*(2), 168–188.

Markus, H. R., & Kitayama, S. (1991). Culture and the self—Implications for cognition, emotion, and motivation. *Psychological Review, 98*(2), 224–253.

Markus, H. R., & Kitayama, S. (2003). Culture, self, and the reality of the social. *Psychological Inquiry, 14*(3–4), 277–283.

Maurer, T. J., Barbeite, F. G., Weiss, E. M., & Lippstreu, M. (2008). New measures of stereotypical beliefs about older workers' ability and desire for development: Exploration among employees age 40 and over. *Journal of Managerial Psychology, 23*(4), 395–418.

Minichiello, V., Browne, J., & Kendig, H. (2000). Perceptions and consequences of ageism: Views of older people. *Ageing and Society, 20*(3), 253–278.

Morris, M., & Minichiello, V. (1992). Why choose to work in geriatrics? Factors which affect physiotherapists' decisions to work with older people. *Australian Journal of Physiotherapy, 38*(1), 21–27.

National Seniors Australia & COTA. (2012a). *Ageism in travel insurance: 2012 survey report*. Retrieved from http://www.nationalseniors.com.au.

National Seniors Australia & COTA. (2012b). *Ageism in travel insurance: Older people's stories, July 2012*. Retrieved from http://www.nationalseniors.com.au.

Nelson, T. D. (2005). Ageism: Prejudice against our feared future self. *Journal of Social Issues, 61*(2), 207–221.

Noone, J., O'Loughlin, K., & Kendig, H. (2013). Australian baby boomers retiring early: Understanding the benefits of retirement preparation for involuntary and voluntary retirees. *Journal of Aging Studies, 27*(3), 207–217.

O'Loughlin, K., Humpel, N., & Kendig, H. (2010). Impact of the global financial crisis on employed Australian baby boomers: A national survey. *Australasian Journal on Ageing, 29*(2), 88–91.

Palmore, E. B. (1977). Facts on aging: A short quiz. *The Gerontologist, 17*(4), 315–320.

Palmore, E. B. (1988). *The facts on aging quiz: A review of findings*. New York: Springer.

Patrickson, M., & Ranzijn, R. (2005). Older workers: The challenges for 21st century management. *International Journal of Organisational Behaviour, 10*(4), 729–739.

Posthuma, R., & Campion, M. (2009). Age stereotypes in the workplace: Common stereotypes, moderators, and future research directions. *Journal of Management, 35*(1), 158–188.

Posthuma, R., & Guerrero, L. (2013). Chapter 14: Age stereotypes in the workplace: Multidimensionality, cross-cultural applications, and directions for future research. In J. Field, R. J. Burke, & C. Cooper (Eds.), *The SAGE handbook of aging, work and society*. Thousand Oaks: Sage.

Ranzijn, R., Carson, E., & Winefield, A. (2004). Barriers to mature aged re-employment: Perceptions about desirable work-related attributes held by job seekers and employers. *International Journal of Organisational Behaviour, 8*(7), 559–570.

Rosencranz, H. A., & McNevin, T. E. (1969). A factor analysis of attitudes towards the aged. *The Gerontologist, 9*(1), 55–59.

Ryan, The Hon Susan, Age Discrimination Commissioner (2012). Age discrimination: a waste of human resources. Address to Australian Human Resources Institute, HR Directors Networking Forum, 6 March 2012. https://www.humanrights.gov.au/news/speeches/age-discrimination-waste-human-resources-2012.

Shah, P., & Kleiner, B. (2005). New developments concerning age discrimination in the workplace. *Equal Opportunities International, 24*(5/6), 15–23.

Stewart, L., Giles, L., Paterson, J., & Butler, S. (2005). Knowledge and attitudes towards older People: New Zealand students entering health professional degrees. *Physical & Occupational Therapy In Geriatrics, 23*(4), 25–36.

Syed, J. (2006). Older workers in Australia: A policy perspective. *Journal of Economic and Social Policy, 11*(1), 21–43.

Tadd, W., Hillman, A., Calnan, S., Calnan, M., Bayer, T., & Read, S. (2011). Right place—Wrong person: Dignity in the acute care of older people. *Quality in Ageing and Older Adults, 12*(1), 33–43.

Tam, K. L., Chandran, K., Yu, S., Nair, S., & Visvanathan, R. (2014). Geriatric medicine course to senior undergraduate medical students improves attitude and self-perceived competency scores. *Australasian Journal on Ageing, 33*(4), E6–E11.

Tavener, M., Byles, J., & Loxton, D. (2014). Expert perceptions of the popular baby boomer image. *Australasian Journal on Ageing, 33*(4), E31–E35.

Taylor, P., McLoughlin, C., Meyer, D., & Brooke, E. (2013). Everyday discrimination in the workplace, job satisfaction and psychological wellbeing: Age differences and moderating variables. *Ageing and Society, 33*(7), 1105–1138.

Twigg, J. (2008). Clothing, aging and me—Routes to research. *Journal of Aging Studies, 22*(2), 158–162.

Vaughan, G. M., & Hogg, M. A. (2013). *Social psychology* (7th ed.). French's Forest: Pearson.

Wells, Y., Foreman, P., Gething, L., & Petralia, W. (2004). Nurses' attitudes toward aging and older adults—Examining attitudes and practices among health services providers in Australia. *Journal of Gerontological Nursing, 30*(9), 5–13.

Woolhead, G., Calnan, M., Dieppe, P., & Tadd, W. (2004). Dignity in older age: What do older people in the United Kingdom think? *Age and Ageing, 33*(2), 165–170.

Chapter 4
Population Ageing: A Demographic Perspective

Peter McDonald

4.1 Introduction

Population ageing is the outcome of the history of a country's fertility, mortality and migration rates, but especially its fertility trend. Australia's population profile has been ageing continuously for the past 150 years, but the rate at which this will occur will be greater in the period 2010–2020 because, in these years, the baby boom generation born in the late 1940s and early 1950s will pass age 65. The extent of future ageing in Australia will be contingent on future fertility levels but also notably upon the level of international migration. A substantial increase in migration from 2005 onwards has greatly reduced the extent of future ageing in Australia and contributed positively to future growth of Gross Domestic Product (GDP) per capita. Comparatively, Australia is in a more favourable position than most other developed countries, and some former developing countries, such as China and Korea, which are also ageing more rapidly than Australia. This chapter provides an overview and explanation of population ageing, including the key elements that contribute to demographic change, the trends in population ageing evident in developed countries such as Australia and elsewhere, and the challenges and opportunities associated with population ageing as a policy issue for governments.

P. McDonald (✉)
Crawford School of Public Policy, Australian National University,
Canberra, ACT 0200, Australia
e-mail: peter.mcdonald@anu.edu.au

K. O'Loughlin et al. (eds.), *Ageing in Australia*, International
Perspectives on Aging 16, DOI 10.1007/978-1-4939-6466-6_4

4.2 The Demography of Ageing

Whether a population's profile is 'young' or 'old' depends on its history, over a period of about 100 years, of the three elements of demographic change: births, deaths and migration. Of the three elements, it is the birth rate (fertility) that has the largest impact. Changes in the death rate (mortality) and the level of migration can also affect the age distribution of a population but to a much lesser extent than changes in the birth rate.

The impact of changes in the birth rate and net migration rate on the size and age structure of a population over a period of 100 years are shown in Table 4.1. Four of the scenarios included in this table are illustrated graphically in the age pyramids in Fig. 4.1.

Scenario 1 shows the age distribution of a population of 10 million people that has had zero migration and a constant fertility rate of 2.08 births per woman over a long period of time. The rate of 2.08 births per woman is the fertility rate that leads to exact replacement of the existing population when the level of migration is zero and when the level of mortality is equivalent to recent mortality in Australia. This results in a zero rate of population growth and a population size and age distribution which remains constant across time. In demography, such a population is referred to as a *stationary population* because nothing ever changes: the number of individuals at each age remains constant, and the number of deaths each year remains constant and equal to the number of births.

Scenarios 2 and 3 illustrate what happens to the size and age distribution of the abovementioned population after 100 years if the fertility rate increases to 6.0 or decreases to 1.3 from Year 0, respectively, whilst net migration remains at zero.

Under Scenario 2, the age structure changes dramatically from the flat-sided distribution in Scenario 1 to a 'pagoda' shape (refer Fig. 4.1). Under Scenario 1, the proportions of the population in the youngest and oldest age brackets are roughly matched at around 20 % but, under Scenario 2, 45 % of individuals are aged less than 15 and only 4 % aged 65 and over (refer Table 4.1). This is the power of 100

Table 4.1 Changes in population size and age structure after 100 years under varying assumptions of fertility and migration

Scenario	Total fertility rate	Annual net migration	Size of population Year 0 (millions)	Size of population Year 100 (millions)	Percentage of total population in age range Year 100		
					<15	15–64	65+
1	2.08	0	10	10	18.5	60.1	21.4
2	6.0	0	10	211.7	44.8	51.0	4.3
3	1.3	0	10	3.5	10.3	54.7	34.9
4	2.08	100,000	10	24.2	19.1	61.7	19.2
5	6.0	100,000	10	305.0	44.5	51.1	4.4
6	1.3	100,000	10	13.4	12.1	61.7	26.2

Source: Author's calculations

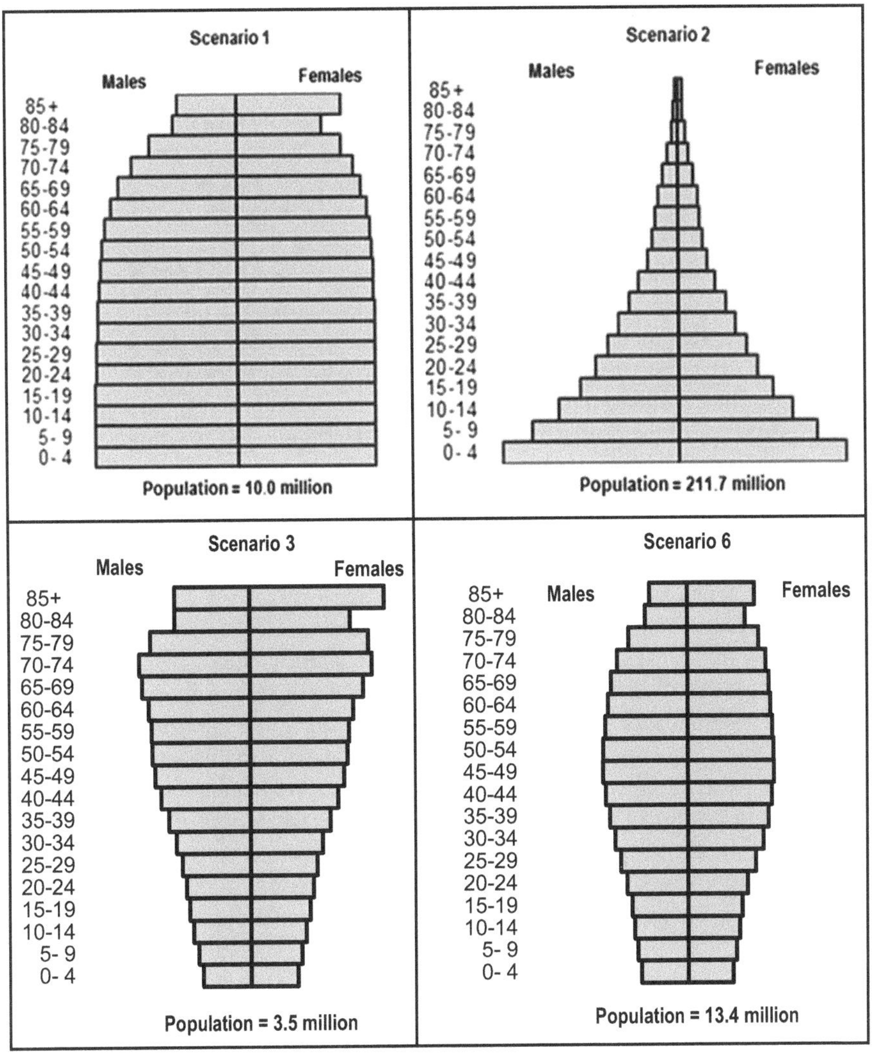

Fig. 4.1 Scenarios showing the impact of changing levels of fertility and migration on population size and age structure after 100 years. *Scenario 1* Age structure of a 'stationary' population in Year 100. Assumes Year 0 population of 10 million, zero migration and birth rate of 2.08. *Scenario 2* Age structure in Year 100 assuming Year 0 population size and structure as per Scenario 1, zero migration and fertility rate of 6.0. *Scenario 3* Age structure in Year 100 assuming Year 0 population size and structure as per Scenario 1, zero migration and fertility rate of 1.3. *Scenario 6* Age structure in Year 100 assuming Year 0 population size and structure as per Scenario 1, fertility rate of 1.3 with 100,000 migrants per annum. *Source*: Author's calculations

years of higher fertility. Note also that, after 100 years, the population which has the higher fertility rate has expanded dramatically, from 10 million to 212 million, while the Scenario 1 population has remained at 10 million.

Under Scenario 3, where fertility falls to a low level, after 100 years the population falls in size to just 3.5 million, and the age distribution is an inverted pyramid with many people at older ages and few at younger ages. Again, this demonstrates the substantial impact that changes in fertility have upon age distributions.

Scenarios 4, 5 and 6 in Table 4.1 repeat the projections shown in Scenarios 1, 2 and 3 respectively, but now include annual net migration of 100,000. Does migration affect the age structure of the population? The answer is a qualified 'yes'. The qualification is that the impact of migration upon age structure is dependent on the level of fertility. Comparing Scenarios 1 and 4, in which fertility is set at 2.08 births per woman, the addition of 100,000 migrants per annum has very little impact on the age structure of the population even though the population size more than doubles. Comparing Scenarios 2 and 5, in which fertility is at the higher rate of 6.0 births per woman, the addition of 100,000 migrants per annum similarly has virtually no impact on the age structure of the population while adding almost 100 million people to the population over 100 years.

However, when the fertility rate is low, as in Scenarios 3, and 6, migration has a notable impact on both the size of the population and the age structure. The populations shown in Scenarios 3 and 6 have the same starting age distribution and the same (low) level of fertility for 100 years, but the population in Scenario 6 has annual net migration of 100,000 compared to the zero migration of Scenario 3.[1] After 100 years, the population in Scenario 3 declines markedly and ends up with 35 % of its population aged 65 years and over, while the population in Scenario 6 increases and has 26 % aged 65 years and over, a very sizeable difference. It also has a considerably higher percentage of its population in the working ages (15–64 years) than the population in Scenario 3. Thus, a country facing the prospect of rapid ageing and population decline, as in Scenario 3, could overcome this problem through a migration programme. However, comparing the population size outcomes of Scenarios 3 and 6 in Table 4.1, it is evident that after 100 years migrants, and their descendants, would numerically dominate the population.

While these scenarios are models, they are not purely theoretical. The strategy of using migration to compensate for sustained low fertility (refer Scenario 6) is being followed at present by Singapore (National Population and Talent Division 2013). Because Singapore's population is relatively small, the number of migrants required to compensate for the low birth rate is also relatively small. The situation would be very different if Japan, for example, opted to follow the same strategy as Singapore

[1] Of course, the age distribution of the migrants also makes a difference. In these projections, the age distribution of the migrants is assumed to be the same as the current age distribution of migrants to Australia. Most of these migrants are young, being concentrated in the age range of 15–34 years. There is a 'double' effect of young migrants on the future age distribution in that young migrants have births after they arrive. In these scenarios, migrants are assumed to have the same fertility rate as the non-migrants.

because the number of migrants required to compensate for Japan's low fertility would be extremely large. As a consequence, official projections of the future population of Japan (National Institute of Population and Social Security Research 2014) show an outcome rather like that of Scenario 3 while official projections for Singapore show an outcome more like that of Scenario 6.

4.3 The Emergence of Population Ageing as an Australian Policy Issue

In Australia, fertility and mortality rates have been falling for 150 years. In 1870, 42 % of the population was aged less than 15 years and only 2 % was aged 65 years and over. However, by 2013, those aged less than 15 years had fallen to 19 % whilst those aged 65 years and over had grown to 14 %.

While population ageing has a long history, it has come into policy prominence only since the 1980s. In the Borrie Report (National Population Inquiry 1975), the most comprehensive report on Australia's population ever undertaken, population ageing received only passing reference and no mention at all in relation to policy. Indicative population projections from 1970 to 2070 made in the Borrie Report forecast that the proportion of the population aged 65 years and over would peak at between 9 % and 15 %, with this peak being reached by 2030 (National Population Inquiry 1975: 1: 294). In only 2 of the 6 indicative projections did the proportion aged 65 years and over peak at 12 % or more, a level that was in fact reached in 1996.

After the Borrie report was published, both fertility and mortality rates in Australia fell much more rapidly than had been assumed in the projections for the report. The fertility rate in Australia in 1972, the last year for which the authors of the report had data, was 2.74 births per woman. By 1980, the rate had fallen to 1.89 births per woman and it has remained around that level ever since. Also from around 1972 onwards, death rates at older ages in Australia fell rapidly after having been almost constant in the previous 50 years. For example, in 1920–1922 the expectation of life for Australian men at age 65 was 12.0 years. This was little changed at 12.2 years in 1970–1972. However, by 2010–2012, it had increased significantly to 19.1 years. Thus, while it is often said that population ageing today is driven by the baby boom generation reaching the older ages, this is only the case because of the massive reductions in both fertility and mortality that occurred from the early 1970s onwards.

Since the late 1990s in Australia, the *Charter of Budget Honesty Act 1998* (Commonwealth of Australia 1998) has required the government of the day to produce an Intergenerational Report (IGR), at least once every 5 years, which addresses the financial implications of demographic change (see Chaps. 8 and 15). So far, four such reports have been published (Commonwealth of Australia 2002, 2007, 2010, 2015). The reports project future population and age structure over a 40 year period. In the making of population projections, assumptions about future levels of fertility, mortality and migration tend to be influenced very heavily by the most recent trends. When the projections were made for the 2003 IGR, fertility had hit a low point of

Table 4.2 Long-term demographic assumptions in the 2003 and 2015 intergenerational reports (IGR) and the 2003 and 2013 projections of the Australian Bureau of Statistics (ABS)

	Total fertility rate (births per woman)	Annual net migration (000's)	Expectation of life at birth in 2051 (years)	
			Males	Females
IGR 2003	1.6	90	83.2	88.2
IGR 2015	1.9	215	87.5	90.1
ABS 2003	1.6	100	84.2	87.7
ABS 2013	1.8	240	84.2	87.7

1.73 and appeared to be still heading downwards and migration had been constant at around 90,000 net per annum. Soon afterwards, however, the trend in the fertility rate turned upwards and migration increased very substantially even reaching 300,000 in 2008. Official agencies making population projections also have a tendency to be conservative about future falls in mortality rates. Accordingly, they tend to underestimate future expectations of life. By the time of the 2015 IGR, forecasters had adjusted to the new demographic realities and moved their assumptions of future levels of fertility, expectation of life and migration upwards. The changes in the projection assumptions between the 2003 and 2015 Intergenerational Reports are shown in Table 4.2. Also shown in the table are the assumptions made by the Australian Bureau of Statistics (ABS) for the official projections made around the same time as the 2003 and 2015 IGR reports were produced.

These changes of assumption have had a considerable impact on the projected future populations in different age groups as illustrated in Fig. 4.2. This figure shows the percentage changes in the projected numbers in selected age groups in 2051 that result from moving from the 2003 ABS assumptions to the 2013 ABS assumptions. By simply changing the assumptions, the population of children (0–9 years) in 2051 is 80 % larger in the 2013 ABS projection than it is in the 2003 projection. At the other end of the age distribution, the population aged 80 and over is only 3 % larger in 2051 according to the 2013 ABS projection compared with the 2003 projection. The 2013 projection is a lot younger than the 2003 projection for the year 2051 because of the higher level of fertility that was assumed in the 2013 projection but, to a much larger extent, because of the higher level of migration that was assumed in the 2013 projection. This is because the migrant cohort will be much younger than the general Australian population when they arrive, and these individuals will go on to have children and grandchildren[2] before they reach the very old age group.

[2] Note the projections assume that, once in Australia, migrants have the same fertility and mortality rates as the general Australian population.

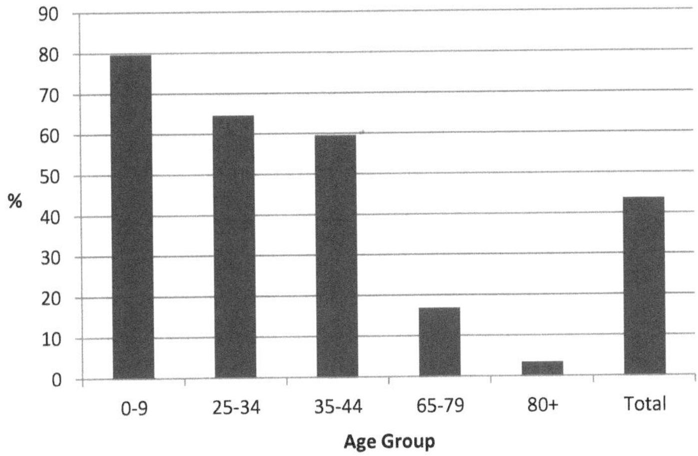

Fig. 4.2 Percentages by which the 2051 population projected by ABS in 2013 exceeds the 2051 population projected by ABS in 2003. *Source*: Authors calculations from Australian Bureau of Statistics. (2003). *Population Projections Australia, 2002–2101*. Canberra: ABS; Australian Bureau of Statistics. (2013). *Population Projections Australia, 2012–2101*. Canberra. ABS

This example shows clearly how the extent of future ageing of the Australian population can be modified by future demography.

While as just described, projections of fertility and mortality can change in unexpected ways, the fertility rate in 2013 (1.95 births per woman) was in fact the same as it was in 1978, and between these years the (downward) movements have been relatively small. It can also be presumed that expectation of life in Australia will continue to improve into the future. Given these assumptions, alternative population futures for Australia will be contingent primarily upon future levels of migration.

4.4 Two Alternative Population Futures for Australia

Two alternative population projections for Australia for the year 2053 are shown in Figs. 4.3 and 4.4. Both projections assume the fertility rate remains constant at an average of 1.9 births per woman and the expected lifespan for men rises from 80.3 years in 2013 to 87.6 years in 2053 and for women from 84.3 years in 2013 to 90.4 years in 2053.

The difference between the two projections lies in the assumed levels of annual net migration in the future. In Fig. 4.3, annual net migration is assumed to fall from 240,000 in 2013 to zero in 2018 and then remains at zero from 2018 to 2053. In Fig. 4.4, annual net migration remains constant at 240,000 from 2013 to 2053. The age distribution of migrants is assumed to be the same as it was in Australia in the year 2011–2012.

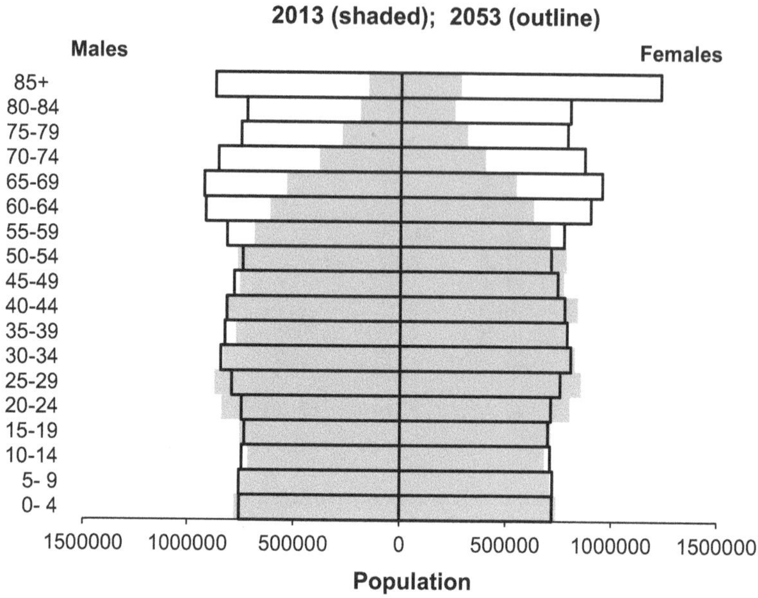

Fig. 4.3 Australian age distributions, 2013 and 2053: zero annual migration assumption. *Source*: Author's calculations

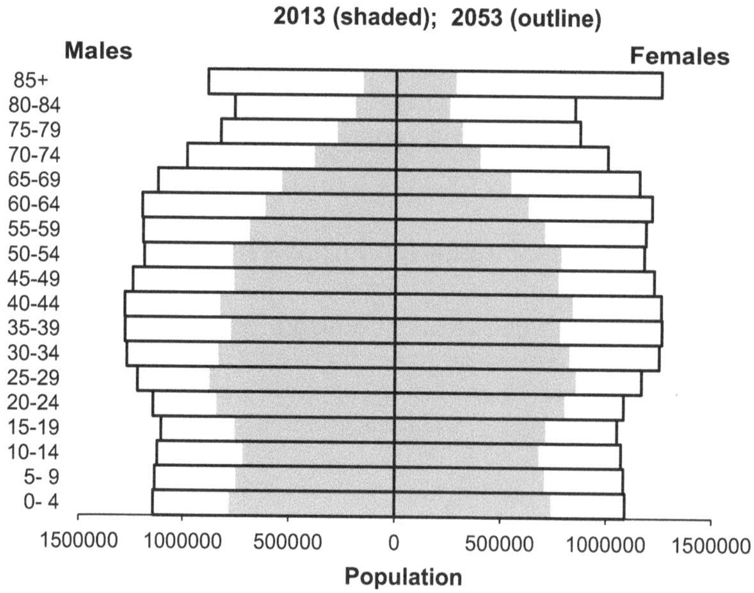

Fig. 4.4 Australian age distributions, 2013 and 2053: 240,000 annual migration assumption. *Source*: Author's calculations

In Fig. 4.3, the zero migration projection, the total population increases by 5.7 million, to 28.8 million between 2013 and 2053, but all of the increase is at ages 60 years and over. Consequently, the population is considerably older by 2053, with the percentage of population aged 65 years and over increasing from 14.4 % to 30.4 %. With no increase in the population under age 60, there would be no increase in the labour supply in the prime working ages of 25–59 years. While still supporting the same number of children as in 2013, the working age population would be faced with supporting twice as many older persons in 2053 than in 2013.

In Fig. 4.4, migration is assumed to remain constant at the 2013 level of 240,000 per annum. The population gets much larger by 2053, exceeding 40 million, but is younger with only 24.1 % being aged 65 years and over. Thus, large-scale immigration has a meaningful impact on the extent of ageing in Australia's population but results in very large increases in the total population.

A comparison of the results of the two projections shows that there is a trade-off between the size of Australia's future population and its age distribution. To what extent is it worthwhile to use migration as a means of changing the age distribution? This question is addressed in the next section using an economic criterion.

4.5 Population Ageing and Immigration Policy

Gross domestic product per capita (GDP per capita) is an outcome of the age distribution of a population interacting with age-specific labour force participation rates and output per worker (productivity). This relationship can be used to assess the long-term impact of different levels of migration upon the growth rate of GDP per capita. The more the population is concentrated in the working ages, the higher will be the rate of growth of GDP per capita, all else being equal. GDP per capita is a frequently used index of living standard: it is an economic criterion that can be used to assess the relative merits of different levels of net overseas migration.

Of course, the merits of migration can also be assessed based on a range of other criteria. For example, on the positive side, increased population through migration may increase the supply of labour; reduce wage inflation; provide economies of scale; allow higher levels of specialisation; and promote mobility, productivity and innovation. On the negative side, increased population through migration could also create price inflation due to relative shortages of commodities or housing, diseconomies of scale related to congestion or crowding, and environmental damage as population expansion competes with biodiversity.

Two recent studies have investigated the impact of immigration on GDP per capita in Australia (McDonald and Temple 2013; Migration Council Australia 2015). Figure 4.5 is taken from the 2013 study. It shows that population ageing in Australia will have a strong negative effect upon the rate of growth of GDP per capita in the decade 2013–2023. With zero net migration, the rate of growth of GDP per capita would fall from just above 1.8 % per annum to just over 1.3 % in this decade, simply due to the ageing of the population. This fall would be mitigated to

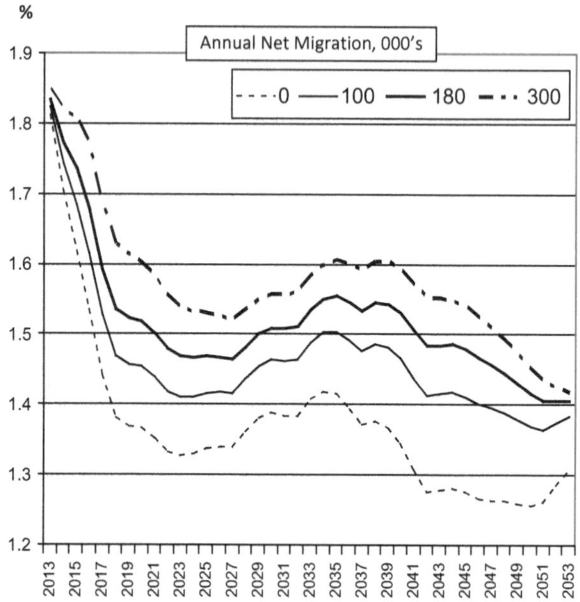

Fig. 4.5 The effects of varying levels of net overseas migration (NOM) upon the percentage growth rate of GDP per capita, Australia, 2013–2053. Assumes migrants have the same labour productivity growth as non-migrants (1.6 % per annum). *Source*: McDonald and Temple (2013)

a meaningful extent if net migration were to remain at around its 2013 level of 240,000. On this rationale, migration in part offsets the negative effect of ageing on the Australian economy in both the short term and the long term. In addition to meeting short-term labour shortages, Australian migration policy is influenced by this mitigating effect of migration on the age structure of the population.

4.6 Local and Regional Ageing

Much of the attention in policy for an ageing population is in the provision of aged care services to older people. In this regard, it matters where older people are located. There is considerable variation in the degree of population ageing across Australia.

The simple story is that younger people tend to move to the large cities. International migrants are also attracted to the large cities. This very long-term Australian trend means that the proportion of older people is generally much higher in country and coastal areas than it is in the cities.

This pattern of cities being younger is illustrated in Fig. 4.6 which shows that, with the exception of Western Australia and Northern Territory, the capital city is always younger than the rest of the state.

The Western Australian exception is due to the young population in the mining areas (refer Table 4.3). In addition, some remote areas of Australia do have quite young populations due to having a higher proportion of indigenous people in their

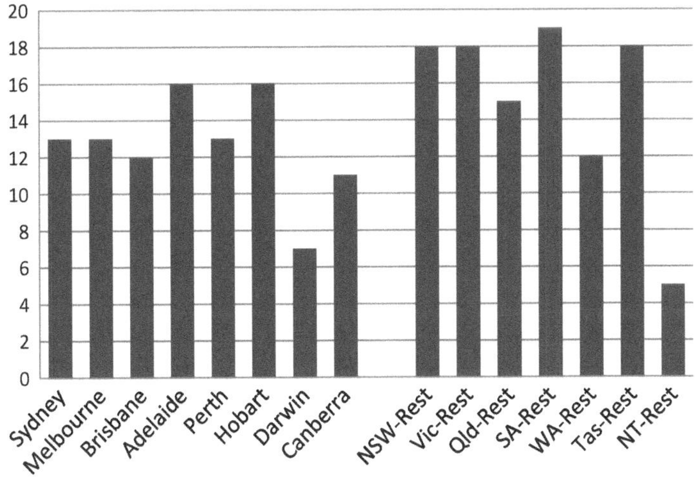

Fig. 4.6 Percentage of population aged 65 years and over by capital city and rest of state. *Source*: Australian Bureau of Statistics Census of Population and Housing 2011. Accessed using Tablebuilder

populations (see Chap. 7). Examples of this are Outback North and Far North of Queensland and East Arnhem in the Northern Territory. There are also some areas of Australia, particularly coastal areas, to which people tend to move when they retire, while young people from the same area move to the city. These coastal areas have the oldest population profiles in Australia because of this double effect. On the other hand, the youngest areas in Australia tend to be outer suburbs in the big cities populated by families with children. An interesting exception is Melbourne City, which has a young population due to the high proportion of students living in the centre of the city.

Table 4.3 shows the percentage of the population aged 65 years and over at the time of the 2011 Census for the youngest and oldest areas of each state and territory, based on the Australia Bureau of Statistics SA3 areas.[3] The youngest area in Australia, with a low 1.9 % of population aged 65 and over, is the Pilbara, the iron ore mining area of Western Australia, while the oldest area, with 29.6 % of individuals aged 65 and over, is Great Lakes, a coastal area in New South Wales north of Newcastle.

As expected, variations in the extent of ageing also occur within SA3 areas. For example, within the SA3 area of Dubbo, in western New South Wales, the proportion of the population aged 65 years and over ranges from 21.7 % in Coonabarabran to 11.2 % in East Dubbo. This high degree of geographic variation presents challenges to providers of aged care services.

[3]The ABS SA3s provide a standardised regional breakup of Australia: the aim is to create a standard framework for the analysis of ABS data at the regional level (ABS 2011b).

Table 4.3 Oldest and youngest areas of Australia showing (in brackets) the percentage of the population aged 65 and over at the 2011 census (ABS 2011a)

State or territory	Five oldest areas	Five youngest areas
NSW	Great Lakes (29.6), Port Macquarie (24.7), South Coast (23.7), Shoalhaven (23.3), Upper Murray (23.1)	Blacktown North (6.0), Rouse Hill (8.1), Mt Druitt (8.5), Bringelly-Green Valley (8.5), Auburn (8.6)
VIC	Gippsland East (23.3), Maryborough-Pyrenees (21.8), Gippsland South West (21.7), Moira (21.6), Mornington Peninsula (21.5)	Melbourne City (6.6), Wyndham (6.9), Melton-Bacchus Marsh (7.3), Casey South (7.4), Nillumbik-King Lake (9.4)
QLD	Bribie-Beachmere (29.3), Hervey Bay (22.7), Caloundra (21.3), Noosa (20.5), Bundaberg (19.7)	Springfield-Redbank (5.6), Central Highlands (6.1), Jimboomba (7.0), Outback North (7.1), Far North (7.2)
SA	Fleurieu-Kangaroo Island (26.4), Yorke Peninsula (25.9), Holdfast Bay (21.6), Burnside (20.3), Mid North (20.1)	Outback North and East (11.0), Adelaide City (11.2), Salisbury (12.0), Playford (12.0), Adelaide Hills (12.8)
WA	Mandurah (20.3), Manjimup (16.8), Albany (16.7), Cottesloe-Claremont (16.3), Wheat-belt North (16.0)	Pilbara (1.9), Kimberley (4.8), Goldfields (5.3), Wanneroo (8.8), Serpentine-Jarrahdale (9.0)
TAS	South East Coast (24.7), Burnie-Ulverstone (18.2), North East (18.0), Devonport (17.7), Meander Valley-West Tamar (17.7)	Brighton (9.4), West Coast (13.6), Central Highlands (13.8), Hobart North West (14.7), Hobart South and West (14.7)
NT	Darwin Suburbs (7.4), Darwin City (6.7), Litchfield (6.6), Barkly (6.0)	East Arnhem (2.1), Palmerston (3.9), Daly-Tiwi-East Arnhem (4.2), Katherine (5.3), Alice Springs (5.4)
ACT	Woden (17.9), Weston Creek (17.0), South Canberra (14.9)	Gungahlin (4.7), Tuggeranong (7.9), North Canberra (10.9), Belconnen (11.1)

Note: Area level used is the Australian Bureau of Statistics Statistical Area Level 3 (SA3) (ABS 2011b)

4.7 An International Perspective on Population Ageing

Differences in population ageing across countries are due to the past histories of fertility rates in each country. Figure 4.7 shows the percentages of population aged 65 years and over for a selection of countries in 2015 and in 2050, as estimated by the United Nations Population Division (2012).

The countries with the oldest population profiles, both in 2015 and 2050, are those that have a sustained history of very low fertility such as Japan, Italy and Germany. Other Southern European and German-speaking countries and wealthy East Asian countries have a similar pattern of ageing.

Among the developed countries, there is another group for which the extent of ageing is much more modest because their fertility rates have never been very low (i.e. below 1.5 births per woman on average). These include all the English-speaking countries, all the French and Dutch-speaking countries and all the Nordic countries.

China will inevitably experience rapid ageing of its population between 2015 and 2050 as a result of the speed of fertility decline in that country. From being

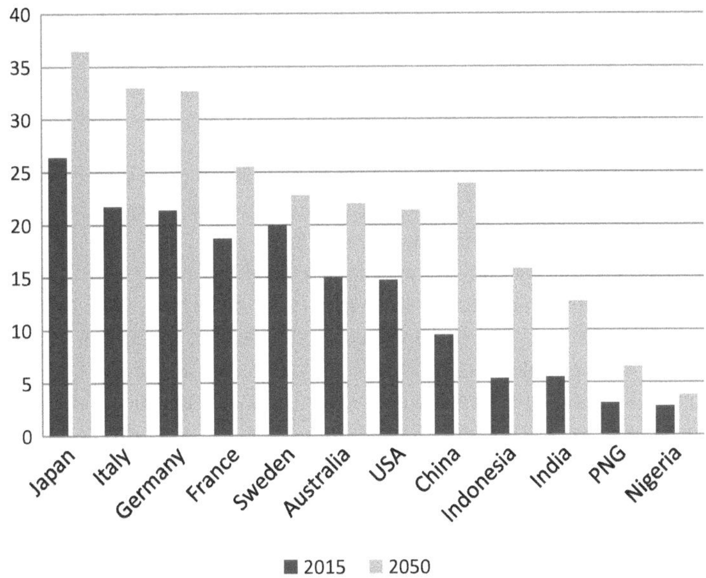

Fig. 4.7 Percentage of population aged 65 years and over in 2015 and 2050, by country. *Source*: United Nations Population Division. *World Population Prospects 2012 Revision*. http://esa.un.org/unpd/wpp/index.htm

much younger than Australia and the USA in 2015, China will become older than these two countries by 2050.

Because ageing has transpired relatively slowly in today's developed countries, these countries have been able to develop systems to address population ageing. However, countries like China and Indonesia, where population ageing will be rapid, face a substantial challenge to develop systems to accommodate their rapidly ageing populations in a much shorter time frame. On the other hand, some developing countries, like Nigeria and Papua-New Guinea, will remain very young even in 2050 because of the slowness of fertility decline.

4.8 Policy Responses

While many countries or localities may prefer that population ageing was not as severe, or was occurring more slowly, usually the demographic trends that contribute to the level or speed of ageing are beyond the control of policy makers. Virtually all countries with fertility rates below 1.5 births per woman have pursued policy approaches to raise the birth rate so that ageing is less severe and less rapid. However, only a few countries in East and Central Europe have been successful in this regard.

In Australia, the long-term trend of inter-generational relocation between city and country/coastal towns and the preference of migrants to settle in cities has also been an important factor in the variation of population ageing between regions. This has policy implications for the provision of aged care services. Towns in country Australia have tried to stem the outflow of young people with a range of policy measures such as the establishment of regional universities and colleges but, for the most part and especially for smaller towns, these attempts have been unsuccessful. The states of Tasmania and South Australia would prefer that their populations were not ageing quite so rapidly but, again, they have had little success to date in slowing the trend. Thus, in most situations, countries and localities must find ways to adjust to the reality of population ageing and its effects.

In a few cases, such as Australia and Singapore, where total population levels are comparatively low, international migration policies have been used relatively successfully to slow the rate of population ageing. However, this is not a solution for countries with large populations and has its own inherent advantages and disadvantages which need to be considered outside of the issue of an ageing population.

4.9 Conclusion

Population ageing is a consequence of changes in the demography of a population, principally falling fertility and mortality rates. It is of concern to many countries because of the potentially strong negative effect it is likely to have upon the rate of growth of GDP per capita, a frequently used index of living standard. While it is often said that population ageing in Australia is driven by the baby boomer generation reaching the older ages, this is only the case because of the substantial changes in fertility and mortality that occurred from the early 1970s onwards.

In comparative terms, the ageing challenge faced by Australia is less than that faced by most other countries. By the time the baby-boom generation reaches ages 80 and over where the expenses associated with ageing increase sharply, there will be a new generation entering the labour force that is larger than was previously expected because of the increase in the number of births between 2005 and 2015. This gives Australia the opportunity to plan for future ageing without the need for short-term crisis measures. For the longer term, Australia needs to encourage employment at older ages, ensure that its retirement incomes policy is fair and effective, and examine ways to control health costs and the financing of aged care. These issues are taken up and discussed in detail in the following chapters.

References

Australian Bureau of Statistics. (2003). *Population projections Australia, 2002–2101* (Cat. No. 3222.0). Canberra: ABS.

Australian Bureau of Statistics. (2011a). *Census of population and housing 2011* (Cat. No. 2011.0). Tablebuilder. Retrieved from http://www.abs.gov.au/ausstats/abs@.nsf/lookup/2011.0Chapter5402011.

Australian Bureau of Statistics. (2011b). *Australian statistical geography standard (ASGS): Volume 1—Main structure and greater capital city statistical areas, July 2011* (Cat. No.

1270.0.55.001). Retrieved from http://www.abs.gov.au/ausstats/abs@.nsf/Latestproducts/E7369D1FCE596315CA257801000C64E5.

Australian Bureau of Statistics. (2013). *Population projections Australia, 2012–2101* (Cat. No. 3222.0). Canberra: ABS.

Commonwealth of Australia. (1998). *Charter of budget honesty act 1998*, Commonwealth of Australia No. 22. Australia: Commonwealth of Australia.

Commonwealth of Australia. (2002). *Intergenerational report: 2002–03*. Canberra: Department of the Treasury. Retrieved from http://www.budget.gov.au/2002-03/bp5/html/01_bp5prelim.html.

Commonwealth of Australia. (2007). *Intergenerational report: 2007*. Canberra: Department of the Treasury. Retrieved from http://archive.treasury.gov.au/igr/igr2007/report/PDF/IGR_2007_final_report.pdf.

Commonwealth of Australia. (2010). *Australia to 2050: Future challenges*. Canberra: Department of the Treasury.

Commonwealth of Australia. (2015). *2015 intergenerational report: Australia in 2055*. Canberra: Department of the Treasury. Retrieved from http://www.treasury.gov.au/PublicationsAndMedia/Publications/2015/2015-Intergenerational-Report.

McDonald, P., and Temple, J. (2013). *The long term effects of ageing and immigration upon labour supply and per capita gross domestic product: Australia 2012–2062*. Canberra: Demographic Insight P/L.

Migration Council Australia. (2015). *The economic impact of migration*. Canberra: Migration Council of Australia.

National Institute of Population and Social Security Research. (2014). *Population and household projections*. Tokyo: National Institute of Population and Social Security Research. Retrieved August 21, 2014, from www.ipss.go.jp/pr-ad/e/eng/04.html.

National Population and Talent Division. (2013). *A sustainable population for a dynamic Singapore: Population white paper*. Singapore: Prime Minister's Office. Retrieved from population.sg/whitepaper/downloads/population-white-paper.pdf.

National Population Inquiry. (1975). *Population and Australia: A demographic analysis and projection* (Vol. 2). Canberra: AGPS.

United Nations Population Division. (2012). World population prospects 2012 revision. Retrieved from http://esa.un.org/unpd/wpp/index.htm.

Chapter 5
Health and Ageing

Julie Byles

5.1 Introduction

Chapter 4 outlined population ageing in Australia. When considering the impact of this demographic change, it is also important to consider the health of the population and how people can be supported to age well. As people age, there is an increased risk of many chronic conditions and disabilities. Consequently, as populations age, the prevalence of these conditions will increase, even if there are great improvements in age-specific disease rates. There is therefore an ever increasing need to support the health of people as they age, including preventive and restorative approaches. In this chapter, we consider the health of older people in Australia, as well as opportunities to optimise well-being.

5.2 Health in Older Age

Health in older age can be measured in various ways including mortality rates and life expectancy, disease and disability-free life expectancy, quality of life, and self-rated health.

J. Byles (✉)
Research Centre for Generational Health and Ageing, The University of Newcastle, Newcastle, NSW, Australia

Hunter Medical Research Institute, Newcastle, NSW, Australia
e-mail: julie.byles@newcastle.edu.au; https://www.newcastle.edu.au/research-and-innovation/centre/gha/

K. O'Loughlin et al. (eds.), *Ageing in Australia*, International Perspectives on Aging 16, DOI 10.1007/978-1-4939-6466-6_5

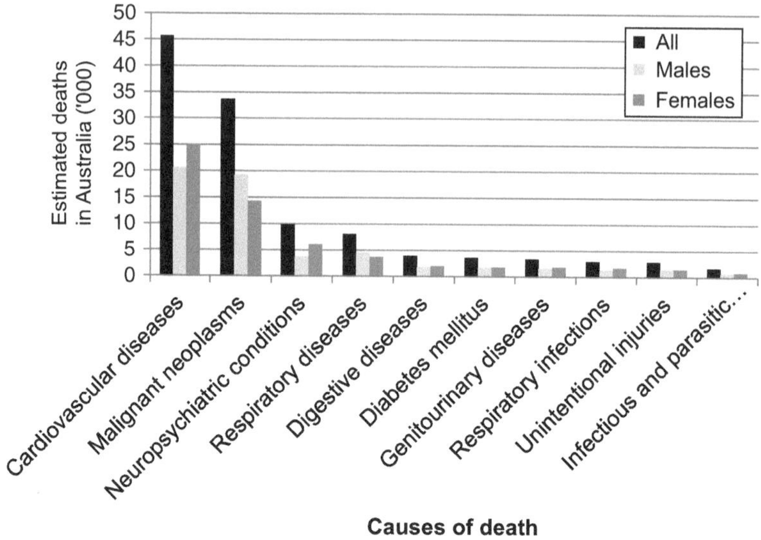

Fig. 5.1 Estimated total deaths of people aged 60 years and over in Australia, by cause and sex (top ten major disease groups), 2008. *Data source*: World Health Organization (2008)

5.2.1 Life Expectancy and Causes of Death at Older Ages

Currently, a baby girl born in Australia can expect to live to age 84.4 years, and a boy can be expected to live to age 80.3 years (Australian Bureau of Statistics 2016a). Australia has a high life expectancy relative to other countries and has seen rapid increase in life expectancy at birth over the last century. This increase is largely attributable to improvements in infant mortality rates and prevention of deaths at younger ages.

Life expectancy at age 60 years has not increased as rapidly as life expectancy at birth. In 2013, a 60 year old woman was expected to live another 27 years, and a man could expect to live another 24 years (World Health Organization 2015).

In Australia, around 90 % of deaths among people aged 60 years and over are due to non-communicable diseases. Figure 5.1 shows common causes of death for people in this age group, including cardiovascular diseases (mostly ischaemic heart disease and cerebrovascular disease), cancers, neuropsychiatric disorders (mostly Alzheimer's disease and other dementias), and chronic obstructive pulmonary disease. The majority of accidental deaths among people aged 60 years and over are due to falls.

5.2.2 Healthy Life Expectancy

Health-adjusted life expectancy (HALE) estimates the years that a person can be expected to live free of disease or disability (Mathers et al. 2004). In 2010, HALE in Australia was 68.4 years for men and 71.8 years for women (Salomon et al. 2012).

If healthy life expectancy increases relative to life expectancy, people will spend more of their life in good health with 'compression of morbidity' (Fries et al. 1989). If people live longer due to reduction in fatal illness, but with more disability due to chronic disease, there will be 'expansion of morbidity'. A 'dynamic equilibrium' could occur if the prevalence of some chronic diseases increases with population ageing, but progression of degenerative diseases reduces. Some data suggest Australia is beginning to see a compression of morbidity (Australian Institute of Health and Welfare 2012b). Other OECD countries have also reported compression of morbidity but some have reported an expansion of morbidity (Lafortune and Balestat 2007).

However, for people aged 65 years or over, the increase in life expectancy between 1998 and 2012 was around twice the increase in healthy life expectancy. The increase in life expectancy for men aged 65 years was 3 years, with an extra 1.6 years spent with no disability and 2.3 years without severe or profound disability. For women the corresponding increases were 2.2 years increase in life expectancy, 0.8 additional years with no disability, and 2.0 years without severe or profound disability (Australian Institute of Health and Welfare 2014b). This means that while increases in life expectancy at older ages were not disability free, they were mostly without severe or profound disability.

5.2.3 *Self-rated Health*

Self-rated health reflects overall well-being and is strongly associated with objective measures of illness and disability, and survival (McCallum et al. 1994; DeSalvo et al. 2006; Simons et al. 2011). The Sax Institute 45 and Up Study provides the largest sample estimate of self-rated health among older people in Australia (see Fig. 5.2). Ratings of health as good, very good, or excellent decrease with age and women are less likely to rate their health highly compared to men.

5.2.4 *Functioning, Disability, and Health*

As a person ages, their health and functional capacity depend on the inter-relationship between age-related changes to their body, changes occurring due to disease or injury, their own personal situation and resources, activity and participation in life and community, and the extent to which they are supported within their social and physical environments (Üstun et al. 2003).

Table 5.1 shows some normal physiological changes that occur with ageing. These normal changes limit bodily reserves, reduce the ability to maintain homeostasis, and increase the potential for illness.

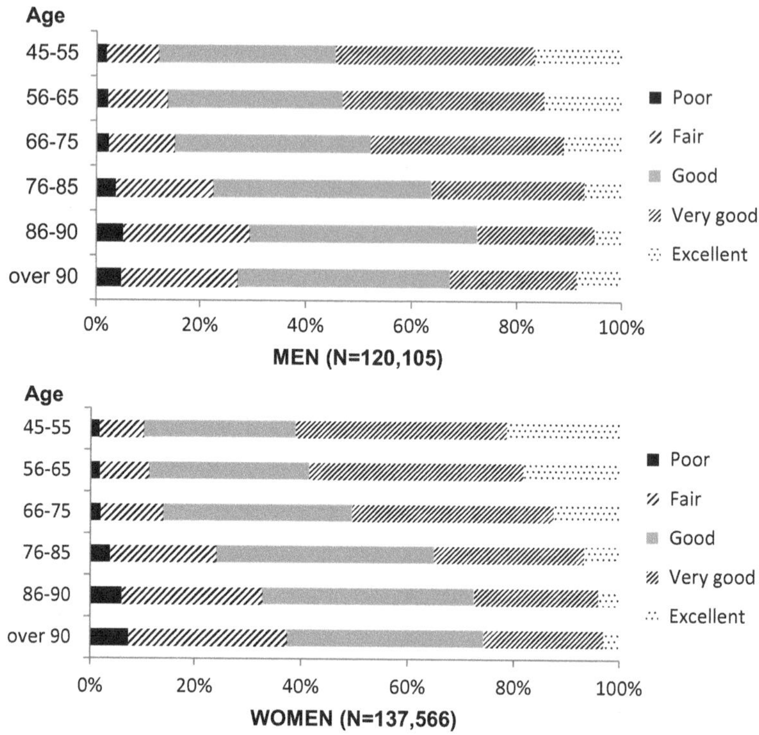

Fig. 5.2 Self-rated health, men and women aged 45 years and over. *Source*: Sax Institute 45 and Up Study, https://www.saxinstitute.org.au/our-work/45-up-study

5.2.5 Successful Ageing

A number of longitudinal studies have been conducted in an attempt to disentangle the effects of ageing and disease. These studies describe 'successful ageing' as

> avoidance of disease and disability, the maintenance of high physical and cognitive function, and sustained engagement in social and productive activities (Rowe and Kahn 1997, p. 439).

In the Australian Longitudinal Study of Ageing, successful ageing was associated with lower age at baseline, being male, and having more education and financial assets. Chronic conditions such as diabetes, arthritis, stroke and asthma, and hip fracture and low physical activity, were risk factors for poorer levels of functioning. Psychological resources (reflecting control and self-esteem) were also important for successful ageing (Andrews et al. 2002). Similarly, in the Melbourne Collaborative Cohort Study, healthy weight, not smoking, physical activity, and not having arthritis, asthma, hypertension, or gall stones were associated with successful ageing (Hodge et al. 2013).

Table 5.1 Age-associated changes to body structure and function

Skin:	Digestive system:
• Thinner and more fragile	• Less efficient absorption of nutrients (especially vitamins)
• Slower healing	• Changes in smooth muscle—increased constipation
• Less efficient production of Vitamin D	• Smaller, less efficient liver—increasing the risk of drug toxicity
• Increased vulnerability to extremes of temperature	Kidney and bladder:
• Reduced sensitivity to touch	• Decreased renal function
Muscle and bone:	• Difficulty maintaining fluid and electrolyte balance
• loss of muscle mass	• Vulnerable to dehydration
• Loss of bone mass and demineralisation (osteoporosis)	• Reduced bladder capacity
• Tendon shortening	• Weak and unstable bladder muscles (incontinence)
• Stiffer cartilages	• Urethral fragility
Cardiovascular system:	• Enlarged prostate leading to bladder obstruction
• Stiffer arteries	
• Slowed electrical activity	
• Postural hypotension	
• Reduced right ventricular function	
Respiratory system:	
• Stiffer lungs	
• Larger alveoli	
• Collapsed bronchioles	
• Weaker chest muscles	
• Reduced respiratory capacity	
• Reduced cough reflex	

5.3 Disease and Disability in Later Life

5.3.1 Major Causes of Disease Among Older Australians

Major causes of disease affecting Australians aged 60 years and over, measured as disability-adjusted life years (DALYs), are shown in Fig. 5.3. Major disease groups contributing to disability include malignant neoplasms (cancers), cardiovascular disease, neuropsychiatric disorders (including dementias), respiratory diseases, sense organ diseases, and musculoskeletal conditions.

Prostate cancer is the most common malignant cancer in men, and breast cancer is the most common cancer in women, and the incidence of these cancers increases

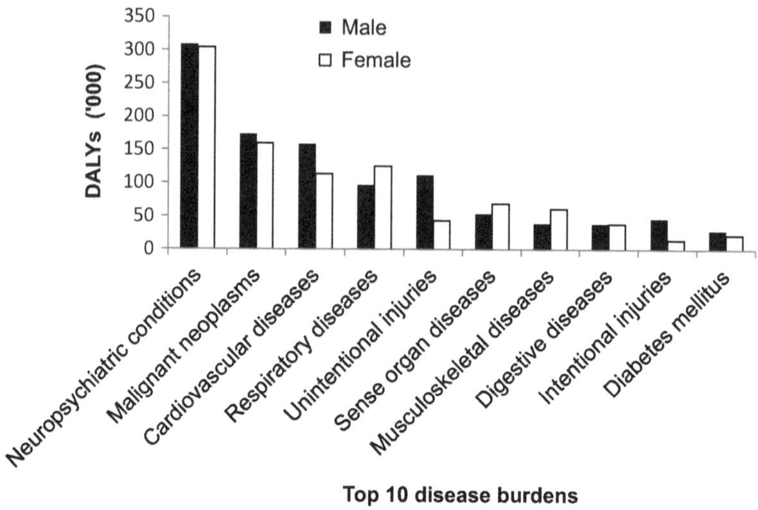

Fig. 5.3 Disability-adjusted life years (representing disease burden) for top ten causes in Australia in 2004, by sex. *Data source*: World Health Organization (2008)

with age. The prevalence of cancer is increasing in Australia. This appears to be due in part to population ageing. It is also related to better screening and early detection of prostate cancer, early bowel cancer, and small-diameter breast cancers, and partly due to longer survival. In 2011, a man in Australia had a one in three (36 %) chance of being diagnosed with cancer by age 75, and a one in two (54 %) chance of being diagnosed with cancer by age 85. Corresponding risks for women were a one in four (27 %) chance by age 75, and a one in three (39 %) chance by age 85 (Australian Institute of Health and Welfare 2015). The higher risk of cancer in men was largely due to smoking-related cancers, but women are catching up as result of their greater uptake of smoking.

Coronary heart disease is the most common form of heart disease affecting older Australians, with higher incidence among people who smoke, who are overweight, and who have high serum cholesterol (Australian Institute of Health and Welfare 2014a). Coronary heart disease affects men and women equally although women have a later age of onset. Case-fatality rates for people who have acute myocardial infarctions (heart attacks) have improved over recent decades due to better emergency treatments. However, angina and cardiac failure remain significant health problems for many older people.

Cerebrovascular disease causes significant disability among older Australians, with around 3–4 % of men and women aged 75 years and over having had a stroke. Stroke is frequently fatal, and survivors are often left with long-term severe or profound disability (Australian Institute of Health and Welfare 2007). A recent Australian study showed that 38 % of women surviving the acute effects of an incident stroke lived for at least another 12 years (Byles et al. 2015), with many experiencing high levels of disability over this period.

The most common and significant neuropsychiatric condition affecting older Australians is dementia, which is the highest cause of disease burden among people aged 75 years or over (Australian Institute of Health and Welfare 2012a). Around 298,000 people in Australia had dementia in 2011; one-third of these were aged 75–84, and 41 % were aged 85 and over (Australian Institute of Health and Welfare 2012c). Among people over 85 years of age the prevalence of dementia is estimated at around 30 % (Australian Institute of Health and Welfare 2012c). However, the Dynamic Analyses to Optimise Ageing (DYNOPTA) study, which combines nine different population-based studies across Australia, suggests that the age-specific prevalence of dementia may be even higher than these estimates (Anstey et al. 2010).

It is projected that the number of people with dementia will triple by 2050 (Australian Institute of Health and Welfare 2012c). Dementia is also more common among women than among men, partly due to women's longer life expectancy, but also due to higher age-specific prevalence (Australian Institute of Health and Welfare 2012c). There have been many studies assessing risk factors for dementia (see e.g. Launer et al., 1999, Anstey et al. 2007). Most of these risk factors are also risk factors for cardiovascular disease and stroke (Tolppanen et al. 2012).

Respiratory diseases affecting older Australians include chronic obstructive pulmonary disease (COPD) and asthma. The major risk factor for COPD is smoking, but other occupational and environmental exposures can also predispose to this condition (Australian Centre for Asthma Monitoring 2011). Asthma is commonly under-diagnosed in older age, even though many people have a reversible component to their airway restriction. It was estimated that in 2004–2005, 9.4 % of people aged 70 or over had asthma and the average age of death from asthma was 79 years (Australian Institute of Health and Welfare 2010a).

Sense organ conditions include hearing and vision problems. In 2007–2008, the prevalence of vision problems among Australians aged 65 years or over was 94 % (Australian Institute of Health and Welfare 2012a). Cataract is the most common eye disease and affects over 70 % of people aged 80 years and over (Australian Institute of Health and Welfare 2007). Cataract can be surgically removed. Glaucoma is another common disease affecting vision and while treatment can slow its progress, lost vision cannot be restored. Similarly, for the progressive vision loss condition known as age-related macular degeneration, treatment can delay degeneration but there is no cure (Australian Institute of Health and Welfare 2007).

Hearing loss affects 75 % of Australians aged over 70 years (Kiely et al. 2012). In one Australian study, 89 % of men and 87 % of women aged over 85 years had audiometric hearing loss (Kiely et al. 2012). Age-related hearing loss has been linked with reduced quality of life (Hogan et al. 2009), poor mental health (Gopinath et al. 2009), cognitive impairment (Tay et al. 2006), reduced social engagement (Kiely et al. 2012), increased use of community services (Schneider et al. 2010), and shorter survival (Karpa et al. 2010). Among participants aged 55 years and over in the Blue Mountains Eye Study, the prevalence of combined hearing and vision impairment was 26.8 % for ages 80 years and over (Schneider et al. 2012).

Musculoskeletal problems include arthritis, falls, and other injuries. Arthritis is one of the most common causes of activity limitation and disability among older people in Australia. In the 2014–2015 National Health Survey, 51.4 % of women aged 55 years or over had arthritis, compared with 35.2 % of men (Australian Bureau of Statistics 2016b). Nearly one-third of Australians with arthritis report restrictions in their daily activities (Australian Institute of Health and Welfare 2010c).

Each year, around 16 % of people aged 60 years and over, and around one-third of people aged 65 years and over report falls; with over 30 % of these people experiencing injuries (such as hip or pelvic fracture and/or head injury) and requiring medical attention (Morris et al. 2004). The rate of falls resulting in hospitalisation in 2009–2010 for the over 65 age group in Australia was 2,663 per 100,000 (Bradley 2013). Fall-related injuries accounted for 72 % of all hospitalised injury cases for the 65 and over age group in Australia in 2009–2010 (Australian Institute of Health and Welfare 2012a). The cost of acute care due to falls among older people in the years 2007–2008 was estimated to be $648.2 million (Bradley 2012).

The rate of falls and associated injuries is even higher for older people in residential aged care and acute care settings (Gibson et al. 2008). In one study in aged care, the rate of falls was 171 falls for every 1,000 bed-months of observation. Around 5 % of these falls required hospital admission, 1.4 % resulted in fractured neck of femur, and 1.5 % resulted in some other form of fracture (Gibson et al. 2008). In 2009–2010, 22 % of hospital admissions for falls were from aged care facilities (Bradley 2013).

Table 5.2 shows the estimated prevalence of major conditions among men aged 60 years and older from the DYNOPTA study. In these data, hypertension and arthritis were the most prevalent diseases (Bielak et al. 2012).

Older people are also likely to have more than one condition affecting their health at any one time. In one Australian study of men and women aged 70 years and over, the median number of conditions per person was 7.0 (Byles et al. 2005). Multimorbidity is associated with many important health outcomes such as quality of life, activities of daily living, health service utilisation, and mortality.

5.3.2 Disability Rates Among Older Australians

The Australian Bureau of Statistics conducts large national surveys to determine the prevalence of disability in Australia. In these surveys, a person is deemed to have a disability if they have any one of a range of health problems that has lasted at least 6 months and that has restricted everyday activities (Australian Bureau of Statistics 2013). In 2012, using this broad definition, 18.5 % of Australians had a disability. The prevalence of disability rose from 3.6 % in young children to 40 % for people aged between 65 and 69, and to over 85 % for people aged 90 years and over. Among people aged 90 and over, 18 % of people had mild or moderate core activity limitation (indicating mild levels of dependency) and 67 % reported a profound or severe core activity limitation (indicating more serious dependency) (Australian Bureau of Statistics 2013) (see Fig. 5.4).

While age-specific rates of disability and dependency are expected to remain stable in the future, the absolute numbers of people aged 65 years and over with a

Table 5.2 Prevalence of chronic disorders per 100 males by age group

	60–64	65–69	70–74	75–79	80–84	85+	Total 60+
Disease	Prevalence (95 % CI)	Prevalence (95 % CI)	Prevalence (95 % CI)	Prevalence (95 % CI)	Prevalence (95 % CI)	Prevalence (95 % CI)	Prevalence (95 % CI)
Diabetes	9.1 (7.4–10.7)	10.5 (5.5–15.5)	12.0 (5.0–19.1)	8.8 (4.6–13.1)	11.3 (6.9–15.6)	7.2 (3.8–10.6)	9.3 (8.0–10.6)
Cataracts	9.2 (7.6–10.9)	11.9 (6.6–17.2)	25.3 (15.9–34.7)	22.4 (16.1–28.6)	11.7 (7.3–16.1)	14.0 (9.4–18.5)	12.0 (10.6–13.4)
Hypertension	39.4 (36.5–42.2)	44.9 (36.5–53.2)	–	0	3.4 (0.1–6.6)	0	31.9 (29.6–34.2)
Stroke	4.8 (3.6–6.0)	3.5 (0.5–6.5)	–	0	0	0	3.7 (2.8–4.6)
Emphysema or Bronchitis	–	–	–	0	0.8* (–0.8–2.5)	0	0.3* (–0.2–0.9)
Asthma	7.9 (6.4–9.5)	9.1 (4.4–13.8)	–	2.9* (–1.1–7.0)	8.0 (3.8–12.1)	7.5 (3.9–11.2)	7.8 (6.5–9.1)
Arthritis	26.7 (24.1–29.2)	23.8 (16.8–30.8)	60.2 (49.7–70.8)	44.1 (36.7–51.6)	32.7 (26.3–39.1)	26.0 (20.3–31.8)	29.9 (27.9–31.9)
Osteoporosis	–	–	3.6* (–0.4–7.6)	4.7 (1.5–7.9)	5.0 (1.6–8.3)	4.1 (1.1–7.0)	4.4 (2.8–6.1)
Any cancer	7.1 (5.6–8.6)	7.0 (2.8–11.1)	30.1 (20.2–40.0)	28.2 (21.5–35.0)	20.5 (15.0–26.0)	22.2 (16.7–27.6)	12.9 (11.4–14.4)
Skin cancer	–	–	24.1 (14.9–33.3)	20.0 (14.0–26.0)	9.9 (5.3–14.6)	6.4 (2.7–10.1)	13.8 (11.0–16.6)

Source: Bielak AAM, Byles JE, Luszcz MA, Anstey KJ. Combining longitudinal studies showed prevalence of disease differed throughout older adulthood. Journal of Clinical Epidemiology, 2012; 65(3): 317–24 (included with permission)

Notes: *CI* confidence interval, **CI* includes null result. Empty cells (–) indicate there was insufficient data to estimate prevalence. 0=no observed cases of condition. Not all participants were asked about every medical condition. All cell $n>6$

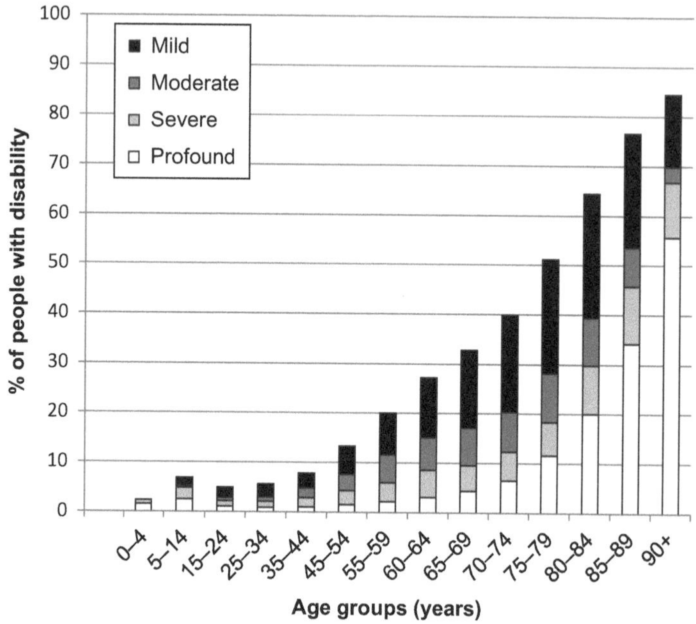

Fig. 5.4 Rates of disability in Australia by age and stacked according to disability status (profound, severe, moderate, or mild core activity limitation). *Data source*: Australian Bureau of Statistics (2013)

core activity restriction can be expected to increase as the population ages. It is this disabled portion of older people that drives health, residential care, and community service requirements within the older population.

5.3.3 Mental Health in Older Age

People may experience poorer mental health in late life, due partly to psychosocial stressors and loss, but also due to increasing frailty and physical illness. 'Terminal decline' occurs when people with relatively stable mental health throughout life experience sudden onset of psychological morbidity towards the end of life (Gerstorf et al. 2010). In the Australian Longitudinal Study on Women's Health, over 70 % of women maintained good or excellent levels of mental health related quality of life as they aged from their 70s to their 90s, and there was no evidence of terminal decline (Leigh et al. 2015). There was however a strong association between chronic disease and poor mental health. As populations age and chronic disease rates increase, mental health conditions will also rise, with projections that depression will be the second leading cause of global disease burden by the year 2020 (Moussavi et al. 2007).

5.4 Opportunities for Prevention of Illness and Promotion of Health

Three stages in promoting healthy old age have been described (Kalache and Kickbusch 1997). The first is increasing capacity for health in early life and includes good maternal nutrition and building resources (such as increasing peak bone density to protect against osteoporosis). The second stage occurs in adult life and involves preventing damage (such as not smoking), protecting against damage (healthy diet), and preventing atrophy (physical activity). The third stage is in late life and involves minimising progression of disease and disability, supportive environments, and compensating for lost capacity. Examples of these approaches to health promotion include secondary prevention of stroke, rehabilitation, exercise and strength training, social support, correction of deficits in vision and hearing, and modifications to domestic and outdoor environments.

5.4.1 Healthy Behaviours

Several key health behaviours are associated with diseases that are common in older age. For example, smoking is a risk factor for cardiovascular disease, stroke, respiratory disease and many cancers, and smoking prevention is important in reducing the incidence of these conditions, even at older ages. The British Doctor's Study shows that over seven years of life expectancy are gained by avoiding smoking after the age of 35 years (Doll et al. 2004). Smoking is the biggest cause of preventable ill health and mortality in Australia; however, in the age group 65 years and over, only 7 % smoke daily (Australian Institute of Health and Welfare 2012a). Even among these people who have continued to smoke well into later life, smoking remains a risk for disease and reduced longevity (Dobson et al. 2012). It is never too late to gain health benefits from quitting smoking.

In Australia, a lack of physical activity accounts for 6.6 % of the total burden of disease and injury in Australia (Begg et al. 2007) and is associated with increased all-cause mortality in both older men and older women (Brown et al. 2012). However, older people may find it difficult to maintain high levels of exercise, and encouragement of exercise requires attention to physical environments as well as education of the individual (Saelens et al. 2003). An Australian study has found that the strongest factor associated with physical activity was the ability to travel alone (Lim and Taylor 2005) and improving this ability could potentially improve the physical activity levels of older people. Older Australian men have been found to have significantly higher levels of physical activity than older women (Booth et al. 2000).

A healthy diet is important for promoting health in older people; however, a third of Australians over 65 years eat less than the recommended intake of fruit and two-thirds eat less than the recommended intake of vegetables (Australian Institute of Health and Welfare 2010b).

Obesity and overweight are increasing in prevalence and are associated with increased risk of stroke, coronary artery disease, diabetes, arthritis, and some cancers (Popkin 2011). However, in older age, a slightly higher Body Mass Index (BMI) measure of around 26.5 is associated with greater health benefits (Flicker et al. 2010).

5.4.2 *Prevention of Disease and Disability*

Many clinical trials provide evidence in support of drug therapy in primary and secondary prevention of cardiovascular disease (O'Keefe et al. 2009), and it is estimated that the combined effects of six preventative drugs in combination (three low-dose antihypertensives, a statin to lower cholesterol, aspirin, and folic acid) could prevent up to 88 % of coronary heart disease and 80 % of stroke (Wald and Law 2003). In the 1921–1926 cohort of the Australian Longitudinal Study on Women's Health, around 70 % of women had prescription claims for statins or antihypertensive medications (alone or in combination) in 2007 (Stewart Williams et al. 2013).

Treatment of diabetes in older patients must consider not only glucose control, but also other health risks and priorities, patient preferences, and life expectancy (Rosenstock 2001). For people with shorter life expectancy, long-term control of hyperglycaemia may be less important than in younger patients (Durso 2006). Control of other risk factors such as lipids and hypertension is also important and may have considerable short-term advantages. Given that a recent Australian cross-sectional study of diabetics over 65 years of age found that the median number of comorbidities was five, it is also particularly important that comorbid conditions are examined to reduce or prevent treatment conflict (Caughey et al. 2010).

A number of interventions have been shown to reduce the risk of falls, including health/environment risk factor screening/intervention programmes, muscle strengthening and balance retraining, and home hazard assessment and modification (Gillespie et al. 2012). Multi-factorial falls prevention programmes have been shown to be effective in reducing falls in community-dwelling older people, and for older people with a history of falling (Gillespie et al. 2012). Important components include group or home-based exercise (including Tai Chi), individual risk assessment, home safety assessment, and modification (Gillespie et al. 2012).

Vitamin D has not been shown to reduce falls among community-dwelling people, but may be important among people with lower vitamin D levels, and may prevent falls and fractures (including hip fracture) among people living in institutional care (Avenell et al. 2009). Bisphosphonates improve bone density and reduce osteoporotic fractures (Cranney et al. 2002); however, the level of adherence to bisphosphonate treatment by older Australian women with osteoporosis is poor (Berecki-Gisolf et al. 2008). Hormone Replacement Therapy (HRT) may also reduce fractures; however, studies show increased risk of cardiovascular disease and breast cancer and so HRT is not currently recommended for fracture prevention (Rossouw et al. 2002).

Vaccination of community-dwelling older people may be a cost-effective way to prevent hospitalisation and deaths due to influenza or pneumonia in Australia (Vu et al. 2002). Influenza and pneumococcal vaccines are available in Australia free of charge to people aged 65 and over (Australian Institute of Health and Welfare 2012a).

Some studies have shown medication review to be beneficial for older people living in the community (Zermansky et al. 2002). One large Australian randomised controlled trial demonstrated a positive effect on medication use and a reduced incidence of falls 12 months after medication review (Pit et al. 2007). Likewise, evidence for prevention-based assessment programmes suggests that assessments have some positive effects on reducing disability burden among older Australians (Byles et al. 2004).

Many cancer screening programmes for younger people are not promoted for older people. While BreastScreen Australia provides free mammographic screening to women over 40, it particularly targets women in the 50–69-year age group (Department of Health and Ageing 2010) and women over 70 years may not be screened. Evidence for ceasing screening after age 70 is not clear since few mammography screening trials enrolled women older than 69 and no trials enrolled women older than 74. The incidence of breast cancer increases with age, but the benefits of screening may be outweighed by competing morbidities among older women. Screening for cervical cancer is not recommended for women over the age of 70 years if they have had two normal Pap smears in the past 5 years (Department of Health and Ageing 2011). Also over 25 % of women in Australia have had a hysterectomy and do not require cervical screening; however, hysterectomy rates in Australia have declined (Hill et al. 2010).

The Australian Cancer Network recommends screening for colorectal cancer from the age of 50 (Australian Cancer Network Colorectal Cancer Guidelines Revision Committee 2005). While there is limited evidence of benefit of screening for oral cancer, assessment of the mouth, teeth, and gums is helpful for older people with cognitive impairment and who may have conditions that limit chewing, swallowing, and communication (National Guideline Clearinghouse 2011).

5.4.3 Restorative Care

Even among frail older people with chronic illness, it is possible to improve independence and health status through restorative approaches to care that help people relearn or develop new skills to manage illness and participate in daily activities. These approaches benefit the older person and their families and can reduce needs for aged care (Ryburn et al. 2009).

5.4.4 Social Determinants of Health

While it is useful to highlight individual behaviours as risk factors for ill health at later ages, addressing these problems requires understanding and improvement, where possible, of the social circumstances that underpin these behaviours. Social disadvantage is a major determinant of ill health at all ages (Marmot 2005). Poor circumstances have their first impact during pregnancy, and poor foetal development is a risk factor for future health and can create predispositions for chronic diseases in later life (Barker 1998). The impact of disadvantage continues throughout childhood and adult life and manifests in terms of psychosocial stress (which can have both physical and mental health impacts), poor education, poor nutrition, social exclusion, underemployment and poor job security, and work-related stress (Power and Matthews 1997). Lower occupational status is associated with higher prevalence of tobacco use and smoking is much more common among the unemployed than the employed in Australia (Smith and Leggat 2007). People living in the lowest socio-economic areas of Australia have been found to be twice as likely to smoke as people in the highest socio-economic areas (Australian Institute of Health and Welfare 2012a). Trends in alcohol and drug use also follow a pattern of increased use with greater socio-economic disadvantage (Redonnet et al. 2012). Disadvantaged people also tend to live in poorer neighbourhoods which have social, economic, and environmental issues negatively impacting on residents (Power 2012).

An Australian longitudinal cohort study found that strong social networks may improve survival of older Australians (Giles et al. 2005). It is thought that these health effects are mediated through the influence of social support on health behaviour, improved access to care, and possibly through physiological pathways. A systematic review of 30 studies of social support interventions for older people identified some evidence that educational and social activity group interventions that target specific groups can alleviate social isolation and loneliness among older people (Cattan et al. 2005). However, more work is required to develop both theoretically sound social support interventions and better methods for evaluating the outcomes of these approaches.

5.5 Conclusion

Australians experience very long life expectancy, with most of these years lived in good health with little disability. There is an expectation for compression of morbidity as our population ages, although it is important to note that increases in longevity are currently greater than increases in healthy life expectancy. Many older people will have one or more chronic conditions, with limitations on their activities of daily living, and will need ongoing health care. Prevention of these conditions is important for healthy ageing and extends across the life course. Appropriate management of chronic conditions, and support for older people with needs for assistance in activities of daily living is also important for enabling Australians to age well. It is also important to consider how built and social environments increase the risk of poor health in older age.

References

Andrews, G., Clark, M., & Luszcz, M. (2002). Successful aging in the Australian Longitudinal Study of Aging: Applying the MacArthur model cross-nationally. *Journal of Social Issues, 58*, 749–765.

Anstey, K. J., Burns, R. A., Birrell, C. L., Steel, D., Kiely, K. M., & Luszcz, M. A. (2010). Estimates of probable dementia prevalence from population-based surveys compared with dementia prevalence estimates based on meta-analyses. *BMC Neurology, 10*, 62–73.

Anstey, K. J., Sanden, C. V., Salim, A., & O'Kearney, R. (2007). Smoking as a risk factor for dementia and cognitive decline: A meta-analysis of prospective studies. *American Journal of Epidemiology, 166*, 367–378.

Australian Bureau of Statistics. (2012). *Australian Health Survey. First Results, 2011-12, Arthritis and osteoporosis [Online]*. Canberra, Australia: ABS.

Australian Bureau of Statistics. (2013). *4430.0—Disability, ageing and carers, Australia: Summary of findings, 2012*. Canberra, Australia: ABS.

Australian Bureau of Statistics. (2016a). *4125.0—Gender indicators, Australia*. Canberra, Australia: ABS.

Australian Bureau of Statistics. (2016b). *National health survey: first results, 2014–15*. Canberra, Australia: ABS.

Australian Cancer Network Colorectal Cancer Guidelines Revision Committee. (2005). *Guidelines for the prevention, early detection and management of colorectal cancer*. Sydney, Australia: Australian Cancer Network.

Australian Centre for Asthma Monitoring. (2011). *Asthma in Australia 2011: AIHW Asthma Series No. 4*. Canberra, Australia: AIHW.

Australian Institute of Health and Welfare. (2007). *Older Australia at a glance* (4th ed.). Canberra, Australia: AIHW.

Australian Institute of Health and Welfare. (2010a). *Asthma among older people in Australia*. Canberra, Australia: AIHW.

Australian Institute of Health and Welfare. (2010b). *Australia's health 2010*. Canberra, Australia: AIHW.

Australian Institute of Health and Welfare. (2010c). *A snapshot of arthritis in Australia. Arthritis Series No. 13*. Canberra, Australia: AIHW.

Australian Institute of Health and Welfare. (2012a). *Australia's health 2012. Australia's Health Series No.13*. Canberra, Australia: AIHW.

Australian Institute of Health and Welfare. (2012b). *Changes in life expectancy and disability in Australia 1998 to 2009. Bulletin No. 111*. Canberra, Australia: AIHW.

Australian Institute of Health and Welfare. (2012c). *Dementia in Australia*. Canberra, Australia: AIHW.

Australian Institute of Health and Welfare. (2014a). *Cardiovascular disease, diabetes and chronic kidney disease: Australian facts: Prevalence and incidence. Cardiovascular, diabetes and chronic kidney disease, Series No. 2. Cat. No. CDK 2*. Canberra, Australia: AIHW.

Australian Institute of Health and Welfare. (2014b). *Healthy life expectancy in Australia: Patterns and trends 1998 to 2012. Bulletin No. 126*. Canberra, Australia: AIHW.

Australian Institute of Health and Welfare. (2015). *Australian cancer incidence and mortality (ACIM) books [Online]*. Canberra, Australia: AIHW. Retrieved 11 July, 2015, from http://www.aihw.gov.au/acim-books/

Avenell, A., Gillespie, W., Gillespie, L., & O'Connell, D. (2009). Vitamin D and vitamin D analogues for preventing fractures associated with involutional and post-menopausal osteoporosis. *Cochrane Database of Systematic Reviews, 2*, CD000227.
Barker, D. (1998). In utero programming of chronic disease. *Clinical Science, 95*, 115–128.
Begg, S., Vos, T., Barker, B., Stevenson, C., Stanley, L., & Lopez, A. D. (2007). *The burden of disease and injury in Australia 2003*. Canberra, Australia: AIHW.
Berecki-Gisolf, J., Hockey, R., & Dobson, A. (2008). Adherence to bisphosphonate treatment by elderly women. *Menopause: The Journal of the North American Menopause Society, 15*, 984–990.
Bielak, A. A. M., Byles, J. E., Luszcz, M. A., & Anstey, K. J. (2012). Combining longitudinal studies showed prevalence of disease differed throughout older adulthood. *Journal of Clinical Epidemiology, 65*, 317–324.
Booth, M. L., Owen, N., Bauman, A., Clavisi, O., & Leslie, E. (2000). Social–cognitive and perceived environment influences associated with physical activity in older Australians. *Preventive Medicine, 31*, 15–22.
Bradley, C. (2012). *Hospitalisations due to falls by older people, Australia 2007–08. Injury Research and Statistics Series No. 61. Cat. No. INJCAT 137*. Canberra, Australia: AIHW.
Bradley, C. (2013). *Hospitalisations due to falls by older people, Australia 2009–10. Injury Research and Statistics Series No. 70. Cat. No. INJCAT 146*. Canberra, Australia: AIHW.
Brown, W., McLaughlin, D., Leung, J., McCaul, K., Flicker, L., Almeida, O., et al. (2012). Physical activity and all-cause mortality in older women and men. *British Journal of Sports Medicine, 46*, 664–668.
Byles, J., Taverner, M., O'Connell, R., Nair, B., Higginbotham, N., Jackson, C., et al. (2004). Randomised controlled trial of health assessments for older Australian veterans and war widows. *Medical Journal of Australia, 181*, 186–190.
Byles, J. E., D'Este, C., Parkinson, L., O'Connell, R., & Treloar, C. (2005). Single index of multimorbidity did not predict multiple outcomes. *Journal of Clinical Epidemiology, 58*, 997–1005.
Byles, J. E., Francis, J. L., Chojenta, C. L., & Hubbard, I. J. (2015). Long-term survival of older Australian women with a history of stroke. *Journal of Stroke and Cerebrovascular Diseases, 24*, 53–60.
Cattan, M., White, M., Bond, J., & Learmouth, A. (2005). Preventing social isolation and loneliness among older people: A systematic review of health promotion interventions. *Ageing and Society, 25*, 41–67.
Caughey, G. E., Roughead, E. E., Vitry, A. I., McDermott, R. A., Shakib, S., & Gilbert, A. L. (2010). Comorbidity in the elderly with diabetes: Identification of areas of potential treatment conflicts. *Diabetes Research and Clinical Practice, 87*, 385–393.
Cranney, A., Wells, G., WillanI, A., Griffith, L., Zytaruk, N., Robinson, V., et al. (2002). II. Meta-analysis of alendronate for the treatment of postmenopausal women. *Endocrine Reviews, 23*, 508–516.
Department of Health and Ageing. (2010). *Breast Screen Australia National Policy* [Online]. DoHA. Retrieved June 26, 2013, from http://www.cancerscreening.gov.au/internet/screening/publishing.nsf/Content/national-policy
Department of Health and Ageing. (2011). *National Cervical Screening Program: NCSP policies* [Online]. DoHA. Retrieved June 19, 2013, from http://www.cancerscreening.gov.au/internet/screening/publishing.nsf/Content/NCSP-Policies-1
Desalvo, K. B., Bloser, N., Reynolds, K., He, J., & Muntner, P. (2006). Mortality prediction with a single general self-rated health question: A meta analysis. *Journal of General Internal Medicine, 21*, 267–275.
Dobson, A., McLaughlin, D., Almeida, J., Brown, O., Byles, W., Flicker, J., et al. (2012). Impact of behavioural risk factors on death within 10 years for women and men in their 70s: Absolute risk charts. *BMC Public Health, 12*, 669.
Doll, R., Peto, R., Boreham, J., & Sutherland, I. (2004). Mortality in relation to smoking: 50 years' observations on male British doctors. *BMJ, 328*, 1519.

Durso, S. (2006). Using clinical guidelines designed for older adults with diabetes mellitus and complex health status. *JAMA, 295*, 1935–1940.
Flicker, L., McCaul, K. A., Hankey, G., Jamrozik, K., Brown, W., Byles, J. E., & Almeida, O. P. (2010). Body mass index and survival in men and women aged 70 to 75. *Journal of the American Geriatrics Society, 58*, 234–241.
Fries, J. F., Green, L. W., & Levine, S. (1989). Health promotion and the compression of morbidity. *The Lancet, 333*, 481–483.
Gerstorf, D., Ram, N., Mayraz, G., Hidajat, M., Lindenberger, U., Wagner, G., & Schupp, J. (2010). Late-life decline in well-being across adulthood in Germany, the United Kingdom, and the United States: Something is seriously wrong at the end of life. *Psychology and Aging, 25*, 477–485.
Gibson, R. E., Harden, M., Byles, J., & Ward, J. (2008). Incidence of falls and fall-related outcomes among people in aged-care facilities in the Lower Hunter region, NSW. *NSW Public Health Bulletin, 19*, 166–169.
Giles, L., Glonek, G., Luszcz, M., & Andrews, G. (2005). Effect of social networks on 10 year survival in very old Australians: The Australian longitudinal study of ageing. *Journal of Epidemiology and Community Health, 59*(7), 574–579.
Gillespie, L., Gillespie, W., Sherrington, C., Gates, S., Clemson, L., & Lamb, S. (2012). Interventions for preventing falls in older people living in the community (review). *Cochrane Database of Systematic Reviews, 9*, CD007146.
Gopinath, B., Wang, J. J., Schneider, J., BurlutskyU, G., Snowdon, J., McMahon, C., et al. (2009). Depressive symptoms in older adults with hearing impairments: The Blue Mountains study. *Journal of the American Geriatrics Society, 57*, 1306–1308.
Hill, E., Graham, M., & Shelley, J. (2010). Hysterectomy trends in Australia—between 2000/01 and 2004/05. *Australian and New Zealand Journal of Obstetrics and Gynaecology, 50*, 153–158.
Hodge, A., English, D., Giles, G., & Flicker, L. (2013). Social connectedness and predictors of successful ageing. *Maturitas, 75*, 361–366.
Hogan, A., O'Loughlin, K., Miller, P., & Kendig, H. (2009). The health impact of a hearing disability on older people in Australia. *Journal of Aging and Health, 21*, 1098–1111.
Kalache, A., & Kickbusch, I. (1997). A global strategy for healthy ageing. *World Health, 50*, 4–5.
Karpa, M. J., Gopinath, B., Beath, K., Rochtchina, E., Cumming, R. G., Wang, J. J., & Mitchell, P. (2010). Associations between hearing impairment and mortality risk in older persons: The Blue Mountains hearing study. *Annals of Epidemiology, 20*, 452–459.
Kiely, K. M., Gopinath, B., Mitchell, P., Luszcz, M., & Anstey, K. J. (2012). Cognitive, health, and sociodemographic predictors of longitudinal decline in hearing acuity among older adults. *The Journals of Gerontology Series A: Biological Sciences and Medical Sciences, 67*, 997–1003.
Lafortune, G. & Balestat, G. (2007). *Trends in severe disability among elderly people: Assessing the evidence in 12 OECD countries and the future implications* (OECD health working paper, No. 26). Paris: Organisation for Economic Cooperation and Development.
Launer, L. J., Andersen, K., Dewey, M. E., Lettenneur, L., Ott, A., Amaducci, L. A., et al. (1999). Rates and risk factors for dementia and Alzheimer's disease: Results from EURODEM pooled analyses. *Neurology, 52*, 78.
Leigh, L., Byles, J. E., Chojenta, C. & Pachana, N. A. (2015). Late life changes in mental health: A longitudinal study of 9683 women. *Aging & Mental Health*, 1-11
Lim, K., & Taylor, L. (2005). Factors associated with physical activity among older people—A population-based study. *Preventive Medicine, 40*, 33–40.
Marmot, M. (2005). Social determinants of health inequalities. *Lancet, 365*, 1099–1104.
Mathers, C. D., Iburg, K. M., Salomon, J. A., Tandon, A., Chatterji, S., Ustün, B., & Murray, C. J. (2004). Global patterns of healthy life expectancy in the year 2002. *BMC Public Health, 4*, 66.
McCallum, J., Shadbolt, B., & Wang, D. (1994). Self-rated health and survival: A 7-year follow-up study of Australian elderly. *American Journal of Public Health, 84*, 1100–1105.
Morris, M., Osborne, D., Hill, K., Kendig, H., Lundgren-Lindquist, B., Browning, C., & Reid, J. (2004). Predisposing factors for occasional and multiple falls in older Australians who live at home. *Australian Journal of Physiotherapy, 50*, 153–159.

Moussavi, S., Chatterji, S., Verdes, E., Tandon, A., Patel, V., & Ustun, B. (2007). Depression, chronic diseases, and decrements in health: Results from the World Health Surveys. *The Lancet, 370*, 851–858.

National Guideline Clearinghouse. (2011). *Oral hygiene care for functionally dependent and cognitively impaired older adults* [Online]. Rockville MD: Agency for Healthcare Research and Quality (AHRQ). Retrieved July 25, 2013, from http://www.guideline.gov/content.aspx?id=34447

O'Keefe, J., Carter, M., & Lavie, C. (2009). Primary and secondary prevention of cardiovascular diseases: A practical evidence-based approach. *Mayo Clinic Proceedings, 84*, 741–757.

Pit, S. W., Byles, J. E., Henry, D. A., Holt, L., Hansen, V., & Bowman, D. A. (2007). A Quality Use of Medicines program for general practitioners and older people: A cluster randomised controlled trial. *Medical Journal of Australia, 187*, 23–30.

Popkin, B. M. (2011). Does global obesity represent a global public health challenge? *The American Journal of Clinical Nutrition, 93*, 232–233.

Power, A. (2012). Social inequality, disadvantaged neighbourhoods and transport deprivation: An assessment of the historical influence of housing policies. *Journal of Transport Geography, 21*, 39–48.

Power, C., & Matthews, S. (1997). Origins of health inequalities in a national population sample. *The Lancet, 350*, 1584–1589.

Redonnet, B., Chollet, A., Fombonne, E., Bowes, L., & Melchior, M. (2012). Tobacco, alcohol, cannabis and other illegal drug use among young adults: The socioeconomic context. *Drug and Alcohol Dependence, 121*, 231–239.

Rosenstock, J. (2001). Management of type 2 diabetes mellitus in the elderly-special considerations. *Drugs & Aging, 18*, 31–44.

Rossouw, J. E., Anderson, G. L., Prentice, R. L., Lacroix, A. Z., Kooperberg, C., Stefanick, M. L., et al. (2002). Writing Group for the Women's Health Initiative. Risks and benefits of estrogen plus progestin in healthy postmenopausal women: Principal results from the Women's Health Initiative randomized controlled trial. *JAMA, 288*, 321–333.

Rowe, J. W., & Kahn, R. L. (1997). Successful aging. *The Gerontologist, 37*, 433–440.

Ryburn, B., Wells, Y., & Foreman, P. (2009). Enabling independence: Restorative approaches to home care provision for frail older adults. *Health & Social Care in the Community, 17*, 225–234.

Saelens, B., Sallis, J., Black, J., & Chen, D. (2003). Neighbourhood-based differences in physical activity: An environment scale evaluation. *American Journal of Public Health, 93*, 1552–1558.

Salomon, J. A., Wang, H., Freeman, M. K., Vos, T., Flaxman, A. D., Lopez, A. D., & Murray, C. J. L. (2012). Healthy life expectancy for 187 countries, 1990–2010: A systematic analysis for the Global Burden Disease Study 2010. *The Lancet, 380*, 2144–2162.

Schneider, J., Gopinath, B., Karpa, M. J., McMahon, C. M., Rochtchina, E., Leeder, S. R., & Mitchell, P. (2010). Hearing loss impacts on the use of community and informal supports. *Age and Ageing, 39*, 458–464.

Schneider, J., Gopinath, B., McMahon, C., Teber, E., Leeder, S. R., Wang, J. J., & Mitchell, P. (2012). Prevalence and 5-year incidence of dual sensory impairment in an older Australian population. *Annals of Epidemiology, 22*, 295–301.

Simons, L. A., Simons, J., Friedlander, Y., & McCallum, J. (2011). Predictors of long-term mortality in the elderly: The Dubbo study. *Internal Medicine Journal, 41*, 555–560.

Smith, D., & Leggat, P. (2007). Tobacco smoking by occupation in Australia: Results from the 2004 to 2005 National Health Survey. *Journal of Occupational and Environmental Medicine, 49*, 437–445.

Stewart Williams, J. A., Wallick, C. J., Byles, J. E., & Doran, C. M. (2013). Assessing patterns of use of cardio-protective polypill component medicines in Australian women. *Drugs & Aging, 30*, 193–203.

Tay, T., Wang, J. J., Kifley, A., Lindley, R., Newall, P., & Mitchell, P. (2006). Sensory and cognitive association in older persons: Findings from an older Australian population. *Gerontology, 52*, 386–394.

Tolppanen, A.-M., Solomon, A., Soininen, H., & Kivipelto, M. (2012). Midlife vascular risk factors and Alzheimer's disease: Evidence from epidemiological studies. *Journal of Alzheimer's Disease, 32*, 531–540.

Üstun, T. B., Chatterji, S., Bickenbach, J., Kostanjsek, N., & Schneider, M. (2003). The International Classification of Functioning, Disability and Health: A new tool for understanding disability and health. *Disability and Rehabilitation, 25*, 565–571.

Vu, T., Farish, S., Jenkins, M., & Kelly, H. (2002). A meta-analysis of effectiveness of influenza vaccine in persons aged 65 years and over living in the community. *Vaccine, 20*, 1831–1836.

Wald, N., & Law, M. (2003). A strategy to reduce cardiovascular disease by more than 80%. *British Medical Journal, 326*, 1419.

World Health Organization. (2008). *The global burden of disease: 2004 update*. Geneva: WHO.

World Health Organization. (2015). *World health statistics*. Geneva, Switzerland: WHO.

Zermansky, A., Petty, D., Raynor, D., Lowe, C., Freemantle, N., & Vail, A. (2002). Clinical medication review by a pharmacist of patients on repeat prescriptions in general practice: A randomised controlled trial. *Health Technology Assessment, 6*, 1–86.

Chapter 6
Cultural Diversity, Health and Ageing

Harriet Radermacher and Susan Feldman

6.1 Introduction

Ethnic diversity is a hallmark of the Australian population and the aim of this chapter is to provide an overview of the research evidence examining influences of ethnicity and migration on health and ageing. The chapter takes a particular focus on the challenges facing individuals, families and health providers in our culturally and linguistically diverse nation. We acknowledge that many older people in Australia live in a range of settings including residential aged care facilities and supported accommodation, as well as receiving care and support in hospitals, clinics and in their own homes. However, this chapter will focus specifically on the experiences of those people who live in a community context, especially given the widely accepted preference for older people to live in their own homes as they age. We will draw on the national research literature, which includes a range of our previous research studies, specifically those investigations which were concerned with understanding the quality of life and the health and well-being of older community dwelling people from ethnic minority backgrounds. While the focus of this chapter is on the Australian experience, we do of course acknowledge that migration is a global phenomenon (Zimmerman et al. 2011). This means that policies and practices to promote the health and well-being of migrants who have come to Australia may also need to consider the continuing and dynamic relationships people have with their home country, considerations which transcend national borders (Torres 2004).

To set the scene for the chapter, a brief history of post-World War II migration to Australia will be presented. This will be followed by a demographic profile of Australia's ethnic diversity, which will acknowledge the various ways in which cultural and linguistic diversity has been and is currently defined in Australia.

H. Radermacher (✉) • S. Feldman
Monash University, Melbourne, Victoria, Australia
e-mail: harriet.radermacher@monash.edu; susan.feldman@monash.edu

K. O'Loughlin et al. (eds.), *Ageing in Australia*, International Perspectives on Aging 16, DOI 10.1007/978-1-4939-6466-6_6

Consideration will be given to concepts of healthy ageing, disparities in health and service usage amongst ethnic minority Australians. Based both on the Australian research evidence and our experience in the field, we will then discuss additional factors that significantly influence the health and well-being of older ethnic minority Australians. These factors include communication, English proficiency and the use of interpreters, health literacy, the role of family and community, and technological advances. Finally, we will summarise the important considerations for service providers in their work to promote the health and well-being of older Australians from ethnic minority backgrounds.

6.2 Context

6.2.1 Multicultural Australia

With 27 % of its population born overseas, Australia is one of the most culturally and linguistically diverse countries in the world (Australian Bureau of Statistics (ABS) 2012a). The proportion of Australia's population born overseas is more than double that found in the United States of America (13 %) and the United Kingdom (12.3 %), and compares even more starkly with Japan which has only 1.3 % of its total population born overseas (Rienzo and Vargas-Silva 2012; Statistics Bureau of Japan 2011; US Census Bureau 2012). A range of immigration policies and programmes have determined the cultural profile of Australia over the years. People have migrated primarily on account of either their 'skills' or to seek asylum. First generation Australians, people living in Australia who were born overseas, are a diverse group including Australian citizens, permanent residents and long-term temporary residents. The older cohort is no exception, currently comprising people who came to Australia when they were young and have now turned 65 years, as well as a small percentage who have come in older age for family reunification or retirement (Ip et al. 2007). Australian immigration history is characterised by the intake of people from specific groups or regions over time. For example, Italian immigrants arrived primarily for economic reasons in the 1950s, and now constitute a well-established community group. In contrast, Vietnamese immigrants were more likely to be traumatised political refugees and represent a relatively new community group, arriving in Australia in the 1970s (Thomas 1999, 2007). While recent arrivals are still coming from Europe, larger numbers are now arriving from the Middle East, Asian and African countries.

In 2011, the top ten countries from which overseas-born Australians 65 years and over originated were as follows: United Kingdom, Italy, Greece, Scotland, New Zealand, Germany, Netherlands, China, Croatia and India (ABS 2012a). While Europeans dominate the current older overseas-born cohort, much greater numbers of people aged between 50 and 64 are from Asian countries (Federation of Ethnic Communities Council of Australia (FECCA) 2015).

In 2006, 35 % (953,702 people) of people over 65 years were born overseas, with 61 % of these coming from non-English-speaking countries (Australian Institute of Health and Welfare (AIHW) 2007). As the cohort of post-war immigrants is ageing, the proportion of older people from non-English speaking backgrounds is increasing as compared to those from English-speaking countries. Despite the large proportion of older people born in non-English speaking countries, 83 % of older people spoke only English at home, and only 6 % reported to speak another language at home and speak English poorly. With only 24 % of those currently under 65 born overseas (ABS 2012a), together with the changing profile of immigration, it is likely that new issues and challenges will emerge for the older cohort in the coming decades.

Dispersal of migrants across Australia is not uniform, with numbers varying across states. In 2006, the largest proportion of migrants was in Western Australia (30 %), followed by New South Wales (27 %) and Victoria (26 %), and the lowest proportion was found in Tasmania (11 %) (ABS 2012b). There is further variation in dispersal within states. Migrants predominantly reside in urban locations. In 2011, 82 % of the overseas-born population lived in capital cities compared with 66 % of the total population (ABS 2012a). Even within Metropolitan regions migrants from particular communities cluster in particular areas. This is well illustrated in Howe's (2006) analysis of the trends, profiles and distribution of Victoria's culturally diverse community.

6.2.2 *Defining Cultural Diversity*

Talking about and defining cultural diversity is challenging, primarily because it is a multi-faceted concept (Radermacher and Feldman 2014). The Australian Bureau of Statistics identified a set of variables in order to develop consistency across data collection (McLennan 1999). These variables also illustrate the comprehensive range of factors that comprise cultural diversity. The core variables included Country of Birth of Person, Main Language Other Than English Spoken at Home and Proficiency in Spoken English. The full standard set also includes Ancestry, Country of Birth of Father, Country of Birth of Mother, First Language Spoken, Languages Spoken at Home, Main Language Spoken at Home, Religious Affiliation and Year of Arrival in Australia. It is likely that the specific context will determine the most important indicators. For example, languages spoken at home and English language proficiency have been proposed as the most useful indicators for identifying the needs of ethnic minority older people in relation to the provision of Home and Community Care (HACC) services (Howe 2006).

This multi-faceted concept may explain why there is no consensus or universal term to describe and distinguish people who come from different ethnic and cultural backgrounds. In Australia, Government and policy makers use the term 'culturally and linguistically diverse' (CALD), to generally refer to people who are born overseas. 'CALD' replaced 'Non-English Speaking Background' (NESB) in 1996 to acknowledge that people from English-speaking nations can still have cultural identities distinct from the mainstream population. Equivalent terminology is 'Black and Minority Ethnic' (BME) in the UK, and 'Racial and Ethnic Minorities' in the US.

In the context of this chapter, we use a range of terms depending on the context. In alignment with Sawrikar and Katz (2009), we find the term 'CALD' problematic. All people are from culturally and linguistically diverse backgrounds, and hence the term's correct application is in reference to a population of people from many different backgrounds, such as the whole Australian population. Our preferred term, although not without its own limitations, is 'ethnic minority'. We use the word 'minority' both to describe the generally smaller critical mass of people from specific ethnic and cultural groups, but also in recognition that many ethnic and cultural groups have significantly less power and status than the Anglo-Celtic 'majority'. And, after all, it is due to this disadvantage, which manifests in a number of ways, that specific attention is given to these groups.

6.2.3 Acknowledging Difference

Rowland's (1999) distinction between different levels of need amongst the ethnic minority aged population supports the widely held assertion that, like other ageing populations, they are not a homogenous group. Assumptions and popular conceptions concerning a lack of diversity in the experience of ageing may mirror wider social images of homogeneity of ageing, influencing policy and service practitioners in their work with, and relationships to, older persons (Holstein and Minkler 2007). The differences within any population can be attributed to a range of factors including age, gender, ethnicity, geographic location, occupation, family structure and marital and socioeconomic status. These factors must also be considered when looking at populations of older people (Arber et al. 2003; Arber and Ginn 1995; Feldman et al. 2012; Radermacher and Feldman 2014). Indeed, there is an extensive body of literature which highlights the heterogeneity of experience and needs amongst people from different ethnic groups (e.g. Mackenzie 1999; Matsuoka and Sorenson 1991; Patel 1999) and it is well recognised that heterogeneity has obvious implications and challenges for developing effective and responsive service delivery models.

Throughout this discussion we talk generally about older people. However, we would strongly suggest that gender should be considered central to understanding life experiences, changes and transitions, specifically the complex nature of growing old for both men and women. In addition, integrating concepts of gender in any study about ageing, regardless of cultural background, builds a foundation for understanding inequalities, not only within specific communities of older people, but that also potentially exist between men and women. A gendered perspective also highlights the interrelationships between all aspects of life, particularly an individual's social and economic circumstances as well as cultural values, beliefs and practices (Calasanti and Slevin 2006). As Arber and Ginn (1995) remind us, age and gender are inextricably linked in social life (p. 11).

6.3 Cultural Diversity, Health and Well-being

Striving for a long, active and engaged life, surrounded by family and community and in optimal health and well-being is the ideal of most people as they grow older. The intersection between social, physical and psychological health is complex and challenging (Browning and Thomas 2007; Feldman and Radermacher 2011) but it is this intersection that will be highlighted in our discussion. Many writers, including Baltes and Baltes (1990), Rowe and Kahn (1997, 1998) and Jorm et al. (1998) propose that successful ageing is both an important and multi-dimensional concept. They elaborate that successful ageing must be seen as 'encompassing three distinct domains: avoidance of disease and disability, maintenance of high physical and cognitive function, and sustained engagement in social and productive activities' (Rowe and Kahn 1997, p. 439). In addition, the World Health Organization (WHO 2002) also advocates 'Active Ageing' which is 'the process of optimising opportunities for health, participation and security in order to enhance quality of life as people age' (p. 12). This definition includes a strong focus on collaboration and a collective approach to ageing, specifically through the inclusion of older people's views about their own experiences of growing older, but also pertaining to policy development, planning and service delivery. Importantly, the later work of Browning and Thomas (2007) extended the indicators of successful or healthy ageing to include cultural and gender conceptualisations. They defined successful ageing as '… a process whereby people can achieve or maintain the best possible state of physical, cognitive and mental health and wellbeing, meaningful and positive engagement with people, community and institutions, and a personal sense of security, choice and autonomy with active adaptation to ageing processes from the individual, familial and society perspectives' (Browning and Thomas, p. 48).

While current conceptualisations of healthy ageing are extending previous predominantly biomedical definitions, they are still rather narrow in that they are largely based on Western perspectives. A small but growing literature is investigating healthy ageing from non-Western perspectives (e.g. Chou and Chi 2002; Fernández-Ballesteros et al. 2008). While many of the same dimensions are being identified as important (e.g. physical and mental health, family and community engagement), some aspects appear to hold greater significance for some communities; for example, the role of spirituality for Malays, both in Australia and in Malaysia (Tohit et al. 2012).

A range of physical, social and support circumstances of people from ethnic minority communities in Australia have been identified (Nimri 2007; Orb 2002; Rao et al. 2006). For example, in Orb's review of the literature, three areas of concern for ageing migrants were identified: physical health, mental and psychological well-being, and socioeconomic welfare. A Queensland scoping project identified similar major issues as economic and financial needs; social needs, social isolation and quality of life; transport; housing; health needs; and aged care (Bartlett et al. 2006).

Furthermore, where a person is born, and the language they speak, remains fundamental to their health and well-being, particularly as they grow older. The importance of these aspects of cultural identity has been identified in various research studies conducted by the authors involving older ethnic minority people in a range of community and geographical settings (e.g. Bird et al. 2009; Feldman and Radermacher 2011; Karunarathna et al. 2010; Radermacher et al. 2008). One study (Feldman and Radermacher 2011) investigating the health and well-being of older men from four ethnic groups living in rural Victoria reported that 'men in all the groups reinforced the important role of their home country's traditions and culture, and emphasised that it was integral to their identity and wellbeing' (Feldman et al. 2012, p. 92). In addition, a number of the men (and other family members) indicated that returning to visit their villages and towns over the years had enabled them to maintain strong ties with their home country. They also talked about how these ties were strengthened through regular communication with family and friends (Torres 2004; Feldman et al. 2012), thus reinforcing their sense of identity and connection with the 'old' way of life. For many of the participants in a range of these studies, their ethnic and cultural background and traditions remained critical to their identities, regardless of how 'Australianised' they perceived themselves to be or whether they had ever returned to their home country. As one older Italian male said:

> *I have no problem to integrate with the Australian community but my blood is Italian blood running around... I came here at 15 years of age and I've been naturalised since about 17 with my parents, but deep inside I feel Italian (Italian male, 60+; unpublished data).*

In this regard, we agree with Warburton et al. (2009, p. 176) that 'culture can provide a foundation for positive ageing'.

6.3.1 Disparities in Health

People born overseas in countries where English is not the first language spoken, as compared to their Australian-born and English speaking peers, are disadvantaged on a number of levels, as evidenced by their poorer health status, health outcomes and access to services (FECCA 2015; Migliorino 2010; Rao et al. 2006; Rowell 2005; Warburton et al. 2009). Additional demographic factors, which could also contribute to migrants being further disadvantaged, include gender, age, English proficiency, education, income and geographic location. For example, older people from ethnic minority backgrounds are more likely than Anglo-Australians to be poorly educated, to work in unskilled occupations and to face socio-economic disadvantage (Hassett and George 2002).

Disparities are also reported to exist in how health and health support agencies are conceptualised across different cultures (Kleinman 1988). This may lead to under-reporting of health and medical conditions that would ordinarily receive treatment in Anglo-Australian communities. For example, there is evidence of under-recognition of depression and other affective disorders among non-English-speaking communities, and an associated reluctance to utilise psycho-geriatric services (Hassett et al. 1999).

Further disparities, this time in favour of being born overseas, are evident in the 'healthy migrant effect', whereby when migrants enter the country they generally enjoy better health than their Australian-born counterparts. This can be attributed to the strict health requirements and eligibility criteria imposed when they migrate. However, this initial health advantage has been shown to decrease with increasing length of residence in Australia (Young 1992; cited in AIHW 2002).

6.3.2 Service Usage by Older Ethnic Minority People

Looking at the patterns of service usage can be a useful way to gain insight into the health status of people from ethnic minority backgrounds. However, it is important to be mindful when interpreting service usage patterns. While high service usage may be indicative of poorer health, low service usage does not necessarily indicate good health. Rather, low service usage may be indicative of several factors, such as ignorance about availability, dissatisfaction and inability to access services (due to communication barriers, lack of transport, etc.).

Using HACC services as an example, while there is evidence to suggest under-utilisation of services by ethnic minority communities, there is also evidence that the rate of use of HACC basic services by ethnic minority people across metropolitan Melbourne is increasing (Australian Healthcare Associates 2007). Similarly, a review of HACC social support services for ethnic minority people found services were well attended and provided important opportunities for socialisation, physical activity and access to support (Haralambous et al. 2007).

The pattern of HACC service usage by people from ethnic minority backgrounds in rural areas differs, however, from those with Anglo-Australian backgrounds. Ward et al. (2005) investigated the use of HACC services by overseas-born people compared to Australian-born people in rural areas of Victoria. Whilst they found that the proportion of overseas-born people who were HACC users was consistent with the demographic profiles, this was not the case for their extent of service usage (they received 35 % less hours than Australian-born counterparts).

In summary, it is evident that there are some differences in the way older ethnic minority Australians think about their health, and subsequently engage with health services. In the next section, we identify four key factors that need to be considered when thinking about a health service system that is inclusive of all Australians.

6.4 Factors That Influence Health and Well-being

6.4.1 Communication, English Language Proficiency and Use of Interpreters

Of particular importance, regardless of age, language spoken or cultural background, is the desire of most people to have the capacity to communicate clearly with others, including family, friends, community and service providers.

The perceived barriers to accessing health and support services for older ethnic minority people include a range of factors such as not knowing the services exist, cultural inappropriateness of services, language barriers, cultural barriers (e.g. food/religious requirements), lack of links between organisations and ethno-specific groups, as well as a lack of available bilingual staff (Gallagher and Truglio-Londrigan 2004; Karunarathna et al. 2010; Kruger et al. 2007; Migrant Information Centre and Yooralla 2006). Above all, studies consistently indicate that language and communication are the primary barriers to health and support service utilisation (Bartlett et al. 2006; Bird et al. 2009; FECCA 2015; Feldman and Radermacher 2011; Karunarathna et al. 2010; Radermacher et al. 2008; Thomas 2007; Warburton et al. 2009).

As a number of studies have found, older individuals from ethnic minority backgrounds recognise their limitations with English language proficiency, with most people accepting that someone who they can trust and can rely on to interpret for them must accompany them to medical appointments (Alexander et al. 2004; Feldman and Radermacher 2011). Unfortunately in many cases their choice is not necessarily the most appropriately trained individual and often determined by availability alone, as identified in a study previously described (Feldman and Radermacher 2011):

> *You have to ask for help. Because even when we go to the doctor have to come someone who to translate to us because we cannot explain properly... Yesterday we called another lady with us who understand properly to translate to us and back from us to the doctor....It's hard...All the younger generation they work. It's a bit hard for them. That lady came yesterday. She left her job to come with us to translate (Macedonian male, in a group of 8 men aged between 63 and 82 years; unpublished data).*

Children are also commonly and inappropriately used as interpreters for family members; inappropriate not only due to issues of privacy and confidentiality, but also because children do not necessarily understand the difference between interpreting and translating, as the following quote from the same study describes:

> *If kids is talking English here they couldn't understand interpreting. Interpreting is different. They can talk the English they can talk the Turkish, but interpreting and explaining that is different (Turkish male, in a group of 8 men aged between 44 and 70 years; unpublished data).*

Even when an appropriately trained interpreter is available to assist communication with a doctor, there are additional issues that need to be addressed. For example, some people will feel uncomfortable talking about personal issues with people present from the opposite sex, as indicated by the observations of an older Muslim Albanian woman in the same study:

> *Well to have somebody, you know, an interpreter or something ...Sometimes I feel embarrassed even to talk to him ...And I believe men should have men interpreters. Because some of the men won't speak to a woman about their problems. They won't speak to a woman. I mean they'll speak to a doctor woman ...But not an interpreter (Albanian female, 57 years; unpublished data).*

The inability to communicate well in English may also contribute to limited health-seeking behaviours and a reluctance on behalf of older ethnic minority people to access mainstream community and service organisations (Feldman and Radermacher 2011). It therefore follows that fundamental to any successful service system is the availability of a pool of accredited interpreters and translators. There is also a potential role for trained bilingual community-based navigators that the community trusts and respects (Henderson and Kendall 2011).

6.4.2 Health Literacy

Health literacy has been defined as 'the degree to which individuals have the capacity to obtain, process and understand basic health information and services needed to make appropriate health decisions' (Ratzan and Parker 2000; para 7). In 2006, 41 % of Australians were assessed as having adequate health literacy (ABS 2009). This proportion is low, but it is lower still for older adults, people whose first language is not English, and those born overseas. A good description of poor health literacy was provided by a service provider working in a rural Victorian community, in a study about older ethnic minority men's health (Radermacher and Feldman 2014). His observation illustrates how older ethnic minority men are not aware of the more complex dimensions to health, in particular the psychosocial elements:

> *They're not aware [older ethnic minority men] because they've never been in that sphere, no one has ever told them... their health is their heart, their legs, their back (Service provider).* (p. 11)

Poor health literacy can lead to people being less likely to engage in healthy behaviours such as regular exercise, healthy eating, not smoking, etc. This in turn places people at greater risk of developing chronic conditions such as heart disease, high blood pressure and diabetes. Once diagnosed, individuals are less likely to manage the disease well if they have limited comprehension of the nature and process of the disease, don't understand written health information or don't fully apply what health professionals advise them.

Health literacy therefore is a useful lens through which to consider the impact of ethnicity and migration on health and ageing, particularly because of the link between low health literacy and poor health outcomes (e.g. Schillinger and Davis 2005; Weiss 2005). There are of course many factors contributing to low health literacy rates in older ethnic minority Australians, including limited English language proficiency, literacy levels in an individual's primary language, access to formal education and income. Furthermore, cultural variations in beliefs about health and subsequent treatment may not be supported by a predominantly Western approach to health and health care, thus further decreasing an individual's health literacy level. So, what strategies might therefore be useful in improving health literacy in ethnic minority communities?

Interestingly, strategies to address health literacy have historically focused on improving the skills and knowledge of individuals (Australian Commission on Safety and Quality in Health Care (ACSQHC) 2011). While this approach places the onus on people with low health literacy to change, more recent strategies have been expanded to acknowledge the additional role and responsibility of the health care system, and the dynamic nature of an individual within the health care system and society more broadly (Baker 2006). This approach clearly places responsibility on the health system and service providers themselves to be more user friendly and culturally appropriate (ACSQHC 2011; Ethnic Communities' Council of Victoria (ECCV) 2012). This latter strategy also demands that health care professionals are supported in their workplaces via adequate training and resourcing, as well as creating organisational cultures that are committed to providing high-quality care for all (ECCV 2012). Such training is undertaken to promote the 'cultural competence' of the workforce.

Cultural competence broadly refers to 'the relationship between the helper and the person being helped, in a cross-cultural context and … focuses on the capacity of the health worker to improve health status by integrating culture into the clinical context' (Royal Australasian College Physicians 2004, p. 1). Training for staff and managers is an important component of this strategy. Its central aim is to increase staff understanding of why organisations have a responsibility to provide an inclusive service, as well as asking questions about how this can best be achieved in practice. 'Cultural competence in health care entails understanding the importance of social and cultural influences on patients' health beliefs and behaviours; considering how these factors interact at multiple levels of the health care delivery system' (Betancourt et al. 2003, p. 293).

6.4.3 The Role of Family

It is difficult to predict the nature of the support that family members may need (whether young or old) but family members are typically reported as a primary support resource for each other (Feldman et al. 2012; Feldman and Seedsman 2005; Harper 2000; Radermacher et al. 2010). In 2003, 78 % of older Australians requiring assistance received informal care from a family member or friend (ABS 2003). Despite the changes in family structure through single parenting, divorce or geographic separation some writers would argue that 'the reports of the demise of the extended family has been exaggerated' (Hatton-Yeo 2007, p. 115).

Family interactions, obligations and exchange remain strong, particularly within ethnic minority communities, and are expressed through informal emotional, physical or financial care and support within the extended family (Bengston 2000; De Vaus 1996). In ethnic minority communities extended family relationships are described as particularly central to health and well-being, a point emphasised by many participants across all of our studies and regardless of cultural background. Within many cultural groups the extended family remains a support network of some significance and is accompanied by the expectation 'that older people may rely solely upon family to care for them in their old age' (Feldman and Seedsman 2005, p. 189), the connections between people being fundamental to the maintenance of health and well-being in their community. A study of older Chinese migrants in Australia identified dependency on children as a key issue which was associated with lack of English proficiency and difficulties accessing language support and interpreter services (Ip et al. 2007). This was especially so for women. The authors suggest that a common belief held by non-Chinese is that Chinese older parents are usually well looked after by their families, this assumption potentially leading to feelings of loneliness and isolation going unnoticed. A British study of black and minority older people also confirmed that assuming the extended family of people from these backgrounds will look after their elders was unwarranted (Butt and O'Neil 2004). While older people from culturally diverse backgrounds may be more likely to live with their families for a range of reasons (family reunification schemes, financial hardship, cultural inappropriateness of public care, etc.), it appears that the support they receive from their families may be more about containment than care.

It must be emphasised however that not all ethnic minority older people have family members living close by and consequently it is not feasible for these older people to rely on family for support even if they wanted or needed to do so. Family members may have to move to another location leaving older parents behind. In one instance, in a study about food security and whether older people were getting enough good food to eat (Radermacher et al. 2010), an older Maltese man was talking about a friend who had moved to be closer to his children, and then his children moved on:

> *He came next to his kids here and then they shifted. They moved somewhere else...There's been people who have come up here from [name of town] at our age to be next to their kids and then their kids say 'we need another house and we want to move' (Maltese Male, in a group of 8 men and women aged between 61 and 80).* (p. 362)

The research findings of this study also indicate that many older people do not want to be a burden on family members and prefer to remain living independently as best they can. In addition, the research data indicates that even if family are close by, they often lead their own lives and have to work and look after their own families and simply don't have the time or resources to assist the older members of their family. In the same study we undertook about food security, a 72-year-old Serbian man described changing family values, particularly in relation to the perceived independence that the Australian law and welfare system affords younger people (Radermacher et al. 2010); an independence which in his view contributes to the erosion of family piety and support networks, noting that 'while the family stays together, young and old together, we have a chance to help one another more' (p. 364).

The assistance received however, by older people from family members, is not necessarily a one-way exchange. As Feldman and Seedsman (2005) remind us 'Any understanding of family life would be in serious deficit if it did not include reciprocation within an exchange relationship' (p. 185). Older people provide a diverse range of support to their families, be it in the form of financial, emotional or material support and this was also evident in our study exploring food security in culturally diverse communities (Radermacher et al. 2010). Older people from all cultural groups included in the study (Macedonian, Serbian, Maltese and Anglo-

Australians) indicated that they may contribute to the household through cooking, caring for children and grandchildren or providing financial support to their extended family members. In addition, family members might have explicitly or implicitly negotiated arrangements about the contribution across the generations. For example, a 68-year-old Macedonian man told us that he specifically kept money aside to pay his grandchildren for assisting him with shopping and household chores. Other people in our study also reported that they had negotiated the financial contributions of their adult children who still lived at home. On the other hand, a 67-year-old Macedonian woman reported that her son and his family lived with her and her husband and that she shopped and cooked for 7 people in the household. Although different in emphasis, all of these examples support the view that Australian families are becoming increasingly diverse in response to the changing economic, social and cultural climate, as well as also reflecting changes to family structure, values and configurations (Feldman and Seedsman 2005). In addition, long established cultural traditions of passing on values, culture and property to following generations were described, in a study about older ethnic minority rural men, as being under threat because of some circumstances beyond their control, including economic development, changing government policy, agricultural practices, family structure and different values of the younger generations (Feldman et al. 2012).

6.4.4 Technological Innovations

Technological advancements as applied to daily life, whether it is the latest in biomedicine, nanotechnology or communication, have a vital role to play in maintaining the health and quality of life of older people. We have seen over the past 100 years an unprecedented development in technological advancements in all aspects of life, but perhaps none so profound as in relation to facilitating a sense of self-determination and independence of older people, particularly those who continue to live in their own homes as they age. An Australian government policy document outlines how the implementation, development and promotion of strategies for healthy ageing must include discussions about the potential role of technology (Prime Minister's Science, Engineering Innovation Council 2003). The document advocates for the integration of technology into all aspects of older people's lives, including psychosocial domains. Many of these innovations have well and truly demonstrated their capacity to contribute to the quality of life of older people who live outside of institutional care, often in their own homes and with minimal support. Technology can play a central role in facilitating communication with family and friends, keeping people connected and in touch. Technology can provide the mechanism for ensuring the inclusion of those individuals who may not be able to physically join in activities of family, friends and community outside of their home. Technological innovations can also assist people to access goods and services and are particularly appropriate for older people living in remote locations, as well as those who do not drive or who find public transport a struggle. New developments in technology may provide the key to quality communication and thus a sense of well-being and inclusion.

It is however particularly relevant that any consideration of innovative technological endeavours also includes the perspectives and needs of older ethnic minority people. Our research indicates that it is fundamental that people from diverse cultural backgrounds take an active role in ensuring that they have up to date and relevant health information as this ensures a sense of self-determination, which in turn contributes to quality of life, health and well-being. Both older people and people from ethnic minority backgrounds have the potential to be further excluded as we forge into a new technological era. Therefore, it is essential that technology is user friendly and relevant, and acknowledges and accommodates for the diverse levels of literacy, learning preferences and language preferences that make up the Australian population. It is also timely to reiterate that older people are not an homogenous group and the significance of difference in culture, gender and life experiences are elements that shape both the ageing experience and the uptake of any technologies.

6.5 Conclusion

This chapter elucidates a number of factors that influence the health and well-being of older people from ethnic minority backgrounds, with a view to encouraging the consideration of how the Australian service system may better support this group of Australians. The life experiences, health concerns and expectations of older Australians from ethnic minority backgrounds are in many ways different from those of older people from the Anglo-Australian majority. Therefore, when considering the components of an effective service system, the need to respond to the health and support needs of *all* Australians should be kept in mind. Currently, evidence suggests that this is not the case, particularly for those people with poor English language proficiency and health literacy (Rowland 1999).

The majority of evidence indicates great diversity both between and within different ethnic groups, and in this regard it would be foolish to assume that 'one size fits all' in relation to developing health promotion and information strategies, policies or when planning services for diverse multicultural and multilingual groups (Quine 1999). Furthermore, it is evident that the service system needs to adapt to the changing, culturally diverse profile of the ageing population, and take account of the particular health and care concerns of older overseas-born persons as active participants who have a capacity to maintain their own health and well-being under the right circumstances. In addition, service planning and delivery needs to be innovative, with adequate flexibility to work with the uneven and dynamic distribution of older ethnic minority people across urban and rural Australia.

Strategies have been implemented at all levels of the Australian government to increase cultural competence and cultural responsiveness in an attempt to increase the capacity of the existing service system to respond to the diverse needs of

Australians. The needs of people from ethnic minority communities with greater English language proficiency and better health literacy are more amenable to this approach, being managed within existing, mainstream service provision (Rowland 1999). Renzaho (2008) highlights a limitation of a cultural competence approach in that it fails to recognise the need for partnership and collaboration to maximise service delivery options and reduce service duplication. An effective health and service system for people from ethnic minority backgrounds must comprise both ethno-specific and mainstream services, preferably working in partnership (Radermacher et al. 2009; Warburton et al. 2009).

Overall, existing research with and about communities of older ethnic minority Australians reinforces the position that social policies and programmes have an opportunity and obligation to respond to the disadvantages that this group face, to reflect and respond to the growing number of older ethnic minority people who wish to remain living as healthy, independent, engaged and contributing members of their family and community. As they age, these people, like any community of older people, need access to appropriate medical, health and other support services. At the heart of the matter is the issue of access and equity in relation to ensuring the health and well-being of older ethnic minority people through appropriate health promotion, information, support and service provision. In the words of a service provider working in rural Australia, but applicable to any location: *'You've got to take it to them, I think that's probably the best model'* (Radermacher and Feldman 2014, p. 13).

6.6 Case Study

The following case study is based on findings from research conducted by the authors in rural Victoria in 2009 and is included to encourage readers to consider how the service system may better support older Australians from ethnic minority backgrounds to maintain and promote their own health and well-being.

6.6.1 Older Migrant Man in a Regional Town

Joe is an 80-year-old Italian man, living in rural Victoria who migrated as a 17-year-old boy with little or no schooling and as an outcome is unable to read or write in his own language. Joe met his Italian wife in Australia, established a farm and, as an orchardist over the past 60 years, has continued to work hard to raise and support his family. Joe's extended family live in close proximity to Joe and his wife with two of his three sons working with him in the orchard. Hard work is valued in Joe's community and over the past decades he has missed very few working days due to sickness, despite suffering from musculoskeletal injuries incurred in the workplace over many years.

Like other men of his generation, Joe says that he does not attend regular medical appointments or seek health advice and will only seek medical attention in a crisis situation or at his family's insistence. Because of his poor English language skills Joe relies upon his family to interpret for him at medical appointments. Joe has been diagnosed with high blood pressure and diabetes, and as such his wife cooks what he describes as 'good' food for him while also gently prodding him to seek health information, particularly in relation to prostate and colon care. He doesn't smoke anymore, and only drinks moderately. Joe and his wife actively participate in their community and family life.

Similar to other older farmers in his district, Joe is determined to hold onto his role as the main breadwinner in his family and head of the household, insisting on continuing to undertake the demanding physical work necessary in the fruit industry, despite his diminishing physical capacity and the protests of his family. Joe's determination to keep on working is within the context of the changing nature of farming practices in the agricultural industry, new government regulations and the introduction of new technologies, which together cause him a great deal of angst. In addition, Joe has been confronted with the decision by his sons and their families to leave the business and move away from the area. These changes to Joe's working life have potential financial implications as well as affecting Joe emotionally. These disruptions to well-established patterns of family relationships and work have put his extended family under pressure, led to an increase in Joe's stress levels and resulted in him being prescribed medication for anxiety and high blood pressure. In addition, Joe's family worry that in the future—when and if necessary—Joe will refuse to accept the care or assistance provided by local government or service provider agencies, preferring to turn to his family members for support. Nevertheless, Joe is from a generation of older migrant men and women whose determination, resilience and accomplishment have mostly held them in good stead, enabling most to continue to live within their community and to face their older years with their sense of cultural identity and strong family values intact.

6.6.2 Discussion Points

- What key factors might practitioners consider in relation to providing appropriate and responsive health and service support for older people from ethnic minority backgrounds such as Joe and his family, both now and in the future?
- Against the background of a life of hard work, both in urban and rural Australia, what appropriate strategies might be helpful in assisting Joe and others to maintain their optimal health and well-being as they grow older?
- Given the changing nature of service delivery and information exchange, what is the role of technology for creating new and innovative opportunities for engaging with older ethnic minority Australians to maintain their health and well-being?
- How can understanding the perspectives of older people from ethnic minority backgrounds help us to rethink strategies for assisting them as they grow older, with particular consideration for the role of family and levels of English language proficiency?

Acknowledgments The authors would like to thank the many people (both participants and service providers) who have assisted us in making our research in this area possible. We also acknowledge Clair Cullen for retrieving some literature for this chapter.

References

Alexander, C., Edwards, R., & Temple, B. (2004). *Access to services with interpreters: User views*. London: Joseph Rowntree Foundation and London South Bank University. Retrieved from http://www.jrf.org.uk/system/files/1859352294.pdf

Arber, S., Davidson, K., & Ginn, J. (Eds.). (2003). *Gender and ageing: Changing roles and relationships*. Buckingham: Open University Press.

Arber, S., & Ginn, J. (1995). Women and ageing. *Reviews in Clinical Gerontology, 4*(4), 93–102.

Australian Bureau of Statistics. (2003). *Disability, ageing and carers: Summary of findings* (Cat No. 4430.0). Canberra: Commonwealth of Australia.

Australian Bureau of Statistics. (2012a). *Reflecting a nation: Stories from the 2011 Census, 2012-2013* (Cat No. 2071.0). Canberra: Commonwealth of Australia.

Australian Bureau of Statistics. (2012b). *Year Book Australia 2012* (Cat No. 1301.0). Canberra: Commonwealth of Australia.

Australian Commission on Safety and Quality in Health Care. (2011). *Patient and consumer centred care: Health Literacy*. Retrieved December 31, 2012, from http://www.safetyandquality.gov.au/our-work/patient-and-consumer-centred-care/health-literacy/

Australian Healthcare Associates. (2007). *Evaluation of the culturally equitable gateways strategy: A draft report*. Melbourne: Department of Human Services.

Australian Institute of Health and Welfare. (2002). *Australian health inequalities* (Bulletin 2). Canberra: AIHW.

Australian Institute of Health and Welfare. (2007). *Older Australia at a glance* (4th ed., Cat. No. AGE 52). Canberra: AIHW.

Baker, D. W. (2006). The meaning and the measure of health literacy. *Journal of General Internal Medicine, 21*, 878–883.

Baltes, P. B., & Baltes, M. M. (1990). Psychological perspectives on successful aging: The model of selective optimization with compensation. In P. B. Baltes & M. M. Baltes (Eds.), European network on longitudinal studies on individual development, *Successful aging: Perspectives from the behavioral sciences* (pp. 1–34). Cambridge, UK: Cambridge University Press.

Bartlett, H., Rao, D. V., & Warburton, J. (2006). *Ageing and cultural diversity in Queensland: Working together to make a difference*. Queensland: Australasian Centre on Ageing, University of Queensland.

Bengston, V. L. (2000). Beyond the nuclear family: The increasing importance of multigenerational bonds. *Journal of Marriage and Family, 63*, 1–15.

Betancourt, J. R., Green, A. R., Carrillo, J. E., & Ananeh-Firempong, O. (2003). Defining cultural competence: A practical framework for addressing racial/ethnic disparities in health and health care. *Public Health Reports, 118*(4), 293–302.

Bird, S., Radermacher, H., Feldman, S., Sims, J., Kurowski, W., Browning, C., et al. (2009). Factors influencing the physical activity levels of older people from culturally diverse communities: An Australian experience. *Ageing & Society, 29* (Special Issue 08), 1275–1294.

Browning, C., & Thomas, S. (2007). *Definitions and predictors of successful ageing and related concepts*. Melbourne: Final report prepared for the Victorian Department of Human Services.

Butt, J., & O'Neil, A. (2004). *'Let's move on': Black and minority ethnic older people's views on research findings*. New York, UK: Joseph Rowntree Foundation.

Calasanti, T. M., & Slevin, K. F. (Eds.). (2006). *Age matters: Re-aligning feminist thinking*. New York: Routledge.

Chou, K. L., & Chi, I. (2002). Successful aging among the young-old, old-old, and oldest-old Chinese. *International Journal of Aging and Human Development, 54*(1), 1–14.

De Vaus, D. (1996). Children's responsibilities to elderly parents. *Family Matters, 45*, 16–21.

Ethnic Communities' Council of Victoria. (2012). *An investment not an expense: Enhancing health literacy in culturally and linguistically diverse communities* (ECCV Policy Paper). Melbourne: Ethnic Communities' Council of Victoria.

Federation of Ethnic Communities' Councils of Australia. (2015). *Review of Australian research on older people from culturally and linguistically diverse backgrounds*. A project funded by the Australian Government Department of Social Services. Deakin, ACT: FECCA.

Feldman, S., & Radermacher, H. (2011). *Understanding ageing well in culturally diverse older males living in Greater Shepparton*. Melbourne: Healthy Ageing Research Unit (HARU), Monash University. Unpublished report prepared by HARU as part of a collaboration with Regional Information Advocacy Council (RIAC), Southern Cross Aged Care Services and Violet Town Bush Nursing Centre.

Feldman, S., Radermacher, H., & Petersen, A. (2012). The vicissitudes of 'healthy ageing': The experiences of older migrant men in a rural Australian community. In A. Kampf, B. Marshall, & A. Petersen (Eds.), *Aging men: Masculinities and modern medicine* (pp. 86–104). London: Routledge.

Feldman, S., & Seedsman, T. (2005). Ageing: New choices, new challenges. In M. Poole (Ed.), *Family: Changing families, changing times*. Crows Nest, NSW: Allen & Unwin.

Fernández-Ballesteros, R., García, L., Blanc, D., Efklides, A., Kornfeld, R., Lerma, A. J., et al. (2008). Lay concept of aging well: Cross-cultural comparisons. *Journal of the American Geriatric Society, 56*, 950–952.

Gallagher, L. P., & Truglio-Londrigan, M. (2004). Community support: Older adults' perceptions. *Clinical Nursing Research, 13*(1), 3–23.

Haralambous, B., Moore, K., & Tate, R. (2007). *HACC social support for people from culturally and linguistically diverse (CALD) backgrounds* (p. 101). Melbourne: NARI.

Harper, S. (2000). Ageing 2000: Questions for the 21st century. *Ageing and Society, 20*(1), 111–122.

Hassett, A., & George, K. (2002). Access to a community aged psychiatry service by elderly from non-English-speaking backgrounds. *International Journal of Geriatric Psychiatry, 17*(7), 623–628. doi:10.1002/gps.656.

Hassett, A., George, K., & Harrigan, S. (1999). Admissions of elderly patients from English-speaking and non-English-speaking backgrounds to an inpatient psychogeriatric unit. *Australian and New Zealand Journal of Psychiatry, 33*(4), 576–582.

Hatton-Yeo, A. (2007). *Intergenerational programs, intergenerational solidarity and social cohesion* (Vol. 23). Barcelona: "la Caixa" Foundation.

Henderson, S., & Kendall, E. (2011). Culturally and linguistically diverse peoples' knowledge of accessibility and utilisation of health services: Exploring the need for improvement in health service delivery. *Australian Journal of Primary Health, 17*, 195–201.

Holstein, M. B., & Minkler, M. (2007). Critical gerontology: Reflections for the 21st century. In M. Bernard & T. Scharf (Eds.), *Critical perspectives on ageing societies* (pp. 13–26). Bristol: Policy Press.

Howe, A. (2006). *Cultural diversity, ageing and HACC: Trends in Victoria in the next 15 years*. Melbourne: Department of Human Services.

Ip, D., Lui, C., & Chui, W. (2007). Veiled entrapment: A study of social isolation of older Chinese migrants in Brisbane, Queensland. *Ageing & Society, 27*(5), 719–738.

Jorm, A., Christensen, H., Henderson, A. S., Jacomb, P., Korten, A., & Mackinnon, A. (1998). Factors associated with successful ageing. *Australasian Journal on Ageing, 17*, 33–37.

Karunarathna, Y., Radermacher, H., & Feldman, S. (2010). *Practising positive partnerships in the ethnic and multicultural community: Networking, coordinating, cooperating, collaborating?* (pp. 1–63). Melbourne, Australia: Monash University, Ethnic Communities' Council of Victoria, and the Southern Metropolitan Region HACC Cultural Diversity Network.

Kleinman, A. (1988). *The illness narratives: Suffering, healing, and the human condition*. New York: Basic Books.
Kruger, E., Tennant, M., Smith, K., & Peachey, J. (2007). The oral health and treatment needs of community-dwelling older people in a rural town in Western Australia. *Australasian Journal on Ageing, 26*(1), 15–20. doi:10.1111/j.1741-6612.2007.00187.x.
Mackenzie, E. R. (1999). Cultural competence increases access to healthcare for at-risk elders. *Innovations in Aging, 28*(2), 3–8.
Matsuoka, A., & Sorenson, J. (1991). Ethnic identity and social service delivery: Some models examined in relation to immigrants and refugees from Ethiopia. *Canadian Social Work Review, 8*(2), 255–268.
McLennan, W. (1999). *Standards for statistics on cultural and language diversity* (ABS Catalogue No. 1289.0). Canberra: Australian Bureau of Statistics, Commonwealth of Australia.
Migliorino, P. (2010). The ageing of the post-war migrants: A challenge for health promotion and service delivery. *Health Voices, 6*, 21–22.
Migrant Information Centre and Yooralla. (2006). *Meeting needs: Perceptions of respite care—Findings from respite service providers and Chinese, Indian & Sri Lankan focus groups*. Melbourne.
Nimri, E. (2007). *Social, care and support needs of older people from culturally and linguistically diverse backgrounds in the Gold Coast Region: Needs analysis*. Ashmore, QLD: Multicultural Seniors Program, Multicultural Communities Council (Gold Coast).
Orb, A. (2002). *Health care needs of elderly migrants from culturally and linguistically diverse (CALD) backgrounds: A review of the literature*. Perth, WA: Freemasons Centre for Research Into Aged Care Services, Curtin University of Technology.
Patel, N. (1999). Black and minority ethnic elderly: Perspectives on long-term care. In Royal Commission on Long Term Care for the Elderly (Ed.), *With respect to old age* (Vol. Research Volume 1, pp. 257–304). London: HMSO.
Prime Minister's Science, Engineering Innovation Council (PMSEIC). (2003). *Promoting healthy ageing in Australia*. Retrieved from http://www.industry.gov.au/science/PMSEIC/Documents/PromotingHealthyAgeing.pdf
Quine, S. (1999). Health concerns and expectations of Anglo and ethnic older Australians: A comparative approach. *Journal of Cross-Cultural Gerontology, 14*(2), 97–111.
Radermacher, H. & Feldman, S. (FirstView Article, 2014). 'Health is their heart, their legs, their back': Understanding ageing well in ethnically diverse older men in rural Australia. *Ageing and Society, 35*(5), 1011–1031. doi:10.1017/S0144686X14001226. Available on CJO 2014.
Radermacher, H., Feldman, S., & Browning, C. (2008). *Review of literature concerning the delivery of community aged care services to ethnic groups—Mainstream versus ethno-specific services: It's not an 'either or'* (pp. 1–41). Melbourne, Australia: Ethnic Communities' Council of Victoria and Monash University.
Radermacher, H., Feldman, S., & Browning, C. (2009). Mainstream versus ethno-specific community aged care services: It's not an 'either or'. *Australasian Journal on Ageing, 28*(2), 58–63. doi:10.1111/j.1741-6612.2008.00342.x.
Radermacher, H., Feldman, S., Lorains, F., & Bird, S. (2010). Exploring the role of family and older people's access to food in different cultures: Will the children be there to help? *Journal of Intergenerational Relationships, 8*(4), 354–368.
Rao, D. V., Warburton, J., & Bartlett, H. (2006). Health and social needs of older Australians from culturally and linguistically diverse backgrounds: Issues and implications. *Australasian Journal on Ageing, 25*(4), 174–179.
Ratzan, S. C., & Parker, R. M. (2000). Introduction. In C. R. Selden, M. Zorn, S. C. Ratzan, & R. M. Parker (Eds.), *National library of medicine current bibliographies in medicine: Health literacy [bibliography online]* (NLM Pub. No. CBM 2000-1). Bethesda, MD: National Institutes of Health, U.S. Department of Health and Human Services. Retrieved from http://www.nlm.nih.gov/archive//20061214/pubs/cbm/hliteracy.html
Renzaho, A. (2008). Re-visioning cultural competence in community health services. *Human Resource Management, 32*(2), 223–235.

Rienzo, C., & Vargas-Silva, C. (2012). Migrants in the UK: An overview. *Migration Observatory briefing*. Oxford: COMPAS, University of Oxford.
Rowe, J. W., & Kahn, R. L. (1997). Successful aging. *The Gerontologist, 37*(4), 433–440.
Rowe, J. W., & Kahn, R. L. (1998). *Successful aging*. New York: Dell Publishing.
Rowell, M. (2005). *TasCOSS Home and Community Care Consumer Consultation Project Annual Report 2005*. Hobart: Tasmanian Council of Social Services.
Rowland, D. T. (1999). The ethnic aged population and the likelihood of special needs. *Australasian Journal on Ageing, 18*(3 Suppl), 50–54.
Royal Australasian College Physicians. (2004). *An introduction to cultural competency*. Sydney: Royal Australasian College Physicians.
Sawrikar, P., & Katz, I. (2009). How useful is the term 'Culturally And Linguistically Diverse (CALD)' in the Australian social policy discourse? In *Proceedings from the ASPC, SPRC, UNSW, Presented at Australian Social Policy Conference: An Inclusive Society? Practicalities and Possibilities, Sydney, 8-10 July 2009*.
Schillinger, D., & Davis, T. (2005). A conceptual framework for the relationship between health literacy and health care outcomes: The chronic disease exemplar. In J. G. Schwartzberg, J. B. VanGeest, & C. C. Wang (Eds.), *Understanding health literacy: Implications for medicine and public health* (pp. 181–204). Chicago: American Medical Association, AMA Press.
Statistics Bureau of Japan. (2011). *Population count based on the 2010 census released*. Retrieved September 6, 2012, from www.stat.go.jp/english/
Thomas, T. (1999). Stress, coping and the mental health of older Vietnamese migrants. *Australian Psychologist, 34*, 82–86.
Thomas, T. (2007). Migration and aging: Australian perspectives and research. In S. Carmel, C. A. Morse, & F. M. Torres-Gil (Eds.), *Lessons on aging from three nations* (pp. 155–167). Amityville, NY: Baywood Publishing Company Inc.
Tohit, N., Browning, C. J., & Radermacher, H. (2012). We want a peaceful life here and hereafter: Healthy ageing perspectives of older Malays in Malaysia. *Ageing & Society, 32*(3), 405–424.
Torres, S. (2004). Making sense of the construct of successful ageing: The migrant experience. In S. O. Daatland & S. Biggs (Eds.), *Ageing and diversity: Multiple pathways and cultural migrations* (pp. 125–140). Bristol: Policy Press.
US Census Bureau. (2012). *The foreign-born population in the United States: 2010*. Retrieved September 6, 2012, from http://www.census.gov/prod/2012pubs/acs-19.pdf
Warburton, J., Bartlett, H., & Rao, D. V. (2009). Ageing and cultural diversity: Policy and practical issues. *Australian Social Work, 62*(2), 168–185.
Ward, B. M., Anderson, K. S., & Sheldon, M. S. (2005). Patterns of home and community care service delivery to culturally and linguistically diverse residents of rural Victoria. *Australian Journal of Rural Health, 13*(6), 348–352.
Weiss, B. (2005). Epidemiology of health literacy. In J. G. Schwartzberg, J. B. VanGeest, & C. C. Wang (Eds.), *Understanding health literacy: Implications for medicine and public health* (pp. 17–40). Chicago: American Medical Association, AMA Press.
World Health Organization. (2002). *Active aging: A policy framework*. Geneva: World Health Organization.
Zimmerman, C., Kiss, L., & Hossain, M. (2011). Migration and health: A framework for 21st century policy-making. *PLoS Medicine, 8*(5), e1001034. doi:10.1371/journal.pmed.1001034.

Chapter 7
Indigenous Australians and Ageing: Responding to Diversity in Policy and Practice

Kathleen Clapham and Cathy Duncan

7.1 Introduction

Recent public discussion around the ageing of the Australian population has paid little attention to the Indigenous population. Governments have started to recognise the importance of taking into account the distinct social and cultural attributes of the Indigenous population in programme planning and policy aimed at older people, but there are still many challenges for services provision. The scant research on this topic means that there is a limited evidence base on which to inform policies and programmes. Drawing on published research literature as well as the grey literature,[1] this chapter provides a review of current literature on ageing and aged care in the Indigenous Australian population in order to identify some of the challenges and opportunities for addressing health and other needs as they age.

[1]The grey literature' includes reports, conference papers, theses, bibliographies, and government reports and documents not published commercially (Lawrence et al. 2014; New York Academy of Medicine 2002).

K. Clapham, B.A. (Hons.), Ph.D. (✉) • C. Duncan, B.S.W. Grad. Cert. H.S.R.
Australian Health Services Research Institute, University of Wollongong,
Level 1, Building 234 iC Enterprise 1, Innovation Campus, Wollongong,
NSW 2522, Australia
e-mail: kclapham@uow.edu.au; cduncan@uow.edu.au

K. O'Loughlin et al. (eds.), *Ageing in Australia*, International Perspectives on Aging 16, DOI 10.1007/978-1-4939-6466-6_7

7.2 The Health and Well-being of Older Indigenous Australians

7.2.1 Characteristics of the Older Indigenous Australian Population

The Australian Indigenous population is a broad heterogeneous group containing distinct language groups and subcultures and enormous cultural, social and geographical diversity.[2] At the 2011 Census there were 548,370 Indigenous Australians (Australian Bureau of Statistics (ABS) 2012), constituting around 2.5 % of Australia's population. Indigenous people occupy different environments in urban, rural and remote locations, with corresponding differences in attitudes, cultural identification and needs. Around 60 % of the Aboriginal and Torres Strait Islander population live in capital cities and inner regional areas, and just over 20 % live in remote and very remote areas (ABS 2013).

The Indigenous population is considerably younger than the non-Indigenous population, largely due to higher fertility rates, lower life expectancy as well as higher mortality rates in the middle adult age group (45–65 years). In 2011, only 8.9 % of the Indigenous population was aged 55 years and above compared to 25.8 % of the non-Indigenous population (Australian Government 2013). But despite its relatively young age structure Australia's Indigenous population is gradually ageing. This demographic trend suggests that the coming decades will see a growing demand for culturally appropriate aged care services by Indigenous Australians, so it is likely that ageing will become an increasingly important issue for Indigenous communities.

Health statistics reveal a stark contrast between the health of Indigenous and non-Indigenous Australians. Relative to non-Indigenous Australians, Indigenous people have poorer health, higher rates of disability and higher mortality rates (Vos et al. 2009). Life expectancy is a key measure of the health of populations. Indigenous males can expect to live to 67 years compared to 78.7 years for all non-Indigenous males; Indigenous females can expect to live to 73 years compared to 82.6 years (Australian Institute of Health and Welfare (AIHW) 2011a). In recent years, the unacceptable gap in life expectancy between Indigenous and non-Indigenous Australians has become a rallying point for action to address health inequities; it became the focal point of a bipartisan approach to Indigenous health policy in December 2007 when the Council of Australian Governments (COAG) committed to closing the gap in life expectancy between Indigenous and non-Indigenous Australians within a generation (COAG 2009).

[2]The terms 'Indigenous Australian' and 'Aboriginal and Torres Strait Islanders' are used interchangeably in this chapter to describe the Indigenous peoples of Australia. The term 'indigenous people' is used more broadly to describe the indigenous peoples of various nations. The terms Koori/Murri/Nunga and other such terms are directly derived from Aboriginal languages and are the names used by Aboriginal people in specific areas when referring to themselves.

Chronic diseases and conditions make up around 80 % of the difference in mortality rates between Indigenous and non-Indigenous Australians aged 35–74 years (AIHW 2011a), with cardiovascular disease the leading cause of disease burden among Indigenous people 55 years and over (Vos et al. 2009). Dementia is emerging as a problem for Indigenous people at comparatively young ages (under 75 years) and this is thought to be due to the high rates of chronic disease and other risk factors they experience throughout life (AIHW 2011b; Li et al. 2014; Smith et al. 2008).

It is widely accepted that a very broad range of social and economic factors underlie the poor health status of Indigenous Australians (Aboriginal and Torres Strait Islander Social Justice Commissioner 2005; Carson et al. 2007; Marmot 2011). They include social and historical factors, such as dispossession and the effects of the 'stolen generation' (Human Rights and Equal Opportunity Commission (HREOC) 1997); educational factors, such as lower levels of schooling; economic factors, such as income poverty and high rates of unemployment; much higher levels of incarceration and ongoing discrimination (Dudgeon et al. 2014; HREOC 1991); and environmental factors such as poorly maintained housing, lack of basic public health infrastructure and dangerous roads and unsafe environments. Other factors, such as the lack of access to good quality health care, misuse of alcohol and other drugs, high rates of tobacco smoking, poor nutrition, obesity and physical inactivity, all contribute to health problems. Underlying these issues, and also exacerbating their impact, is the level of stress which Indigenous people report (AIHW 2011c). These factors affect the health of Indigenous people across the life span.

7.2.2 Addressing the Health Needs of Older Indigenous Australians

Indigenous people experience an increase in the rate of ill health and disability with increasing age. This often leads to a need for care services at comparatively younger ages. Due to the greater burden of chronic and other illnesses and the subsequent need for health and support services, Indigenous Australians have been designated a 'special needs group' in Australian Government aged care legislation and policy; Aboriginal and Torres Strait Islander persons over 50 years are considered 'aged', as compared to those 65 years and over in the general population (Steering Committee for the Review of Government Service Provision (SCRGSP) 2015). This age differential is used to plan and allocate aged care services, such as residential care, for the Indigenous population.

While well intentioned, a number of authors have highlighted the negative implications of this policy. Cotter and colleagues, for example, argue that the underlying assumption that the Indigenous population aged over 50 years have the same set of age-associated conditions and care needs as the non-Indigenous population aged over 70, has led to a lack of balance in service provision across geographic areas and between different levels of care. They question the appropriateness of using aged care facilities designed for non-Indigenous populations to provide chronic disease care services to people in their 50s (Cotter et al. 2011). Others suggest that the policy may contribute to the stereotyping of Indigenous people over the age of 50 as dependent and in need of care (Wall 2013). Another possible consequence of the policy is that the inclusion of 50–70-year-olds in an 'aged care' bracket diverts attention from the full range of their health care needs and potentially excludes this group from Indigenous population health strategies; it reinforces an uncritical acceptance of early deaths and only manages their decline rather than offering more positive interventions (Cotter et al. 2007; Cotter et al. 2012). In future it is likely that Indigenous people will access aged care services at older ages than is currently the case and this will have important implications for the delivery of aged care and health, disability and community welfare systems (Productivity Commission 2008).

It is important to recognise that efforts to improve the health of older Aboriginal people must also address the wider issues impacting on the health of the Indigenous population. Indigenous people view health in broad holistic terms. Health and well-being go beyond physical health to include the mental, spiritual, cultural and emotional health of an individual and the well-being of the whole community (National Aboriginal Health Strategy Working Party 1989). This definition highlights the importance of having in place effective preventive health programmes that address social and cultural issues such as employment across the life cycle, nutrition, housing and education in improving health overall, as well as promotion of factors that stimulate healthy brain growth.

7.3 The Cultural and Social Context of Indigenous Ageing

7.3.1 Historical Context: Impact of Past Policies

The history of colonisation and past government policy in Australia has had long-term negative impacts on today's older Indigenous generation, particularly those who were members of the 'stolen generations', a term which refers to Indigenous children separated from their families and communities as part of past government laws, policies and practices. Many older Indigenous people experienced the trauma of the removal of their children, oppressive policies, prejudice and discrimination, inadequate diet, poor medical care, cultural dispossession and displacement from their homelands. The transgenerational effects of growing up in state-run homes, in foster care or adopted out to non-Aboriginal families continues to effect the lives of many Indigenous people, their families and communities, years after. Even those not removed by the state, suffered as a result of being prohibited from speaking their language or practicing their culture (HREOC 1997).

For many older Indigenous people, the history of exclusion, disadvantage and discrimination across multiple domains throughout their lives has resulted in impoverishment, lack of assets and vulnerability to exploitation (Sykes 1988; Bin-Sallik and Ranzijn 2001). A greater proportion of older Aboriginal people, therefore, are likely to be income poor and reliant on government benefits. As they move into old age, therefore, Aboriginal people are less likely to be home owners, have inherited wealth, savings or superannuation as they have not had the same choices for housing, living conditions or security as most non-Aboriginal Australians.

Colonisation had other negative impacts on Indigenous older people. Indigenous identity is strongly linked to extended family, social networks and maintaining links with cultural values and practices. But colonisation disrupted family relationships and led to a lack of socialisation into the role of elder. It impacted on the time and capacity to care for older people in traditional ways and contributed to a loss of intergenerational learning (King et al. 2009). Despite these hardships, a strong sense of identity continues to underpin the health and well-being of Indigenous people; this is positively reinforced by the strong connection to land and the knowledge that their ancestors have lived here for over 50,000 years (Wilkes 2002).

7.3.2 Culture, Gender and Respect in Old Age: The Role of Older People in Maintaining Health

Respect for older people is an integral feature of Indigenous cultures, and Indigenous cultures are frequently distinguished from Western cultures in terms of the respect bestowed to older members of their communities (Berndt and Berndt 1988; Hulko et al. 2010; Reid 1985; Sykes 1988; Wilson et al. 2010). There are significant gender differences in the roles Aboriginal people assume as they age. Traditionally, the political and ceremonial roles of Aboriginal men increase as they age (Berndt and Berndt 1988; Reid 1985). Aboriginal women also assume greater authority as they age, although differently or separately from the male spheres of influence (Kaberry 1939; Bell 1983). Generational knowledge is transmitted along gender lines.

But not all older Indigenous people are held in the same high esteem within their communities. The term 'elder', which is achieved by some older Aboriginal people, bestows a particular status of honour, wisdom and respect. Elders are highly valued members of their communities who hold the status of an elder statesperson (Arkles et al. 2010; Gray et al. 1991) and have some cultural authority (Cotter et al. 2007). The cultural roles and responsibilities of elders differ from community to community.

Older Indigenous people as a group play a pivotal role in maintaining the health of their communities; they are the repositories of the oral traditions, food and medicinal herbs, spiritual matters and knowledge of the land, all of which link to health and well-being. They provide guidance and support to the community, participate in decision making, cultural activities or ceremonial events, share stories, advice and pass on key information about their heritage (Arabena 2007).

The land has an essential central place in all Indigenous cultures; traditional lands or 'Country' embody kin, history, spirituality and identity. Participation in 'caring for Country' activities by those living on traditional lands in a remote area of Northern Australia has been associated with significantly better health outcomes (Kuipers et al. 2011; McDermott et al. 1998). Indigenous people's connection to their country or homelands is important to cultural continuity. Senior members of Indigenous communities are frequently relied upon to substantiate ongoing connection with land in native title claims (Arabena 2007; Cotter et al. 2007).

The cultural identity of Indigenous people in Australia today has a strong collective perspective with the community at the core (Dudgeon et al. 2014). Aboriginal households may include two or more generations and obligations to the extended family are built on concepts of caring and sharing (Wilkes 2002). Kinship ties foster economic, psychological and physical support and stability and teach young men and women about the relationships which connect families to each other, the community and the land (Bessarab 2006). This social system ensures broad familial support but also entails obligations and responsibilities.

Older people play a central role in maintaining family connections and contribute to family and kin cohesion (Waugh and Mackenzie 2011). They are advisors as well as carers of their families and communities. Maintaining the well-being of Aboriginal elders is vital to the maintenance of the well-being of the whole Aboriginal community (Bin-Sallik and Ranzijn 2001). It is important therefore that they are able to remain within communities and interact with other generations in order to be able to continue to fulfil their cultural and familial roles (Aboriginal and Torres Strait Islander Ageing Committee 2010).

Older Aboriginal people have particularly important roles in relation to younger generations (Arkles et al. 2010; Bin-Sallik and Ranzijn 2001). Grandparents often undertake a major share in raising their grandchildren. They provide opportunities for grandchildren to participate in traditional activities such as fishing and hunting and learn practical bush skills, even in urbanised areas (Waugh and Mackenzie 2011). Elders also teach life skills including teaching young people how to deal with the racism they frequently encounter (Warburton and Chambers 2007). Grandchildren are regarded as a great source of happiness and satisfaction which assists them in remaining healthy and having fulfilling lives (Waugh and Mackenzie 2011). However, the burden of caring for others can also contribute to stress and exhaustion in some older Indigenous people who are obliged to provide for the children and the less able, the sick or members of the extended family who are in need (Wilkes 2002). Often the grand-parenting role of older people extends to becoming primary carers or surrogate parents where parents are absent. These obligations can place a heavy burden on older people particularly when extended families are caught up in a cycle of oppression and poverty (Wilkes 2002).

In both remote areas and in contemporary urban contexts, older Indigenous people have a key role within families and communities in passing on cultural knowledge which is transmitted orally from one generation to another. Elders rely on their memory to teach traditional law, land and language (Arkles et al. 2010; Waugh and Mackenzie 2011). Storytelling is an important way of passing on information. Teaching by elders has also been incorporated into school curricula; elders teach in the classroom and bush excursions educate young people about land, art, history and spirituality (Warburton and Chambers 2007). Fulfilling the roles and responsibilities of elders can provide a sense of purpose and pride and build self-identity as an Indigenous Australian (Waugh and Mackenzie 2011). As educators elders are also in a position to help strengthen Indigenous identity and strengthen family and community values.

Apart from their responsibilities within the extended family, many older Indigenous people make an invaluable contribution to their communities as volunteers. In the contemporary context, this includes activities such as attending meetings, sitting on advisory committees and boards, political activism, visiting prisons and hospitals, volunteering their wisdom and cultural knowledge at schools, and locating truanting school children and returning them to school (Warburton and Chambers 2007).

7.3.3 *Caring for Older People*

There are significant differences in the ways in which Indigenous and non-Indigenous people care for their older people. Like most traditional societies, responsibility for care of the aged and frail falls to family and kin in the older person's own community (Smith et al. 2010). Younger people are expected to care for the elderly as part of their reciprocal obligations. In some Indigenous communities there are limitations on who can provide personal care to the elders. It is important that those who deliver care to the older person are trustworthy family members, be the right gender and be in the 'right' relationship according to the kinship system. In Warlpiri culture, for example, old people are the 'keepers of culture' who are recognised as having made a significant contribution to ceremonial and social life. Therefore, those entrusted to look after them must also have the right standing in the community (Smith et al. 2010). For these reasons it may be difficult for professional staff from the community to provide care for certain members of the group (Productivity Commission 2011).

Older Indigenous people in remote areas may expect to receive care from traditional healers or 'Ngangkari' who continue to play an important role in the spiritual aspects of healing in some communities. For example, the Ngangkari healers programme is delivered throughout the Anangu Pitjantjatjara Yankunytjatjara lands of South Australia (SA), Western Australia and the Northern Territory (Panzironi 2013).

Illness and physical limitation can impact negatively on the ability of older Indigenous people to continue to engage in meaningful activities and maintain social and cultural roles, including being positive role models for younger people (Warburton and Chambers 2007; Waugh and Mackenzie 2011). The high rates of mortality, morbidity and disability and breakdown in cultural continuity in many

Indigenous communities have impeded the intergenerational transmission of social and cultural knowledge. They have also placed an unacceptably high burden on those who take on additional family and community responsibilities.

Being unable to fulfil important roles and engage in the meaningful occupations that support identity can adversely affect an individual's physical, emotional and spiritual health, as well as that of families and communities (Waugh and Mackenzie 2011). Social interaction and opportunities to mix with other older Indigenous people therefore provides motivation to engage in activities that are important for health and a sense of identity (Wilcox 1993 cited in Waugh and Mackenzie 2011; Wilkes 2002).

Personal safety has also become a significant concern for older people, both in the community and within aged care services. Studies undertaken over the past 25 years report that traditional reciprocal care for individuals within the family is often lacking (Arabena 2007; Arkles et al. 2010; Gray et al. 1991; Sykes 1988; Waugh and Mackenzie 2011; Wilkes 2002). Sykes, writing in 1988, noted the existence of elder abuse. Others have commented more recently on the lack of respect and uncaring attitudes towards older people by younger people, elder abuse, and control across generations and in families as emerging issues associated with Indigenous ageing. Participants in an urban study by Waugh and Mackenzie (2011) were angered by the uncaring attitudes towards older people, which they saw as being influenced by the media and modern culture. Behaviour such as younger people swearing in front of older people and answering back to them were cited as contributing negatively to family well-being and cohesion.

7.4 Policy and Programmes for Older Indigenous People

7.4.1 Indigenous Aged Care Policy

As discussed earlier, the care of the aged and frail is traditionally the responsibility of family and kin in Indigenous communities, but there are sometimes limitations on who can provide care to the elders. Consequently, providing culturally sensitive and appropriate care to older Indigenous people within the Australian health care system has been and remains a challenge for policy makers and service providers (AIHW 2011b).

Until the late 1980s older Aboriginal people from remote communities who required residential aged care services outside the family had little choice but to move away from their communities and enter nursing homes in larger regional or urban centres. Aged care services for Indigenous older people were delivered through hostel accommodation funded through Aboriginal Hostels Limited—a company owned by the Australian Government. By the 1990s there was considerable concern across many parts of Australia about the large number of older Aboriginal people who had reluctantly relocated to live permanently in nursing homes or had to travel to major towns to attend culturally foreign assessment centres (Goodwin 1993; Cotter et al. 2007; Smith et al. 2010). In 1994 a more flexible way of funding and delivering services was achieved through the National Aboriginal and Torres Strait Islander Aged Care Strategy which established Aboriginal and Torres Strait Islander Flexible Aged Care Services.

Current aged care service delivery for older people from Aboriginal and Torres Strait Islander communities is governed by the *Aged Care Act* 1997 which states that both residential and community aged care services are to be provided for all Australians. As previously discussed, Aboriginal and Torres Strait Islander people over 50 years are recognised as a special needs group under the *Aged Care Act* 1997 and in the most recent aged care reforms (Aged Care (Living Longer Living Better) Act 2013. As a result of the inclusion as a special needs group under the Act, aged care providers are required to provide culturally appropriate care, however Home Care Packages are now allocated to individuals rather than being regionally allocated to aged care providers based on identified populations of special needs groups.

In rural and remote locations that are too small to support the standard systems of aged care provision, the Multipurpose Services Program provides a more workable care and treatment model by bringing a range of local health and aged care services, often including residential aged care, together under one management structure. This programme is funded by pooling Commonwealth and State health and aged care funds which are then distributed across programmes according to the needs of the local community in these locations. Older Indigenous people can also access Aboriginal and Torres Strait Islander-specific aged care services through the National Aboriginal and Torres Strait Islander Flexible Aged Care Program.

From 1988 to 1994 the Commonwealth Government, in partnership with the State and Territory Governments, began to provide community-based services through the Home and Community Care (HACC) programme. This programme was directed at older people who would otherwise be forced to leave their homes and move away to receive care. Although these services were quite basic in Aboriginal communities, the Commonwealth Government, under the Aboriginal and Torres Strait Islander Aged Care Strategy, supported communities to negotiate locally appropriate services (Smith et al. 2010). In some states, such as SA and NSW, Aboriginal Home Care Services were established. As of 1 July 2015, the HACC programme transitioned into the Commonwealth Home Support Program (CHSP) (except in WA). The Commonwealth Home Support Program (Commonwealth Department of Social Services (DSS) 2015) formally recognises Aboriginal and Torres Strait Islander people (over 50 years) as a special needs group. From 2018, the CHSP will merge with the Home Care Packages Program to form a single community aged care program, with a focus on increasing 'choice and control' for consumers. However, due to the lack of culturally competent community care providers in many areas, and the case management required to manage new individualised budgets for the packages, the 'choice and control' available for many Aboriginal and Torres Strait Islander people may in reality be limited.

There are currently around 200 aged care services directly funded by the Australian Government that are Indigenous specific and/or are located in remote areas. Around 70 of these have a residential care component and 30 come under the Flexible Aged Care Program (Productivity Commission 2011). Funding was allocated to ensure culturally appropriate quality aged care services would be available for Aboriginal and Torres Strait Islander people. Current aged care reforms (e.g. *Longer Living Better' Act* 2013) recognise that the demand for access to the National Aboriginal and Torres Strait Islander Flexible Aged Care programme already far outstrips supply and acknowledges that many older Indigenous people continue to have to relocate large distances from family and home to access aged and health care services.

Since the introduction of compliance auditing under the *Aged Care Act* 1997 there have been quality improvements across all areas of the age care sector (Brooke 2011). However, Indigenous people residing in flexible aged-care facilities have only recently seen the same compliance requirements as those in the rest of the aged care system. The development of a quality framework for flexible aged-care facilities followed the recommendations from a coroner's case, in which an elder had died as a result of burns after falling into a pit fire at Docker River in the Northern Territory (Anonymous 2008; Brooke 2011). Following recommendations from the Aged Care Commissioner, in 2008 the Australian Government announced the introduction of a quality framework, which includes standards for health and personal care, safety and physical environment, culturally appropriate lifestyle, and effective management and governance. This was the first time the Indigenous aged care sector had been subject to federal regulation and quality standards (Brooke 2011).

7.4.2 Challenges for Policy Makers and Service Providers

The increasing demand for aged care services by Indigenous Australians in recent years has created the need for more funding and resources to be provided to support the needs of this population. Policy makers face significant challenges in developing flexible models of care that can deliver services to culturally and linguistically diverse populations in urban, regional, rural and remote locations; ensuring there is adequate and easily accessible information available about aged care services; and ensuring aged care workers have appropriate linguistic skills and adequate training in providing culturally appropriate care (Productivity Commission 2011). One significant challenge is the high cost associated with improving aged care for Indigenous people. Consultations conducted by the Productivity Commission indicated that flexible care services can be difficult to establish and are vulnerable because they are small and located in remote areas where staff are hard to attract and retain (Productivity Commission 2011). Many Indigenous aged care providers are small organisations serving small populations, so there can be difficulty achieving economies of scale and sharing infrastructure across a range of services.

As the Indigenous population transitions to an older demographic, policy makers and service providers need to know much more about how older Indigenous people experience ageing and have a much better understanding of their care needs in order to offer appropriate choices to Aboriginal clients and their families. However, currently there are very few studies which focus on how Indigenous Australians experience ageing, aged care services or the aged care system. A small number of qualitative studies which have been referred to in this chapter (Bin-Sallik and Ranzijn 2001; Dance et al. 2004; Waugh and Mackenzie 2011) provide useful insights into the lived experience or needs of older Indigenous Australians in remote and urban areas. Additionally,there is very little available information from the perspective of Indigenous people, their families or carers, to provide insight into how to address the challenges they face.

7.5 Patterns and Challenges in Access to Aged Care

7.5.1 Patterns of Service Utilisation by Indigenous Older People

Based on the scant literature on the use of aged care services by Indigenous people and on their needs and preferences for aged care, a number of important differences can be discerned in the patterns of service utilisation between older Indigenous Australians and non-Indigenous people. Older Indigenous people use both specialised services and those that are available to the general population. They use dementia and aged care services at a younger age than other Australians (AIHW 2011b). An increasing number of Indigenous people are residing in residential aged care facilities or accessing community care, with the number of aged care assessments undertaken on Indigenous persons aged 50 or older almost doubling from 2004 to 2007 (Brooke 2011). However, the use of residential care is around 90 % lower than for either Indigenous or non-Indigenous people 70 years and over (Cotter et al. 2011).

A recent Productivity Commission report found that Indigenous Australians seek aged care services at just over one-third the rate of the wider population (Steering Committee for the Review of Government Service Provision (SCRGSP) 2015). Nationally, 18.5 Indigenous people per 1000 used aged care residential services, compared with 51.1 per 1000 for the wider population. They were also far less likely to access information services on ageing and aged care. Nationally the average rate of Aboriginal and Torres Strait Islanders accessing home-based aged care services was higher than in the wider community, at 22.1 per 1000, compared with 17.2.

7.5.2 Use of Dementia-Specific Services

There is a significant knowledge gap on the prevalence of dementia in the Australian Indigenous population (Li et al. 2014) or Indigenous peoples' use of dementia services (Arkles et al. 2010; Garvey et al. 2011). Studies in the Northern Territory, Western Australia and New South Wales suggest a substantially higher prevalence of dementia and younger onset dementia in the Indigenous population compared to the non-Indigenous population (Li et al. 2014; Smith et al. 2008). Research using a validated cognitive tool for the assessment of Indigenous Australians (the Kimberley Indigenous Cultural Assessment tool) has documented much higher prevalence of dementia in Indigenous people over age 45 compared with non-Indigenous people of the same age in the Kimberley region of Western Australia (Smith et al. 2008).

Relatively few Indigenous people with dementia access formal government support programmes, particularly in remote communities (AIHW 2011b). One reason for these low rates may be that dementia is not always recognised as a medical condition by family and community members, and so may be undiagnosed. Unequal access to specialist services, geographical dispersion and lack of mobility also explain the large gaps in dementia service delivery (Alzheimer's Australia 2011; Bin-Sallik and Ranzijn 2001). Waiting times for assessments are long and access to services is difficult, and expensive, especially in regional areas where there are few geriatricians.

There is an urgent need for interventions to address the emerging impact of dementia in the Australian Indigenous population (Alzheimer's Australia 2011). New approaches to brain health highlight the need to address factors across the life cycle which 'grow' or impair brains before cognitive decline begins (Arkles et al. 2010). The conceptual framework for addressing dementia in Indigenous populations needs to take into account the impact of history, context (language, residential location, socio-economic status) and culture in understanding the potential risk factors that this population faces, in both the development of dementia syndromes, and of the approaches required to ameliorate and address these at a community health, public health and service provision level.

7.5.3 Problems in Accessing Care

A key challenge identified for Australian Indigenous people is the lack of aged care services that meet their specific needs. Aged care policies have historically been based on government rather than community needs, resulting in a lack of aged care services in areas where Indigenous people live, particularly in remote areas (Productivity Commission 2011; Alzheimer's Australia 2011; Aboriginal and Torres Strait Islander Ageing Committee 2010; Bin-Sallik and Ranzijn 2001). Where services do exist, in both remote and urban areas, Indigenous people frequently experience difficulties in accessing them (Arkles et al. 2010; Dance et al. 2004). A fundamental problem is the failure to communicate information about services in ways which are relevant to the diverse needs and cultural understandings in the Indigenous population. In a scoping study into the needs of Indigenous Aged Care in South Australia, Bin-Sallik and Ranzijn (2001) noted the lack of understanding among elders and Aboriginal people working with elders, about the way both the mainstream aged care system and Aboriginal aged care system work. Given the complexity of the Australian aged care system it is not surprising that Indigenous people may fail to understand the multiplicity of programmes and services which have been set up to care for aged people.

Older Indigenous people may also experience problems accessing services because of social isolation, personal and health problems (Wall 2013). Transport to services is an ongoing challenge, particularly for those living in rural and remote areas (Arkles et al. 2010; Bin-Sallik and Ranzijn 2001), while in urban areas, social isolation and difficulty accessing culturally appropriate services also contributes to the lower access to services (AIHW 2011b).

Services are generally not set up to cater for the complexity of needs of Indigenous people, notably the overburden of chronic health conditions such as diabetes and cardiovascular diseases at early ages (Cotter et al. 2007; Dance et al. 2004; Waugh and Mackenzie 2011). The need for mobile medical services for Indigenous people with dementia has also been identified as requiring consideration because of the special connection Indigenous Australians have with Country (Alzheimer's Australia 2011). Another frequently reported problem is lack of services which are able to adapt to meet local needs, circumstances and community preferences (Arkles et al. 2010). There are numerous problems in using mainstream services. Available services are not always perceived as culture friendly (Wall 2013) and some older Aboriginal people are fearful of, or are reluctant to engage with, medical and welfare systems (Arkles et al. 2010). Pollitt (1997) noted the many deficiencies in service provision for individuals manifesting markedly abnormal behaviour in Indigenous communities.

7.5.4 Care Preferences and Needs

There are frequent references in the literature to the desire for older Indigenous people to remain in their communities to receive care (Alzheimer's Australia 2011; Arkles et al. 2010; Cotter et al. 2007; Dance et al. 2004; Smith et al. 2010; Woenne-Green 1995). Participants surveyed by Alzheimer's Australia indicated a desire for quality residential aged care which was located in their community, and which also enabled them to make choices consistent with their culture. They were also concerned about the lack of choices for residents and the restrictions due to safety regulations in residential aged care facilities (Alzheimer's Australia 2011). Many of those surveyed wished to be cared for in their own communities where they were close to family and kin; the importance of being in or near to one's own home country at the time of death is widely supported in the literature (Anderson and Devitt 2004 cited in Cotter et al. 2007; Arkles et al. 2010). As stated previously, the majority of Indigenous people have a deep attachment to their traditional lands or own home country. There is some evidence for the detrimental effects on the health of older people as a result of having to leave their communities (Alzheimer's Australia 2011).

Services run and delivered by Indigenous people are often preferred to mainstream services (Dance et al. 2004; Bin-Sallik and Ranzijn 2001). Bin-Sallik and Ranzijn (2001) observed that Aboriginal people in South Australia were reluctant to deal with the mainstream system, which they described as 'not Nunga-friendly',[3] cold and impersonal. They felt that non-Aboriginal carers had little or no understanding of Aboriginal society and were racist in many subtle ways, for example the prevailing stereotype was that 'every Aboriginal person is a welfare case'. They also reported that Aboriginal people utilising mainstream services were often frightened to complain in case they lost access to services and/or their information was passed on to other service organisations. Gender is also a highly important consideration in the care of Indigenous older people who may have a strong preference for a person of their own gender to assist with personal care, such as washing, dressing and toileting. Western approaches to person-centred care tend to focus on the individuality of the particular person receiving care and of their immediate family members. A person-centred approach to community care practice with Aboriginal and Torres Strait Islander people also needs to take into account the central importance of connection to family, kinship networks, communities and country (McMillan et al. 2010).

Despite these examples it is important not to make assumptions about the types of services preferred by individual Indigenous people (Wall 2013). Some people want their health and/or aged care needs met by Aboriginal controlled community services such as Aboriginal Medical Services or Aboriginal aged care organisations, while others may choose to use mainstream services, or mainstream services that employ Aboriginal and Torres Strait Islander staff. These preferences may be influenced by family, cultural and historical experiences; the complexity of individual health needs; language issues; cost and service availability. Just over half (56 %) of the Indigenous people surveyed by Dance et al. (2004) in the ACT did not have a preferred cultural background for their carer, highlighting the importance of offering older Indigenous people a range of options for their care. In many remote areas, Aboriginal people have no choice about what type of service to access as there may only be one local provider.

Despite the strong preference of many older Indigenous people to remain in their own communities and to receive aged and health care services (Arkles et al. 2010; Cotter et al. 2007; Woenne-Green 1995) the scarcity of such services in remote areas means that older Indigenous people frequently have to leave their communities to access care but continue to have a strong desire to return home as soon as possible (Alzheimer's Australia 2011). The exodus of older people can leave a 'huge void' in communities (Alzheimer's Australia 2011, p. 22). The Productivity Commission Report (2011) suggests that providing aged care services to Indigenous communities can increase the stability of these communities by prolonging the positive influence and impact of the elders on community development, law and governance (Productivity Commission 2011).

[3] The word 'Nunga' or 'Noongar' refers to Aboriginal people in South Australia.

7.6 Responding to the Care Needs of Older Indigenous Australians

7.6.1 Opportunities for Service Delivery

One of the keys to developing appropriate and responsive aged care services for Indigenous people is to take the time to build genuine relationships with Aboriginal and Torres Strait Islander communities (Wall 2013). The starting point for consultation is to seek advice and direction from local elders, community leaders and community organisations about how to access the community and allow additional time to build genuine trusting relationships. Involving Indigenous people in the design, planning and delivery of services, for example through consultation sessions or by including Aboriginal people on decision making committees, can greatly contribute to culturally appropriate care. Wall (2013) recommends an organisational audit to determine the service's strengths and weaknesses and identify where changes need to be made.[4]

7.6.2 Achieving Cultural Competence in the Workforce

One of the most important challenges for both policy makers and service providers in Indigenous aged care is the limited workforce available to deal with the growing need of the diverse Indigenous aged population. Service providers need to be able to respond in a culturally sensitive way to the linguistic, cultural and geographical diversity of the older Indigenous population (AIHW 2011b; Productivity Commission 2011). This means recruiting, training and retaining staff capable of delivering care in a culturally competent way.

There is general agreement in the literature about the urgent need to develop cultural competence as an essential set of skills within the non-Indigenous aged care workforce (Brooke 2011; Cotter et al. 2007; Productivity Commission 2011; Wall 2013). This requires that workers involved in service delivery receive adequate education and training in Indigenous history and culture, specific local knowledge, including understanding of the local community and the necessary communication and other skills required to effectively work in an Aboriginal community and health context. While many writers emphasise the importance of compulsory cultural awareness training for all non-Aboriginal workers having any contact with Aboriginal people (e.g. Bin-Sallik and Ranzijn 2001) there has been little rigorous evaluation of such programmes to determine their effectiveness and impact and, anecdotally, it is suggested that many mandated programmes are a 'token effort'.

[4] There are a number of useful resources which offer practical advice for service providers and practitioners (see e.g. Wall 2013; Brooke 2011).

The development of such programmes needs to be carefully considered and tied to the needs and perspectives of the local population.

A crucial aspect of culturally appropriate service provision is ensuring that services are culturally safe and secure. 'Cultural safety' is a term which refers to practice that respects and supports cultural identity and empowers the individual to express that identity and have their cultural needs met (McGrath and Holewa 2007). In practice it involves a critical examination of the power imbalances in healthcare encounters between Indigenous clients and non-Indigenous healthcare providers (Smith et al. 2010).

Respect for Indigenous people and their histories and culture, is an integral component of culturally appropriate service delivery, particularly beliefs and protocols around gender, traditional healers and utilising local Indigenous staff, in addition to fulfilling the normal requirements of aged care services (Productivity Commission 2011). Respecting the diversity between and within Indigenous communities, and getting to know the local community and what is acceptable in that community, are all important components of effective aged care service delivery.

Language and communication are particularly important and sensitive aspects of culturally competent care. Language is not only a means of communication, but also encapsulates the history of peoples, including their ways of living and ways of thinking (Wuethrich 2000 cited in McGrath and Holewa 2007). Service provision in remote areas requires cross-cultural communication skills and understanding the significance of language for the Indigenous client and their families. Clear communication is fundamental to culturally appropriate aged care, including assessment and diagnosis of dementia and in palliative care. McGrath, for example, highlights that an understanding of the role of language in palliative care is a core dimension of ensuring cultural safety (McGrath and Holewa 2007). The choice of interpreters also needs careful consideration, as interpreters require a good understanding of patients' cultures as well as their languages as they have to be able to skilfully translate the concept from one culture to another. It is important not only that the interpreter understands the local language but also is in the 'right' kinship relationship to the patient (McGrath and Holewa 2007).

It is also important to remember that even though there are important cultural differences between Indigenous and non-Indigenous people, culture does not account for everything. Older Indigenous people and their family members have different life histories, personalities and experiences that shape and influence them and need to be taken into account in the way they interact with caregivers and service providers. Reflective practice by practitioners needs to take into account both the cultural context as well as individual differences between Aboriginal clients.

7.6.3 Enhancing the Skills of the Indigenous Workforce

Equally as important as developing a culturally competent non-Indigenous workforce is recruitment, retention and training of the Aboriginal and Torres Strait Islander workforce (Arabena 2007; Alzheimer's Australia 2011; Aboriginal and

Torres Strait Islander Ageing Committee 2010). The increasing number of Indigenous people using residential or community-based aged care services requires both a workforce with specific skills and adequate resourcing of the sector. However, the sector experiences both a lack of Indigenous workers in aged care and a lack of training opportunities. Services rely heavily on volunteers and the under-resourcing of the sector leads Aboriginal paid workers to become overwhelmed by the work (Bin-Sallik and Ranzijn 2001). Recruiting Aboriginal health and aged care workers helps to address the language barriers as well as trust barriers which Indigenous people may encounter. Aboriginal aged care workers are considered part of the community and therefore have a different kin-based relationship with older clients. A key challenge for workforce development is the ability to attract and retain Indigenous aged care workers, and to invest in training to develop their capacity to provide high-quality service (Productivity Commission 2011).

7.7 Conclusion

This chapter has highlighted the considerable challenges and complexities which need to be taken into consideration in ensuring appropriate support and services for the growing number of older Indigenous Australians requiring care; it has also identified the many opportunities for doing so. A good starting point is to understand that the needs of older Indigenous people are diverse and recognise that the knowledge embedded in the local community is a valuable resource to determine how to deliver aged care services in culturally appropriate ways.

Expanding health services research in the area of Indigenous aged care and addressing the lack of intervention studies for the ageing Indigenous population is fundamental to providing a sound knowledge base for policy makers and providers of care. Much can also be gained by sharing good practice examples. The literature contains a small number of case studies of culturally appropriate models of care for older Indigenous people, such as the Yuendumu Old People's Programme (Smith et al. 2010). There is a small but growing interest in development of culturally effective assessment tools such as the Kimberley Indigenous Cognitive Assessment Tool (Smith et al. 2008; Radford et al. 2014). But there have been very few studies of specific care interventions for the ageing Indigenous population (Brooke 2011) and existing research is largely focused on remote areas, although recent work being undertaken as part of the Koori Growing Older Well Study is starting to address this research gap (Radford et al. 2014).

As Sykes (1988) argued, almost two decades ago, aged care should be part of community development; partnerships with Aboriginal organisations, therefore, have a crucial role to play. In the planning of aged care services, broad community consultation with appropriate stakeholders at the community level can assist in the understanding of the needs and expectations of older Aboriginal people. The inclusion of Aboriginal and Torres Strait Islander people in determining policy and models of care will also assist in the making of culturally appropriate decisions (Goodwin 1993; Stewart et al. 2011). Involving Indigenous people in the delivery of services for the aged can further help to address needs and improve service delivery. An essential component of planning should be the development of the Indigenous aged care workforce through the provision of training and employment opportunities at all levels.

For individual health professionals and service providers it is important to understand that Indigenous older people have diverse cultural experiences and needs which depend on where they live, as well as their individual, family, social and economic circumstances. There is much work to do to achieve a culturally competent workforce in mainstream aged care services, culturally appropriate dementia services and culturally appropriate aged care assessments and an urgent need to increase the capacity of the Indigenous aged care workforce. Health professionals will play a crucial part in this future development.

7.8 Case Studies

The case studies below provide two fictional scenarios which incorporate some of the challenges and opportunities in healthy ageing for Aboriginal and Torres Strait Islander people.

7.8.1 Bob's Story

The news got around quickly. Bob, a 71-year-old Aboriginal elder had been admitted to hospital with burns to his chest and face. The police said that Bob was very confused and disorientated when found by his wife after falling into an open fire in his backyard. His wife Fran had noticed he had become more confused and unsteady on his feet in the last year, but especially in the last few weeks. The family had also noticed that Bob, who had always been a very tidy and well-presented man, was looking quite unkempt and in the past week his clothes smelt. Bob had worked for many years as a stockman on an outback station. He often lit an open fire in his backyard on the outskirts of town; it reminded him of the 'good old days'. Bob was also good with the young men in the community and shared his experiences of traditional life which helped the young men with their sense of cultural identity.

As Bob's family heard the news they made their way to the hospital. The security staff in the Emergency Department watched with obvious unease as the group of Aboriginal people waiting for Bob grew to around 15 people. The waiting room was full and other patients were growing restless with the 5-h wait. Some started mumbling about the 'blacks' getting preferential treatment and echoed politicians' comments about the Aboriginal Welfare industry.

Six weeks later, after many skin grafts, Bob was ready to leave the hospital. He could hardly turn his head due to the grafting on his neck. Many months of therapy lay ahead. Although the hospital had diagnosed and treated the urinary tract infection that caused the delirium and led to the accident, the geriatrician and social worker told the family that Bob had a mild–moderate dementia. This further reduced Bob's already poor chances of being able to return to his own home due to his increasing frailty, unsteadiness on his feet, possibility of falls and memory loss. Hospital staff suggested it may be time for Bob to be admitted to an aged care facility.

Bob, who was quite depressed after the accident, didn't want to go to a nursing home; he was saying he would kill himself if he couldn't return to his home. The family and community knew how important it was to care for Bob in his home environment, especially as Bob was one of the longest living elders from the local community. However, the family needed some home-based help from government-funded aged care services to keep him at home. Bob had an assessment by the local Aged Care Assessment Team (ACAT). The ACAT arranged for Bob to have day leave prior to discharge so they could assess him in his home environment. At the request of Fran, the ACAT also arranged for the Aboriginal Health Worker from the local Aboriginal Health Service to attend the home visit. The ACAT approved a Home Care package including some home modifications, informing Bob he had the choice of providers, but the Aboriginal Home Care Package provider had no capacity for new clients for twelve months. Bob was suspicious of non-Indigenous service providers. Fran, whose capacity to care for Bob was limited by her chronic heart problem, arthritis and diabetes, was very grateful for the help of her relatives in the first few weeks; they rotated as 'live in carers' so Bob could return home. However, everyone had busy lives, so Fran was not sure how things would work out long term.

7.8.2 *Jean's story*

Jean, a 74 year old, was finding it increasingly difficult to manage housework due to worsening arthritis in her hands and knees. For years she had been catching the local bus to the mall to do weekly shopping, but this was also proving to be very challenging due to the breathlessness she frequently experienced. Jean was very private about her own affairs. She didn't talk about her problems and didn't like asking for help. But when her friend Mary commented on how unwell she looked, Jean opened up and agreed to visit her local Aboriginal Health Service.

Jean had been so busy over the past few months and had not stopped to think about her own health. Her daughter, Sharon, and three grandchildren had recently moved in with her. Sharon's partner, George had always been a gentle man and good with the children, but since a car accident 12 months ago he had become violent and unpredictable. Although George recovered well from the physical injuries, the doctors told Sharon that because he had severe head injuries he may never fully recover mentally. Jean observed that Sharon was anxious and not managing at home. She was also worried that the kids were starting to miss school. After another incident when the police had to be called, Sharon decided that she needed to leave George. Jean wanted her daughter and grandchildren to have all the opportunities to get a good education that she never had, so was prepared to do anything to help Sharon through this difficult time. She offered to look after the kids while Sharon tried to find a job and more suitable housing and book the kids into the new school.

Jean felt better about going to the Aboriginal Health Service after Mary offered to accompany her. Jean had to leave school when she was young and struggled to fill out forms. She told Mary that she was 'shamed' that she couldn't read or write properly. She was also grateful that Mary was able to help her out with the three grandchildren while she was in her appointment. By the time it was her turn to see the doctor Jean had relaxed a bit. The doctor took some time to ask Jean about how she was feeling, and as Jean became more comfortable the doctor learnt about the challenges she was facing with the children and gradually found out the story about Sharon and George. The doctor was amazed by Jean's good humour and willingness to take on the problems life has thrown up at her.

When Jean felt comfortable enough to start talking about her own health issues, the doctor was very encouraging, not only assisting with pain relief for her arthritis and finding out what was causing her breathlessness, but also by suggesting that Jean might benefit from attending the local arthritis support group. The doctor also offered Jean some practical help. She told Jean that the Aboriginal Home Care service could help by providing a cleaner once a week, transport her to the shops in the community bus, provide someone to push the trolley for her and carry the shopping bags into her house. The doctor also told Jean that there were Aboriginal health workers at the health service who could help women in Sharon's situation. She also suggested that the local brain injury service might be able to review George and provide some ongoing assistance.

Jean felt that a weight had been lifted from her shoulders. She was relieved to know that support was available to help the family through this difficult time and that Sharon and the kids would have help getting settled into their new home. She was also hopeful that maybe George might return to his gentle caring self.

7.8.3 Discussion Points

- Each of the case studies raises multiple health and welfare issues. What are the issues in each case they and how are they inter-related?

References

Aboriginal and Torres Strait Islander Ageing Committee. (2010). *Aboriginal ageing: Growing old in Aboriginal communities. Linking services and research.* A report of the 2nd National Workshop of the Australian Association of Gerontology (AAG) Aboriginal and Torres Strait Islander Ageing Committee (ATSIAC), held in Darwin, Northern Territory, 11th August 2010. Retrieved from http://www.neura.edu.au/sites/neura.edu.au/files/pagedownloads/ATSIACNational_Workshop_2010Report.pdf

Aboriginal and Torres Strait Islander Social Justice Commissioner. (2005). *Social Justice Report 2005*. Canberra: Human Rights and Equal Opportunity Commission.
Aged Care Act 1997. Canberra: Commonwealth of Australia.
Aged Care (Living Longer Living Better) Act 2013. Canberra: Commonwealth of Australia.
Alzheimer's Australia. (2011). *Report for the Department of Health and Ageing in relation to services for Consumer Engagement in the Aged Care Reform Process*. Canberrra, Australia: Alzheimer's Australia.
Anonymous. (2008). Safety risk visits to Indigenous aged care services. *Australian Nursing Journal, 16*(5), 6–7.
Arabena, K. (2007). 'Before I forget': Dementia in Aboriginal and Torres Strait Islander communities. *Ngoonjook: A Journal of Australian Indigenous Issues, 30*, 41–49.
Arkles, R., Pulver, L. J., Robertson, H., Draper, B., Chalkley, S., & Broe, G. (2010). *Ageing, cognition and dementia in Australian Aboriginal and Torres Strait Islander peoples: A life cycle approach. A review of the literature*. Sydney: Neuroscience Research Australia and Muru Marri Indigenous Health Unit, University of New South Wales.
Australian Bureau of Statistics. (2012). *Census of population and housing—Counts of Aboriginal and Torres Strait Islander Australians* (Cat. No. 2075.0). Canberra, Australia: ABS.
Australian Bureau of Statistics. (2013). *Estimates of Aboriginal and Torres Strait Islander Australians, June 2011* (Cat. No. 3238.0.55.001). Canberra, Australia: ABS.
Australian Government. (2013). *Closing the gap Prime Minister's Report 2013*. Canberra, Australia.
Australian Institute of Health and Welfare. (2010). *National best practice guidelines for collecting Indigenous status in health data sets*. (Cat. No. IHW 29). Canberra, Australia: Australian Institute of Health and Welfare.
Australian Institute of Health and Welfare. (2011a). *The health and welfare of Australia's Aboriginal and Torres Strait Islander People: An overview 2011* (Cat. No. IHW 42). Canberra, Australia: Australian Institute of Health and Welfare.
Australian Institute of Health and Welfare. (2011b). *Older Aboriginal and Torres Strait Islander people* (Cat. No. IHW 44). Canberra, Australia: Australian Institute of Health and Welfare.
Australian Institute of Health and Welfare. (2011c). *Contribution of chronic disease to the gap in mortality between Aboriginal and Torres Strait Islander people and other Australians* (Cat. No. IHW 48). Canberra, Australia: Australian Institute of Health and Welfare.
Bell, D. (1983). *Daughters of the dreaming*. Melbourne: McPhee Gribble/Allen and Unwin.
Berndt, R. M., & Berndt, C. H. (1988). *The world of the first Australians* (5th ed.). Canberra, Australia: Aboriginal Studies Press.
Bessarab, D. (2006). *A study into the meanings of gender by Aboriginal people living in urban (Perth) and regional (Broome) settings*. Unpublished doctoral thesis, Curtin University, Perth, WA.
Bin-Sallik, M. A., & Ranzijn, R. (2001). *Report on a scoping study into the needs of Indigenous Aged Care in South Australia* [Prepared for Aged Care and Housing Group]. Adelaide, SA: College of Indigenous Education and Research, University of South Australia.
Brooke, N. J. (2011). Needs of Aboriginal and Torres Strait Islander clients residing in Australian residential aged care facilities. *Australian Journal of Rural Health, 19*(4), 166–170.
Carson, B., Dunbar, T., Chenhall, R., & Bailie, R. (Eds.). (2007). *Social determinants of indigenous health*. Sydney: Allen and Unwin.
Commonwealth Department of Social Services. (2015). Commonwealth home support program – Program manual. Canberra, Australia: DSS.
Cotter, P., Anderson, I., & Smith, L. (2007). Indigenous Australians: Ageing without longevity? In A. Borowski, S. Encel, & E. Ozanne (Eds.), *Longevity and social change in Australia* (pp. 65–98). Sydney: University of New South Wales Press Ltd.
Cotter, P. R., Condon, J. R., Anderson, I. P. S., Smith, L. R., & Barnes, T. (2011). Indigenous aged care service use and need for assistance: How well is policy matching need? *Australasian Journal on Ageing, 30*(Suppl 2), 38–44.

Cotter, P. R., Condon, J. R., Barnes, T., Anderson, I. P. S., Smith, L. R., & Cunningham, T. (2012). Do Indigenous Australians age prematurely? The implications of life expectancy and health conditions of older Indigenous people for health and aged care policy. *Australian Health Review, 36*(1), 68–74.

Council of Australian Governments. (2009). *National Indigenous Reform Agreement (closing the gap)*. Canberra, Australia: COAG.

Dance, P., Brown, R., Bammer, G., & Sibthorpe, B. (2004). Aged care services for Indigenous people in the Australian Capital Territory and surrounds: Analysing needs and implementing change. *Australian and New Zealand Journal of Public Health, 28*(6), 579–583.

Dudgeon, P., Wright, M., Paradies, Y., Garvey, D., & Walker, I. (2014). Aboriginal social, cultural and historical contexts. In P. Dudgeon, H. Milroy, & R. Walker (Eds.), *Working together: Aboriginal and Torres Strait Islander mental health and wellbeing principles and practice* (pp. 3–25). Canberra, Australia: Office of Aboriginal and Torres Strait Islander Health, Department of Ageing.

Garvey, G., Simmonds, D., Clements, V., O'Rourke, P., Sullivan, K., Gorman, D., et al. (2011). Making sense of dementia: Understanding amongst Indigenous Australians. *International Journal of Geriatric Psychiatry, 26*(6), 649–656.

Goodwin, M. (1993). Aboriginal aged care. *The Queensland Nurse, 12*(6), 22–23.

Gray, A., Trompf, P., & Houston, S. (1991). The decline and rise of Aboriginal families. In J. R. P. Trompf (Ed.), *The health of Aboriginal Australia* (pp. 80–122). Marrickville: Harcourt Brace Jovanovich.

Hulko, W., Camille, E., Antifeau, E., Arnouse, M., Bachynski, N., & Taylor, D. (2010). Views of first nation elders on memory loss and memory care in later life. *Journal of Cross-Cultural Gerontology, 25*(4), 317–342.

Human Rights and Equal Opportunity Commission. (1991). *Report of the Royal Commission into Aboriginal deaths in custody*. Canberra: Australian Government Publishing Service.

Human Rights and Equal Opportunity Commission. (1997). *Bringing them home: A guide to the findings and recommendations of the national inquiry into the separation of Aboriginal and Torres Strait Islander Children from their families*. Sydney: HREOC.

Kaberry, P. (1939). *Aboriginal women: Sacred and profane*. London: George Routledge and Sons.

King, M., Smith, A., & Gracey, M. (2009). Indigenous health part 2: The underlying causes of the health gap. *Lancet, 374*, 76–85.

Kuipers, K., McIntosh, K., Paluch, T., & Oke, L. (2011). Caring for country was associated with positive health outcomes for Indigenous people living in remote areas of Northern Australia. *Australian Occupational Therapy Journal, 58*(1), 56–57.

Lawrence, A., Houghton, J., Thomas, J., & Weldon, P. (2014). Where is the evidence: Realising the value of grey literature for public policy and practice. Retrieved 21 January, 2015, from http://apo.org.au/research/where-evidence-realising-value-grey-literature-public-policy-and-practice

Li, S. Q., Guthridge, S. L., Eswara Aratchige, P., Lowe, M. P., Wang, Z., Zhao, Y., et al. (2014). Dementia prevalence and incidence among the Indigenous and non-Indigenous populations of the Northern Territory. *Medical Journal of Australia, 200*(8), 465–469.

Marmot, M. (2011). Social determinants and the health of Indigenous Australians. *Medical Journal of Australia, 194*(10), 512–513.

McDermott, R. A., O'Dea, K., Rowley, K., Knight, S., & Burgess, P. (1998). Beneficial impact of the homelands movement on health outcomes in central Australian Aborigines. *Australian and New Zealand Journal of Public Health, 22*(6), 653–658.

McGrath, P., & Holewa, H. (2007). End-of-life care of aboriginal peoples in remote locations: Language issues. *Australian Journal of Primary Health, 13*(1), 18–27.

McMillan, F., Kampers, D., Traynor, V., & Dewing, J. (2010). Person-centred care as caring for country: An Indigenous Australian experience. *Dementia, 9*(2), 163–167.

National Aboriginal Health Strategy Working Party. (1989). *A National Aboriginal Health Strategy* (1996 Reprint ed.). Canberra: Australian Government Publishing Service.

New York Academy of Medicine. (2002). Retrieved January 5, 2015, from http://www.greylit.org/

Panzironi, F. (2013). *Hand-in-hand. Report on Aboriginal traditional medicine. An-angu Pitjantjatjara Yankunytjatjara Lands (APY).* Aп-angu Ngangkar-i Tjutaku Aboriginal Corporation. Seaton, South Australia: ANTAC.

Pollitt, P. A. (1997). The problem of dementia in Australian Aboriginal and Torres Strait Islander communities: An overview. *International Journal of Geriatric Psychiatry, 12*(2), 155–163.

Productivity Commission. (2008). *Trends in Aged Care Services: Some implications.* Canberra, Australia: Commission Research Paper.

Productivity Commission. (2011). *Caring for Older Australian* (Report No. 53). *Final Inquiry Report.* Canberra, Australia.

Radford, K., Mack, H. A., Robertson, H., Draper, B., Chalkley, S., Daylight, G., et al. (2014). The Koori Growing Old Well Study: Investigating aging and dementia in urban Aboriginal Australians. *International Psychogeriatrics, 26*(6), 1033–1043.

Reid, J. (1985). 'Going Up' or 'Going Down': The status of old people in an Australian Aboriginal Society. *Ageing and Society, 5*(1), 69–95.

Smith, K., Flicker, L., Lautenschlager, N., Almeida, O. P., Atkinson, D., Dwyer, A., et al. (2008). High prevalence of dementia and cognitive impairment in Indigenous Australians. *Neurology, 71*, 1470–1473.

Smith, K., Grundy, J. J., & Nelson, H. J. (2010). Culture at the centre of community based aged care in a remote Australian Indigenous setting: A case study of the development of Yuendumu Old People's Programme. *Rural and Remote Health, 10*(4), 1422.

Steering Committee for the Review of Government Service Provision. (2015). *Report on Government Services 2015, Indigenous Compendium* (Vol. F). Community Services. Canberra, Australia: Productivity Commission.

Stewart, J., Lohoar, S., & Higgins, D. (2011). *Effective practices for service delivery coordination in Indigenous communities. Resource Sheet 8.* Produced by the closing the gap clearinghouse. Canberra: Australian Institute of Health and Welfare and Melbourne, Australian Institute of Family Studies.

Sykes, R. B. (1988). *Issues affecting older Aboriginal people: A discussion paper.* Canberra, Australia: Commonwealth Office for the Aged.

Vos, T., Barker, B., Begg, S., Stanley, L., & Lopez, A. D. (2009). Burden of disease and injury in Aboriginal and Torres Strait Islander Peoples: The indigenous health gap. *International Journal of Epidemiology, 38*(2), 470–477.

Wall, S. (2013). Working with older Aboriginal and Torres Strait Islander people. *Research to practice briefing* (Vol. 8). Sydney, Australia: The Benevolent Society.

Warburton, J., & Chambers, B. (2007). Older Indigenous Australians: Their integral role in culture and community. *Australasian Journal on Ageing, 26*(1), 3–7.

Waugh, E., & Mackenzie, L. (2011). Ageing well from an urban Indigenous Australian perspective. *Australian Occupational Therapy Journal, 58*(1), 25–33.

Wilkes, T. (2002). *Indigenous matters while we age.* Paper presented at the 6th Global Conference of the International Federation on Ageing, 27-30 October 2022, Perth, Australia.

Wilson, K., Rosenberg, M. W., Abonyi, S., & Lovelace, R. (2010). Aging and health: An examination of differences between older Aboriginal and non-Aboriginal people. *Canadian Journal on Aging, 29*(3), 369–382. doi:10.1017/s0714980810000309.

Woenne-Green, S. (1995). *They might have to drag me like a bullock: The Tjilpi Pampa Tjutaku Project: Final report.* Alice Springs: Ngaanyatjarra Pitjantjatjara Yankunytjatjara Women's Council Aboriginal Corporation.

Chapter 8
Enhancing the Health and Employment Participation of Older Workers

Jack Noone and Philip Bohle

8.1 Introduction

This chapter begins with a brief background to Australia's ageing workforce before identifying the major strengths of older workers compared to younger workers, such as higher productivity, greater experience, psychological resilience and organisational commitment. It will then describe the challenges that may prevent older workers from working longer and discuss existing policy responses to them. The major challenges include age discrimination, early retirement, declining health status, underdeveloped or outdated job skills, care giving responsibilities and unsustainable work design. The aspects of work design covered include the workplace health and safety (WHS) risks faced by older workers, job demands and functional deterioration, working hours and precarious work. The different challenges for men's and women's employment is also discussed.

The final section discusses the concept of 'work ability', which is defined as a worker's capacity to meet the demands of their job. Low work ability is associated with early retirement and each challenge described in this chapter has been identified as a cause of low work ability. Work ability is therefore a potentially useful framework for understanding the determinants of workforce participation and for informing a broad range of employment policies to enhance the health and productivity of older workers.

J. Noone (✉)
Suite 16.01, Level 16, 6 O'Connell Street, Sydney, NSW 2000, Australia

Ageing, Work and Health Research Unit, Faculty of Health Sciences, Cumberland Campus, 75 East St, Lidcombe, NSW 2141, Australia
e-mail: j.noone@unsw.edu.au

P. Bohle
Ageing, Work and Health Research Unit, Faculty of Health Sciences, The University of Sydney, Cumberland Campus, East Street, Lidcombe, NSW 2141, Australia
e-mail: philip.bohle@sydney.edu.au

K. O'Loughlin et al. (eds.), *Ageing in Australia*, International Perspectives on Aging 16, DOI 10.1007/978-1-4939-6466-6_8

Australian workers over the age of 45 are categorised as 'mature-aged'. The term 'older workers' used in this chapter is roughly equivalent. However, the international research literature on older workers does not consistently define the age range to which the term applies. In any case, such categorisations are arbitrary and do not reflect the continuum of ageing across the life course.

8.2 Background

Australia's labour force is both ageing and shrinking in size. In 1999, workers aged over 55 constituted only 10 % of the workforce but will account for 50 % of workforce growth by 2016. In contrast, 20–44-year-olds constituted almost two thirds of the labour force in 1999 but by 2016 will account for only 20 % of labour market growth (Australian Bureau of Statistics 1999). The 'dependency ratio' is also predicted to increase. In 2013, there were 50 dependents, that is, those younger than 15 and older than 65, for every 100 working-age adults (15–64 years). By 2060, it is estimated there will be 65 dependents for every 100 working-age adults (Australian Bureau of Statistics 2013). This change reflects a 30 % increase in the proportion of the population not in paid work.

In 2010, the then Productivity Commissioner considered the ageing workforce, and a greater dependency ratio, problematic because it would mean less tax revenue to support predicted increases in spending on health, age care and pensions (Swan 2010; see also Chap. 9). This sentiment is also echoed in the latest Intergenerational Report (Department of the Treasury (Australia) (2015)). Successive governments have sought to meet the economic costs of population ageing by encouraging older workers to stay in the workforce longer (Australian Government 2011). Crucially, most policies focus on the individual worker rather than organisational factors, such as work organisation. This approach is problematic because elements of work organisation, such as working hours and precarious (insecure) work, are known to significantly affect health and workforce participation (Stansfeld and Candy 2006; Virtanen et al. 2014). Consequently, any policy agenda that does not sufficiently focus on improving working conditions is unlikely to maximise the workforce participation of older workers. Workers are likely to continue to leave the workforce prematurely, with negative effects on participation and quality of life in retirement (Noone et al. 2013), particularly following the detrimental impact of the Global Financial Crisis on retirement finances (see Chap. 2; Kendig et al. 2013). This chapter will examine some of the complex relationships between worker characteristics, organisations and employment policies. A good place to start untangling these relationships is to consider the strengths older employees bring to the workplace.

8.3 Strengths of an Ageing Workforce

Surprisingly to some, research indicates that productivity at work increases from late adolescence to approximately age 55 before dropping off slightly (Göbel and Zwick 2009). In fact, a recent study of car assembly workers indicated that productivity, measured in terms of errors made, did not decrease until age 60 (Börsch-Supan and Weiss 2013). Other research suggests older workers are 'good value for money' compared to younger workers. For example, a Portuguese study found that pay tended to stop increasing at around age 45 but productivity continued to increase well into workers' 50s (Cardoso et al. 2011). Interestingly, one review (Skirbekk 2004) found that older workers were more productive when the job demands concerned verbal abilities, but less productive when problem solving, speed and learning were involved. However, other researchers have argued that any cognitive and physical declines with age can be offset by older workers' greater wisdom and experience (Ilmarinen 2001).

It is now recognised that, despite the benefits of older workers' skills and experience, too little is being done to maximise the utilisation of this human resource (Tikkanen 2011). Brooke and Taylor's (2005) analysis of two Australian and two UK organisations showed that the experience of older workers was highly valued for training and mentoring younger workers. Older workers were also seen to provide stability in the workplace and to maintain quality standards. However, no formal policies were identified within these organisations to capitalise on older workers' strengths. In addition, stereotypes about older workers' inability to utilise new technology appeared to impede productivity.

Older workers tend to be more resilient and satisfied at work than younger workers. A recent meta-analytic review of multiple research studies indicated that increasing age is associated with stronger workplace coping skills and that older workers reported no more stress at work than their younger counterparts (Rauschenbach et al. 2013). Increasing age may also buffer the negative effects of workload on job satisfaction. For example, a study of 3500 service sector employees, nurses and academics found that older workers' workload and work–life balance had no impact on their job satisfaction, whereas a high workload and work–life conflict were associated with decreased job satisfaction for younger workers (Mauno et al. 2013). Finally, a seminal meta-analysis has shown that older workers consistently exhibit greater organisational commitment (Mathieu and Zajac 1990). Greater organisational commitment has positive flow-on effects on staff retention.

The evidence above suggests that employing older workers has many benefits. However, the 'healthy worker effect', the tendency for workers to be healthier than the general population, may contribute to these findings. Poor health is a significant contributor to early retirement. Therefore older workers participating in research are likely to be healthier than those who do not, particularly those who have already left the workforce due to poor health. Nevertheless, the literature highlights important strengths of older workers. Unfortunately, despite these findings, prevailing age-based stereotypes still portray older workers as less productive, less resilient and unwilling to learn (Rauschenbach et al. 2012). As we have illustrated, these stereotypes have been widely debunked in recent research (Ng and Feldman 2012). Age stereotypes and discrimination are therefore a major challenge for the ageing workforce.

8.4 Challenges for an Ageing Workforce

8.4.1 Age Discrimination

As noted in Chap. 3, age discrimination is a determinant of early workforce exit and is also associated with lower recruitment, training and retention and poorer mental health (Snape and Redman 2006; Gringart et al. 2005). For older workers, discrimination may occur directly if they are denied employment based on age-based stereotypes (see Chap. 3 for details), or indirectly if, for example, employment policy is the same for everyone but disadvantages older workers (Australian Human Rights Commission 2012). Indirect or institutionalised age discrimination is present in Australia's employment and insurance systems. For example, there are age limits on workers' compensation and income insurance for older workers is prohibitively expensive due to steep increases in premiums with age (Australian Government 2011). However, policies to reduce discrimination have mainly focused on reducing direct age discrimination.

The 2007–2013 Labor government bolstered existing strategies to reduce age discrimination with various new policies, including the appointment of an Age Discrimination Commissioner, the consolidation of five anti-age discrimination Acts and a review of existing legislation to identify remaining barriers to mature-age employment. Approximately $2 million was committed to fund research into reducing discrimination and an employer incentive scheme was introduced. The Jobs Bonus Scheme provided employers with $1000 if they hired and retained an eligible mature-aged worker for 3 months or more. The financial incentive for employers to hire older workers was expected to offset discrimination at the time of hiring. The subsequent Liberal-National Coalition government renamed the scheme (Restart), increased the maximum subsidy to $10,000 and the time in employment from 3 months to 1 year. The Corporate Champions Program (Department of Employment 2014) also developed a range of resources and services to help employers attract and retain mature-age workers. These resources include tailored support and assistance to employers (valued up to $20,000), in the form of consultation on strategies for age management in the workplace, including training about age discrimination.

8.4.2 *Early Retirement*

A further challenge is that Australians tend to retire earlier than workers in other Western nations. For example, the average retirement age in the United States is 64 for men and 62 for women (Munnell 2011) compared to 61.1 and 59.2 for Australian men and women, respectively, during the 5 years to 2010 (Australian Bureau of Statistics 2013).

To prolong workforce participation, the 2007–2013 Labor government introduced a policy to increase the pension eligibility age from 65 to 67 by 2023. This was expected to lift employment participation rates by limiting retirement income options. Since then, the 2013–2016 Coalition government has sought to increase the eligibility age to 70 (Department of the Treasury (Australia) (2015)) and calls have also been made to increase the superannuation preservation age to match eligibility for the age pension (Also refer Chap. 9—Australia's Retirement Income Policy).

However, a problem with increasing age eligibility for the pension and superannuation is that some people cannot work due to poor health, commitments to care giving or loss of employment due to redundancy. In these instances, increasing the retirement age may make workers' lives harder as they wait for pension eligibility. Policymakers have therefore looked to other ways of promoting workforce participation, such as improving workers' health.

8.4.3 *Increasing Incidence of Chronic Disease*

Another challenge for older workers is the increasing incidence of chronic disease, which is expected to cause a 50% increase in health-induced early retirement (Schofield et al. 2009). Poor health already accounts for 25% of retirements for men and 20% for women (Australian Bureau of Statistics 2011). An increase of 50% would create a large group of workers who cannot work but who could have to wait 10–15 years to be eligible for an age pension. These involuntary retirees would be financially dependent on either sickness benefits or their personal assets and superannuation. The premature depletion of retirement assets and income could have negative implications for their financial resources and quality of life in old age. This is a form of 'longevity risk' and is discussed further in Chap. 9.

The Healthy Workers Initiative (HWI) (Australian Government 2013a) is major recent policy response to improving worker health. The HWI comprises a $294 million investment by government (2009–2018) to fund work-based physical health promotion programmes. It aims to improve the health of working adults by promoting good diet and exercise and a reduction in alcohol and tobacco consumption. The onus is therefore on individuals to improve their health status rather than on organisations to change potentially health-damaging aspects of the work environment.

Early retirement due to mental ill health costs approximately $1.8 billion in lost gross domestic product per annum (Schofield et al. 2011). Early retirements, lost productivity, and individual costs have prompted Safe Work Australia, the country's primary workplace health and safety policy development organisation, to prioritise mental disorders (Safe Work Australia 2012). beyondblue's National Workplace Program (2015) aims to addresses this issue by raising awareness and reducing the stigmatism of mental ill health. However, there is less focus on improving the psychosocial environment to reduce the prevalence of mental disorders. Therefore, mental health promotion through the design of healthy workplaces remains an unmet challenge in employment policy.

8.4.4 Job Skills and Training

Outdated and unrecognised job skills, mismatches between worker skills and job demands, and inadequate training have been identified by the National Seniors Australia (NSA) as major challenges for older workers' employment participation (Temple et al. 2011). The NSA reported that approximately 20 % of 50–69-year-old workers identified a lack of necessary training as having created difficulties in either finding employment or seeking longer hours. Research suggests that older workers have less access to training than younger workers (Fouarge and Schils 2009) indicating age discrimination and disadvantage with respect to job seeking and retainment for older workers.

The 2007–2013 Labor government's Investing in Experience scheme (Australian Government 2013b) was implemented to modernise older workers' skill sets. It provided funding to employers to have employees' skills formally assessed, recognised and updated through training if necessary. However, the 2013–2016 Coalition government discontinued this scheme. This is a concern, as on-the-job training is associated with increased employability for older workers (Picchio and Van Ours 2013). Skill development for older workers is therefore an unmet challenge.

8.4.5 Unpaid Care Giving

There are approximately 2.7 million unpaid caregivers in Australia and the largest proportion is aged between 45 and 65 (Australian Bureau of Statistics 2013). Although unpaid care giving contributes significantly to the nation's economy (Access Economics 2010), it is also associated with poor health, fewer paid working hours and early retirement of the caregiver (Lilly 2011).

The competing demands of work and care and the important contribution carers make to the nation's economy are now being recognised in policy development (Australian Government 2011). The primary policy strategy to support caregivers is the right to request flexible working arrangement from their employer, including changes in work hours, different patterns of work (e.g. job sharing, split shifts) or relocation of work (e.g. working from home). These arrangements are expected to make it easier for caregivers to remain in the workforce while still performing valuable caring roles.

8.4.6 The Design of Work

A key 'action area' in the Australian Work Health and Safety Strategy 2012–2022 is entitled 'healthy and safe by design' (Safe Work Australia 2012). The aim is to eliminate or minimise hazards by improving workplace design, which includes 'work, work processes and systems of work' (p. 9). This section briefly examines the relationship between age and selected elements of this domain, including workplace health and safety risks, work demands, working hours and precarious work.

8.4.6.1 Older Workers and Workplace Health and Safety (WHS) Risks

The limited research on age and WHS has focused on observable physical injury. National labour statistics also concentrate on physical factors, such as falls or fractures, rather than disease. Older workers have a lower overall frequency of work-related injuries than younger workers although the injuries they sustain tend to be more severe and are more likely to be fatal (Crawford et al. 2010; Rogers and Wiatrowski 2005). For example, falls on the same level, often arising from problems with the floor or ground surfaces, are more frequent for older workers and are more likely to cause a fracture or fatality as age increases. This means that workplaces that are relatively safe for younger workers may be more hazardous for older workers. The safety of older workers may be further jeopardised if they are overlooked for relevant WHS training that identifies specific risks for them.

Despite the greater focus on injury, work-related disease causes substantially more deaths than traumatic injuries at work (Quinlan et al. 2010). The incidence of disease is also higher among older workers (Safe Work Australia 2005), which in part reflects longer workplace exposure and the prolonged latency of some diseases. For example, workers may be exposed to asbestos fibres early in their working lives, but malignant mesothelioma typically takes decades to develop. Most research indicates the median age of diagnosis is around 65 years or older, after which median survival time is short (roughly 9–14 months; Chapman et al. 2008). Preventing or minimising hazardous exposures for all workers is the most effective means of mitigating the heavy toll of these long-term disease processes on older workers. Susceptibility to some occupational risks may genuinely increase with age however. Ritz (1999), for example, reported that exposure to ionising radiation before the age of 40 had a marginal effect on mortality, but exposure after age 40 increased mortality by two or three times for radiosensitive cancers, lung cancers and all cancers combined. After systematically reviewing 59 studies on older workers' WHS needs, Crawford et al. (2010) concluded that diminished physical and psychological capacities increased risks for workers aged over 50, but that there were large differences between workers in these effects and many risks could be controlled by appropriate workplace changes. When there is convincing evidence that older workers are more susceptible to specific hazards, redesigning work is often an effective way to avert negative effects.

In summary, older workers face a higher risk of serious injury and experience a greater burden of work-related disease, either because of longer exposure or specific susceptibilities. These greater risks can often be addressed by changes in work design, for example by reassigning heavier physical tasks or other risk-related activities. Nevertheless, it is important to recognise that industry and occupation, rather than worker characteristics, are the strongest predictors of hazard exposure, disease and injury. For example, jobs that require heavy lifting carry the highest risks of physical injury (Zwerling et al. 1996).

8.4.6.2 Work Demands and Functional Deterioration

Ageing is associated with declines in indices of physical capacity, such as muscle strength, bone density and aerobic capacity, and with deteriorating health, such as a greater prevalence of musculoskeletal disorders (Crawford et al. 2010). However, the level of decline depends on individual factors, such as lifestyle, intellectual and physical activity, and genetics (Crawford et al. 2010; McMahan and Sturz 2006), some of which are amenable to interventions in the workplace (Crawford et al. 2010) or even well beyond working age (e.g. Baker et al. 2007). Relationships between age and cognition are complex because some processes decline but others improve. For example, ageing is linked to slower reaction times but greater accuracy and a better capacity to process complex problems (Crawford et al. 2010; Ilmarinen 2001).

Older workers may compensate for diminished physical or mental capacities by drawing on greater job experience and other resources more efficiently (Crawford et al. 2010). However, these strategies are only possible when job demands remain lower than peak work capacity and when job content is flexible (Laflamme and Menckel 1995), which may not be the case for many older workers, especially those in physically demanding jobs. These findings suggest that work demands should be adjusted to suit the strengths and vulnerabilities of older workers, for example by reducing the physical demands of their jobs as they grow older and regularly evaluating individual capacities against work demands.

8.4.6.3 Working Hours

Weekly working hours and shift work, especially night and early morning work, are critical aspects of work design. A large body of research indicates that long hours and shift work have negative effects on health and work performance (Quinlan et al. 2010). Many older workers remain in shift work despite greater susceptibility to these negative effects. European data, for example, show that the prevalence of shift work involving night work drops only slightly after age 45 and that 24 % of men and 12 % of women in this age group work at night (Costa 2003).

For some time there has been evidence that shift work-related sleep disturbances, fatigue and gastrointestinal symptoms increase with age, especially among night workers (Quinlan et al. 2010). Koller (1983), for example, found that physical and psychological symptoms rose steeply during initial exposure to shift work, tapered off, and then again rose steeply after the age of 40. There was no equivalent trend among day workers. More recent evidence confirms that negative effects from shift work and irregular hours increase with age, especially when night work is required (Bohle et al. 2010; Costa 2003). Ageing is associated with changes in daily ('circadian') biological rhythms and slower adjustment to night work, which are associated with shortened sleep after night shift and increased risk of sleep disorders (Quinlan et al. 2010). In addition to circadian changes, Costa (2005) attributed reduced tolerance of shift and night work among older workers to three factors: 'psychophysical conditions', such as physical fitness or sleep efficiency; 'social conditions', such as commuting; and aspects of work design, such as workloads.

Unfortunately, there has been little specific research on the relationship between age and injury in shift work. Folkard (2008) reviewed the evidence and noted a 'complete lack of studies that have directly examined the combined effects of age and shift work on occupational injuries and accidents' (p. 195). However, he did find evidence that, irrespective of the worker's age, injury rates are higher at night than during the day and they increase over successive night shifts. He also noted evidence that injuries are less frequent among older workers but more severe. Finally, he noted that older workers' physical and cognitive performance and sleepiness levels were poorer over the course of a single night shift and over several successive night shifts. On the basis of this indirect evidence, Folkard concluded that older workers may be exposed to greater risks of injury on night shift.

Various interventions may improve the WHS of older shift workers. Costa (2005) recommended the following measures for shift workers over age 45: (1) avoiding or limiting night work; (2) making permanent night work voluntary and facilitating transfer to day work; (3) increasing scope for workers to select preferred shifts; (4) reducing workloads; (5) shortening working hours; (6) allowing more frequent rest breaks and short naps; (7) conducting health checks at least every 2 years; and (8) providing counselling and training on coping strategies. While these recommendations are based on sound research, there has not been enough rigorous intervention research to confirm they have the intended positive effects on older workers' health and safety.

8.4.6.4 Precarious Work

Over recent decades, there have been profound labour market changes in Australia and many other developed countries. In particular, ongoing (or 'permanent') jobs have declined and been replaced by various forms of 'precarious work', such as casual, temporary or subcontract work (Cranford et al. 2003). Precarious work is more dependent on organisations' immediate requirements for labour and does not offer employment security (Louie et al. 2006). It is often associated with poorer

WHS (Quinlan and Bohle 2009; Quinlan et al. 2001) although different forms of precarious work, such as casual work or subcontracting, may have different effects on health (Louie et al. 2006).

Precarious work is expanding among older workers, but its effects on their health have not been thoroughly investigated. In the USA, at least 15 % of workers aged over 55 are in precarious work. Older workers are more likely than younger workers to work in some types of precarious work, such as independent contracting and on-call work, and workers aged over 65 are more likely than any other age group to work precariously (Bureau of Labor Statistics 2005).

Research indicates that the effects of job insecurity on health are greater for older workers than for their younger co-workers. For example, a study of repeated downsizing in retailing found that older workers were more likely to experience long-term symptoms of distress (Isaksson et al. 2000). An extensive meta-analysis of 133 studies (Cheng and Chan 2008) confirmed that the negative effects of job insecurity on physical and mental health were more severe for older workers, irrespective of gender. Payment by the hour, which is a common characteristic of insecure work, is also associated with adverse health outcomes for older workers (Crawford et al. 2010).

Greater work experience is often used to explain why older workers are injured less often than younger workers. However, Breslin and Smith (2005) found the risk of injury associated with short job tenure was at least as high for older workers as younger ones, suggesting that job-specific knowledge and experience, rather than general work experience, is critical. Although many older workers still have longer job tenure than their younger counterparts, the growth of precarious work, which is associated with shorter periods of employment in particular jobs or industries, has reduced this advantage. Unfortunately, there has been little rigorous research on the effects of age and job-specific experience on WHS.

Many older workers prefer flexible or part-time work to regular, full-time work (Shacklock et al. 2007). However, to secure this flexibility they must often relinquish ongoing employment and accept precarious work. Various forms of precariousness, such as casual or agency employment, do tend to have more 'flexible' (irregular) working hours. In many cases, however, this flexibility primarily benefits employers. Costa et al. (2006) distinguished flexibility primarily under employers' control ('variable' hours) from flexibility primarily controlled by employees ('flexible' hours). Research indicates variable hours have negative effects on health and well-being while flexible hours can counteract the negative effects of even highly irregular hours (Costa et al. 2006; Bohle et al. 2011). Bohle et al. (2011), for example, found that employer-controlled, variable hours were associated with greater dissatisfaction with working hours, work–life conflict, fatigue and psychological symptoms. Conversely, flexible hours (with higher employee control) decreased these negative effects, even when hours were very irregular. This research suggests that if older workers are transferred into precarious work with irregular hours, or indeed any work arrangement in which hours are more irregular, it is important that there are mechanisms to ensure their work schedules are genuinely flexible from their point of view.

This Section 8.4.6 highlights the effects of selected aspects of work design on older workers' health and workplace participation. It shows that older workers are at greater risk of serious or fatal injury and also carry a greater burden of work-related disease. Although they are less susceptible to some risks, and there are substantial individual differences in the effects of other risks, older workers are likely to gain benefits from carefully targeted changes to the design of work. Key interventions include those aimed at optimising job demands and eliminating risks that increase with age, reducing or eliminating long working hours and night or early morning work, enhancing worker control over job tasks and working hours, and reducing job insecurity and other elements of precarious work. Appropriate changes of this nature are likely to reduce the risk of early retirement (Crawford et al. 2010). Unfortunately, there has been insufficient rigorous research to conclusively identify the interventions that will be most beneficial in particular circumstances.

8.4.7 Gender and Workforce Participation

Male and female workers face some significantly different challenges as they grow older. Age discrimination is reported across a variety of occupational settings (Roscigno et al. 2007), but women are more likely to report it than men (Ayalon 2014; Duncan and Loretto 2004). Lower levels of training amongst women have also been reported (Evertsson 2004). Australian Bureau of Statistics (ABS) data also suggest that men are more likely to retire due to poor health than women (Australian Bureau of Statistics 2011), arguably due to their over-representation in physically demanding and hazardous occupations. However, women are more likely than men to provide unpaid care and do so with a higher level of intensity (Australian Bureau of Statistics 2010). Yet, it is important to note that these differences are sometimes small, inconsistent and do not apply to all men or all women. While the gendered findings may inform future research and intervention, such endeavours must acknowledge the complexity of men's and women's lives.

8.5 Older Workers, Work Ability and Employment

'Work ability' has been defined as a 'worker's capacity to do their work with respect to the work demands and their health and mental resources' (Ilmarinen and Tuomi 1992: 8). It reflects a worker's capacity to meet the demands of a job. It is the product of interactions between societal, familial, organisational and individual level factors (Lederer et al. 2014) and is closely linked with workforce participation.

Australia's main strategy for addressing the economic costs of population ageing is to keep workers employed for longer. Work ability is a promising conceptual framework for understanding and improving the workforce participation and well-being of an ageing workforce. Promoting workforce participation for older workers is a key objective of work ability research and intervention. It has also been argued that health, caregiving responsibilities, age discrimination, training and the design of work are important components of workforce participation. These challenges are related to workers' capacity to meet job demands. Factors outside work also influence work ability. Social support from family and friends, work–life balance and social structures, such as gender and socioeconomic status, are all associated with workforce participation. Understanding the combined effects of these domains on work ability is likely to provide a strategic direction for workplace policy development and intervention.

8.5.1 Measuring Work Ability

The most common method of assessing an employee's work ability, the Work Ability Index (WAI), comprises three subscales (Tuomi and Oja 1998). The first requires workers to rate their self-perceived work ability compared to their lifetime best and compared to the mental and physical demands of the job. The second subscale assesses physical health (e.g. number of chronic diseases, sick leave) and the third measures mental health. Scores across all the subscales are weighted and summed to give a range between 7 and 49, with higher scores representing higher levels of work ability.

8.5.2 Outcomes of Work Ability

Workers with low work ability tend to retire earlier than workers with higher work ability, due to poor health or disability. A landmark Finnish study of 4255 municipal workers (Ilmarinen et al. 1991) showed that approximately 35 % of workers originally categorised as low work ability had left work due to disability or illness 5 years later. After 11 years almost two thirds of the original low work ability workers had retired with a disability pension and only 2 % remained working full-time. Over the 11-year period between 1981 and 1992, 12 % of the low work ability employees had passed away.

8.5.3 Work Ability and Age

Research suggests that work ability decreases with age, although the evidence is not consistent (see van den Berg et al.'s 2008 review). The extensive literature on age and work ability has important implications nevertheless. Pohjonen's (2001) study of home care workers showed that the oldest group (aged 55–62 years) had work ability scores nearly seven points (7–49 scale) lower than the youngest group (aged

19–34 years). Work ability started to decline by age 35 for some workers and the most marked declines were identified from ages 40–44 and after 55. Pohjonen argued that work ability should therefore be promoted for workers of all ages, particularly by reducing time pressures, improving ergonomics and by providing opportunities for home care workers to control their work conditions.

8.5.4 Dimensions of Work Ability

The following is based on Gould et al.'s (2008) recent conceptualisation of work ability dimensions. There is some confusion in the literature as to whether individual dimensions form part of work ability or are causes of it (Lederer et al. 2014). In line with Gould et al. (2008), we conceptualise them as causes. The dimensions of work ability can be divided into four interrelated areas: the organisation (work, work community and leadership); the individual (health, work motivations, competencies); family and community; and society. All of the challenges that older workers face map onto these dimensions as described below. Work ability is therefore a broadly based framework for promoting workforce participation and informing employment policy for older workers.

8.5.4.1 Work, Work Community and Leadership

The physical and psychosocial work environment is arguably the most significant determinant of work ability for older workers (Ilmarinen et al. 2005). This dimension concerns the design of safe workplaces, interpersonal relationships at work and the way work is organised (e.g. work hours, precariousness). Age discrimination fits within this domain as it concerns the relationships between older workers and those who hire them, between older workers and their managers and between older and younger workers. Flexible working conditions are also relevant as they concern the organisation of work hours to enable workers to juggle work and other responsibilities. Finally, night work was associated with low work ability, but only for precarious workers (Rotenberg et al. 2009).

8.5.4.2 Health and Functional Capacity

Good health, both mental and physical, is the foundation for work ability (Ilmarinen et al. 2005) because workers need the functional capacity (cognitive or physical) to meet job demands. For example, physically demanding roles demand good physical health. However, as older workers' health and functional capacity decline, their work ability may be maintained by altering job demands, for example by reducing physical workloads (Ilmarinen 2001).

8.5.4.3 Motivations, Attitudes and Values

Workers who are motivated to meet job demands and have positive attitudes towards work have the highest work ability (Ilmarinen et al. 2005). However, workers' motivation is also influenced by the work environment. A positive workplace environment promotes motivation and positive attitudes while a damaging workplace environment can have a negative impact (Gould et al. 2008). For older workers, reduced motivation arising from a poor environment may prompt early retirement.

8.5.4.4 Competencies

Competencies arising from improved knowledge and job skills also enhance work ability (Gould et al. 2008). Skills development, through either on-the-job experience or formal training programmes, can improve workers' capacity to meet job demands.

This is problematic in Australia, as the Investing in Experience scheme (Australian Government 2013b), which provided opportunities for older workers to have their skills formally assessed and updated, has been abandoned. Opportunities for precarious workers (e.g. causals, agency workers) to improve skill sets are also diminished, as these workers are often not in the workplace long enough, are overlooked by managers or are not considered a good investment.

8.5.4.5 Family and Close Community

Gould et al. (2008) argue that the social support a worker receives from their family and close community can have a positive influence on their work ability and this is supported in Ilmarinen et al.'s (2005) study. Ilmarinen (2009) also argues that the balance of work and family life is important for work ability. Conflicts between caregiving and work demands may decrease work–life balance, creating a negative effect on work ability.

8.5.4.6 Society or the External Operating Environment

Societal level factors such as government and the labour market create the rules for employment (Gould et al. 2008). One example is governmental policy to prolong working life by increasing age eligibility for the pension. Australia's WHS laws are a further example. These provide protection for workers (and employers) through the provision of safe work environments. It is these policies that can improve or reduce work ability. Safe Work Australia's Strategic direction (2012) for reducing musculoskeletal disorder and the promotion of workers' mental health is a further example.

8.5.5 Work Ability Interventions

Work ability provides a potentially useful framework for workplace intervention to improve the well-being and workforce participation of older workers. However, a recent systematic review (Cloostermans et al. 2014) identified a dearth of high-quality randomised control trials with older workers, particularly in relation to improving work design, skill sets and mental health. Yet work ability interventions that target physical health generally show positive results. Pohjonen and Ranta (2001) found that participation in a 9-month supervised physical exercise programme for home care workers resulted in improved work ability at both the 1-year and 5-year follow-up. However, the scarcity of more comprehensive, multifactor intervention programmes and the widespread focus on the individual rather than the organisation is a significant gap in work ability research.

Silverstein (2008) suggests that most of the challenges facing older workers can be addressed through sustainable business practices that will improve not only work ability but also productivity and staff retention. Silverstein identifies a need for intervention and evaluation research focused on work environment improvements, health promotion, work–life balance and social support. Evidence of this nature is urgently needed for policy development. Without sound data identifying whether interventions focused on the individual or on the design of work can prolong working lives, and whether one focus is more effective than the other, it is unlikely that governments and organisations will commit to designing healthier workplaces.

8.6 Conclusion

This chapter demonstrates that government policy to promote workforce participation should place greater emphasis on improving work organisation. However, more research is required to demonstrate the benefits of more positive workplaces for workers' health and participation. Safe Work Australia's strategic framework (2012) provides a catalyst for appropriate research and intervention, but this alone is unlikely to be enough. For example, reversing the trend toward precarious work and reducing other negative aspects of work design, such as unhealthy or dangerous working hours, or insufficient worker control over job tasks, requires a major change in workplace culture. The economic recession in the 1970s and the Global Financial Crisis have contributed to downsizing and restructuring and to the expansion of precarious work. Instigating a shift back to more secure employment and genuinely flexible working conditions is clearly a major undertaking. Senior executives and managers will need to understand and accept evidence that changes in work design will improve worker health, productivity and participation—and consequently profitability—before they 'buy in' to more sustainable practices.

Some of the concepts and issues identified in this chapter, such as work ability, are also relevant to the health and workforce participation of younger workers. Exposures to harmful physical and psychosocial work environments often begin early in working careers and accumulate over the life course. This is one of the reasons Brooke and Taylor (2005) argue that organisational age management policies and practices should focus on workers of all ages rather than just older workers. The ageing of the workforce demands that such organisational and policy change should happen sooner rather than later.

8.7 Case Study

Peter is 58 years old and has worked in construction for 35 years. Like many in the industry, he was exposed to asbestos until the 1980s although he was fortunate to work in safer conditions than some. Nevertheless, he displays early signs and symptoms of asbestosis and it has become harder for him to do physical work tasks.

Peter has watched colleagues become ill and die from asbestos-related diseases. This experience was a major reason why he developed a keen interest in workplace health and safety (WHS). From 1995 to 2005 he consulted on many workplace investigations into hazardous exposures. He had a senior role in the construction workers' union from 2002 to 2008 and consulted on the safe design of work for small- and large-scale building sites.

Peter is married with three adult children and five grandchildren. His 85-year-old mother lives in a small flat at the back of their modest home. The Global Financial Crisis had a significant economic impact on Peter and his extended family. His consultancy work dried up and consequently he has had to take whatever construction work he could. It has involved a lot of casual agency work over multiple work sites, including evenings and weekends. He feels fortunate if he works on one site for more than a fortnight. He is relieved and grateful his wife has found work at the local supermarket. However, conflicting shifts often mean that they have little time together and care for Peter's ageing mother is a constant concern.

The increasing use of computers for the development and viewing of building plans is a major challenge. Although Peter is enthusiastic about learning to use new technology, he struggles and is often chastised by younger co-workers for what they see as incompetence. He asked a project manager for computer training but was told the investment could not be justified because he was a temporary worker. Peter cited his low rates of absenteeism, safe work practices and quality of work as justification, but to no avail. He has also asked his agency for training opportunities, but has been told he is too old. Peter fears that without training he will struggle to find work but he cannot afford to retire. Sadly, though, his declining health may demand it.

8.7.1 Discussion Points

- What strengths does Peter have as an older worker and what challenges does he face in the workplace?
- What are the strengths and limitations of existing policies designed to support Peter's employment?
- Which determinants of workplace participation for older workers are most important—biological, psychological or social?

References

Access Economics. (2010). *The economic value of informal care in 2010: Report by Access Economics for Carers Australia*. Canberra, Australia: Access Economics.

Australian Bureau of Statistics. (1999). *Labour force projections: 1999-2016* (Cat. No. 6260.0). Canberra, Australia: ABS.

Australian Bureau of Statistics. (2010). *Australian social trends: Older people and the labour market* (Cat. No. 4102.0). Canberra, Australia: ABS.

Australian Bureau of Statistics. (2011). *Retirement and retirement intentions* (Cat. No. 6238.0). Canberra, Australia: ABS.

Australian Bureau of Statistics. (2013). Disability, ageing and carers, Australia: summary of findings, 2012(Cat. No. 4430.0). In Australian Bureau of Statistics (Ed.). Canberra, Australia: ABS.

Australian Bureau of Statistics. (2013). Disability, ageing and carers, Australia: summary of findings, 2012(Cat. No. 4430.0). In Australian Bureau of Statistics (Ed.). Canberra, Australia: ABS.

Australian Government. (2011). *United Nations Economic and Social Commission for Asia and the Pacific, regional survey on ageing 2011: Australia*. Retrieved June 24, 2013, from http://www.health.gov.au/internet/publications/publishing.nsf/Content/ageing-un-report-regional+survey-ageing-2011-toc

Australian Government. (2013a). *Healthy workers initiative*. Retrieved March 21, 2014, from http://www.healthyworkers.gov.au/

Australian Government. (2013b). *Investing in experience: Skills recognition and training*. Retrieved March 31, 2014, from http://skillsconnect.gov.au/

Australian Human Rights Commission. (2012). *Age discrimination: Know your rights*. Retrieved December 12, 2014, from http://www.humanrights.gov.au/know-your-rights-age-discrimination.

Ayalon, L. (2014). Perceived age, gender, and racial/ethnic discrimination in Europe: Results from the European Social Survey. *Educational Gerontology, 40*(7), 499–517.

Baker, M., Kennedy, D., Bohle, P., Campbell, D., Knapman, L., Grady, J., Wiltshire, J., McNamara, M., Evans, W., Atlantis, E., & Fiatarone Singh, M. (2007). Efficacy and feasibility of a novel tri-modal robust exercise prescription in a retirement community: A randomized, controlled trial. *Journal of the American Geriatrics Society, 55*(1), 1–10.

beyondblue. (2015). *National workplace program*. Retrieved from https://www.beyondblue.org.au/about-us/programs/workplace-and-workforce-program/programs-resources-and-tools/national-workplace-program

Bohle, P., Pitts, C., & Quinlan, M. (2010). Time to call it quits? The safety and health of older workers. *International Journal of Health Services, 40*(1), 23–41.

Bohle, P., Willaby, H., Quinlan, M., & Mc Namara, M. (2011). Flexible work in call centres: Working hours, work-life conflict and health. *Applied Ergonomics, 42*, 219–224.

Börsch-Supan, A., & Weiss, M. (2013). *Productivity and age: Evidence from work teams at the assembly line*. The Research Centre for Education and the Labour Market (ROA), Maastricht University School of Business and Economics.

Breslin, F. C., & Smith, P. (2005). Age-related differences in work injuries: A multivariate, population-based study. *American Journal of Industrial Medicine, 48*(1), 50–56.
Brooke, L., & Taylor, P. (2005). Older workers and employment: Managing age relations. *Ageing and Society, 25*(3), 415–429.
Bureau of Labor Statistics. (2005). *Contingent and alternative employment arrangements*. Retrieved May 31, 2008, from http://www.bls.gov/news.release/conemp.toc.htm
Cardoso, A. R., Guimarães, P., & Varejão, J. (2011). Are older workers worthy of their pay? An empirical investigation of age-productivity and age-wage nexuses. *De Economist, 159*(2), 95–111.
Chapman, A., Mulrennan, S., Ladd, B., & Muers, M. F. (2008). Population based epidemiology and prognosis of mesothelioma in Leeds, UK. *Thorax, 63*(5), 435–439.
Cheng, G. H. L., & Chan, D. K. S. (2008). Who suffers more from job insecurity? A meta-analytic review. *Applied Psychology, 57*(2), 272–303.
Cloostermans, L., Bekkers, M. B., Uiters, E., & Proper, K. I. (2014). The effectiveness of interventions for ageing workers on (early) retirement, work ability and productivity: A systematic review. *International Archives of Occupational and Environmental Health, 88*(5), 521–532.
Costa, G. (2003). Shift work and occupational medicine: An overview. *Occupational Medicine, 53*, 83–88.
Costa, G. (2005). Some considerations about aging, shift work and work ability. In G. Costa, W. Goedhard, & J. Ilmarinen (Eds.), *Assessment and promotion of work ability, health and well-being of ageing workers* (pp. 67–72). San Diego, CA: Elsevier.
Costa, G., Sartori, S., & Akerstedt, T. (2006). Influence of the flexibility and variability of working hours on health and well-being. *Chronobiology International, 23*(6), 1125–1137.
Cranford, C. J., Vosko, L. F., & Zukewich, N. (2003). The gender of precarious employment in Canada. *Relations Industrielles, 58*(3), 454–482.
Crawford, J. O., Graveling, R. A., Cowie, H. A., & Dixon, K. (2010). The health, safety and health promotion needs of older workers. *Occupational Medicine, 60*, 184–192.
Department of Employment. (2014). *Corporate champions*. Retrieved March 12, 2014, from http://employment.gov.au/corporate-champions
Department of the Treasury (Australia). (2015). *2015 Intergenerational report: Australia in 2055*. Department of the Treasury (Australia). Retrieved November 9, 2015, from http://apo.org.au/node/53411
Duncan, C., & Loretto, W. (2004). Never the right age? Gender and age-based discrimination in employment. *Gender, Work and Organization, 11*(1), 95–115.
Evertsson, M. (2004). Formal on-the-job training: A gender-typed experience and wage-related advantage? *European Sociological Review, 20*(1), 79–94.
Folkard, S. (2008). Shift work, safety and ageing. *Chronobiology International, 25*, 183–198.
Fouarge, D., & Schils, T. (2009). The effect of early retirement incentives on the training participation of older workers. *Labour, 23*(s1), 85–109.
Göbel, C., & Zwick, T. (2009). *Age and productivity: Evidence from linked employer employee data* (No. 09-020). Mannheim, Germany: ZEW Research. Retrieved from ftp://ftp.zew.de/pub/zew-docs/dp/dp09020.pdf
Gould, R., Ilmarinen, J., Jarvisalo, J., & Koskinen, S. (2008). *New dimensions of work ability: Results of the Health 2000 Survey*. Helsinki: Finnish Institute of Occupational Health.
Gringart, E., Helmes, E., & Speelman, C. P. (2005). Exploring attitudes toward older workers among Australian employers: An empirical study. *Journal of Aging and Social Policy, 17*(3), 85–103.
Ilmarinen, J. E. (2001). Ageing workers. *Occupational and Environmental Medicine, 58*(8), 546–552.
Ilmarinen, J. (2009). Work ability—A comprehensive concept for occupational health research and prevention. *Scandinavian Journal of Work, Environment and Health, 35*(1), 1–5.
Ilmarinen, J., & Tuomi, K. (1992). Work ability of aging workers. *Scandinavian Journal of Work, Environment and Health, 18*(s2), 8–10.
Ilmarinen, J., Tuomi, K., Eskelinen, L., Nygård, C.-H., Huuhtanen, P., & Klockars, M. (1991). Summary and recommendations of a project involving cross-sectional and follow-up studies on

the aging worker in Finnish municipal occupations (1981—1985). *Scandinavian Journal of Work, Environment and Health, 17*(s1), 135–141.
Ilmarinen, J., Tuomi, K., & Seitsamo, J. (2005). New dimensions of work ability. *International Congress Series, 1280*, 3–7.
Isaksson, K., Hellgren, J., & Petterson, P. (2000). Repeated downsizing: Attitudes and wellbeing of surviving personnel in a Swedish retail company. In K. Isaksson, C. Hogstedt, C. Erisksson, & T. Theorell (Eds.), *Health effects of the new labor market*. New York: Kluwer.
Kendig, H., Wells, Y., O'Loughlin, K., & Heese, K. (2013). Australian baby boomers face retirement during the global financial crisis. *Journal of Aging and Social Policy, 25*(3), 264–280.
Koller, M. (1983). Health risks related to shift work. *International Archives of Occupational and Environmental Health, 53*, 59–75.
Laflamme, L., & Menckel, E. (1995). Aging and occupational accidents: A review of the literature of the last three decades. *Safety Science, 21*, 145–161.
Lederer, V., Loisel, P., Rivard, M., & Champagne, F. (2014). Exploring the diversity of conceptualizations of work (dis)ability: A scoping review of published definitions. *Journal of Occupational Rehabilitation, 24*(2), 242–267.
Lilly, M. B. (2011). The hard work of balancing employment and caregiving: What can Canadian employers do to help? *Healthcare Policy, 7*(2), 23.
Louie, A. M., Ostry, A. S., Quinlan, M., Keegel, T., Shoveller, J., & LaMontagne, A. D. (2006). Empirical study of employment arrangements and precariousness in Australia. *Relations Industrielles, 61*(3), 465–486.
Mathieu, J. E., & Zajac, D. M. (1990). A review and meta-analysis of the antecedents, correlates, and consequences of organizational commitment. *Psychological Bulletin, 108*(2), 171–194.
Mauno, S., Ruokolainen, M., & Kinnunen, U. (2013). Does aging make employees more resilient to job stress? Age as a moderator in the job stressor–well-being relationship in three Finnish occupational samples. *Aging and Mental Health, 17*(4), 411–422.
McMahan, S., & Sturz, D. (2006). Implications for an aging workforce. *Journal of Education for Business, 82*(1), 50–55.
Munnell, A. H. (2011). *What is the average retirement age?* Center for Retirement Research at Boston College. Retrieved from http://crr.bc.edu/wp-content/uploads/2011/08/IB_11-11-508.pdf
Ng, T. W. H., & Feldman, D. C. (2012). Evaluating six common stereotypes about older workers with meta-analytical data. *Personnel Psychology, 65*(4), 821–858.
Noone, J., O'Loughlin, K., & Kendig, H. (2013). Australian baby boomers retiring 'early': Understanding the benefits of retirement preparation for involuntary and voluntary retirees. *Journal of Aging Studies, 27*(3), 207–217.
Picchio, M., & Van Ours, J. C. (2013). Retaining through training even for older workers. *Economics of Education Review, 32*, 29–48.
Pohjonen, T. (2001). Perceived work ability of home care workers in relation to individual and work-related factors in different age groups. *Occupational Medicine, 51*(3), 209–217.
Pohjonen, T., & Ranta, R. (2001). Effects of worksite physical exercise intervention on physical fitness, perceived health status, and work ability among home care workers: Five-year follow-up. *Preventive Medicine, 32*(6), 465–475.
Quinlan, M., & Bohle, P. (2009). Overstretched and unreciprocated commitment: reviewing research on the occupational health and safety effects of downsizing and job insecurity. *International Journal of Health Services, 39*(1), 1–44.
Quinlan, M., Bohle, P., & Lamm, F. (2010). *Managing occupational health and safety: A multidisciplinary approach* (3rd ed.). Melbourne: Palgrave Macmillan.
Quinlan, M., Mayhew, C., & Bohle, P. (2001). The global expansion of precarious employment, work disorganization, and consequences for occupational health: A review of research. *International Journal of Health Services, 31*(2), 335–414.
Rauschenbach, C., Göritz, A. S., & Hertel, G. (2012). Age stereotypes about emotional resilience at work. *Educational Gerontology, 38*(8), 511–519.

Rauschenbach, C., Krumm, S., Thielgen, M., & Hertel, G. (2013). Age and work-related stress: A review and meta-analysis. *Journal of Managerial Psychology, 28*(7/8), 781–804.

Ritz, B. (1999). Radiation exposure and cancer mortality in uranium processing workers. *Epidemiology, 10*(5), 531–538.

Rogers, E., & Wiatrowski, W. J. (2005). Injuries, illnesses, and fatalities among older workers. *Monthly Labor Review, 128*(10), 24–30.

Roscigno, V. J., Mong, S., Byron, R., & Tester, G. (2007). Age discrimination, social closure and employment. *Social Forces, 86*(1), 313–334.

Rotenberg, L., Griep, R. H., Fischer, F. M., Fonseca, M. J. M., & Landsbergis, P. (2009). Working at night and work ability among nursing personnel: When precarious employment makes the difference. *International Archives of Occupational and Environmental Health, 82*(7), 877–885.

Safe Work Australia. (2005). *Surveillance alert: OHS and the ageing workforce*. Canberra, Australia: Safe Work Australia.

Safe Work Australia. (2012). *Australian work health and safety strategy 2012-2022*. Canberra, Australia: Safe Work Australia.

Schofield, D., Passey, M., Earnest, A., Percival, R., Kelly, S., Shrestha, R., & Fletcher, S. (2009). Health&WealthMOD: A microsimulation model of the economic impacts of diseases on older workers. *The International Journal of Microsimulation, 2*(2), 58–63.

Schofield, D., Shrestha, R. N., Percival, R., Passey, M. E., Callander, E. J., & Kelly, S. J. (2011). The personal and national costs of mental health conditions: impacts on income, taxes, government support payments due to lost labour force participation. *BMC Psychiatry, 11*, 72–79.

Shacklock, K., Fulop, L., & Hort, L. (2007). Managing older worker exit and re-entry practices: A 'revolving door'? *Asia Pacific Journal of Human Resources, 45*(2), 151–167.

Silverstein, M. (2008). Meeting the challenges of an aging workforce. *American Journal of Industrial Medicine, 51*(4), 269–280.

Skirbekk, V. (2004). Age and individual productivity: A literature survey. *Vienna Yearbook of Population Research, 2*(1), 133–154.

Snape, E., & Redman, T. (2006). Too old or too young? The impact of perceived age discrimination. *Human Resource Management Journal, 13*(1), 78–89.

Stansfeld, S., & Candy, B. (2006). Psychosocial work environment and mental health—A meta-analytic review. *Scandinavian Journal of Work, Environment and Health, 32*(6), 443–462.

Swan, W. (2010). *Australia to 2050: Future challenges*. Commonwealth of Australia.

Temple, J., Adair, T., & Hosseini-Chavoshi, M. (2011). *Ageing and the barriers to labour force participation in Australia*. Canberra: National Seniors Australia.

Tikkanen, T. (2011). From managing a problem to capitalizing on talent and experience of older workers. *The International Journal of Human Resources, 22*(6), 1217–1220.

Tuomi, K., & Oja, G. (1998). *Work ability index*. Helsinki: Finnish Institute of Occupational Health.

van den Berg, T., Elders, L., de Zwart, B., & Burdorf, A. (2008). The effects of work-related and individual factors on the Work Ability Index: A systematic review. *Occupational and Environmental Medicine, 66*, 211–220.

Virtanen, M., Oksanen, T., Batty, G. D., Ala-Mursula, L., Salo, P., et al. (2014). Extending employment beyond the pensionable age: A cohort study of the influence of chronic diseases, health risk factors, and working conditions. *PLoS One, 9*(2), e88695.

Zwerling, C., Sprince, N. L., Wallace, R. B., Davis, C. S., Whitten, P. S., & Herringa, M. A. (1996). Risk factors for occupational injuries among older workers: An analysis of the health and retirement study. *American Journal of Public Health, 86*(9), 1306–1309.

Part II
Living and Lifestyle

Chapter 9
Retirement Income

Hazel Bateman, Rafal Chomik, and John Piggott

9.1 Introduction

Providing a population with some form of income security for the later stages of their lifespan has been a part of the political agenda in developed economies for more than a century. Policy was first formalised at a national level in Germany in the mid-1880s, championed by the Chancellor Otto von Bismarck. Many countries in Western Europe followed. Australia and New Zealand were the first Anglophone countries to adopt such a policy—in the Australian case the Age Pension was introduced in 1909, replacing some pre-existing, state-based schemes. Roosevelt's 'New Deal', in the mid-1930s, saw the birth of a US policy encompassing both unemployment insurance and retirement provision, and the 1942 Beveridge report formed the basis for the UK's National Insurance plan. By the 1950s, all developed economies, and many developing nations, had in place some form of institutionalised retirement provision policy.

It is worth reflecting on why this should have become a social policy imperative at this time, and why it spread so rapidly to become a more or less global phenomenon. Several factors were at work. The first major historical development was the

H. Bateman (✉)
ARC Centre of Excellence in Population Ageing Research, School of Risk and Actuarial Studies, University of New South Wales, 223 Anzac Parade, Kensington, NSW 2033, Australia
e-mail: h.bateman@unsw.edu.au

R. Chomik
ARC Centre of Excellence in Population Ageing Research, Sydney, NSW, Australia

J. Piggott
ARC Centre of Excellence in Population Ageing Research, School of Economics, UNSW Business School, University of New South Wales, Level 3 East, 223 Anzac Parade, Kensington, NSW 2033, Australia
e-mail: j.piggott@unsw.edu.au

K. O'Loughlin et al. (eds.), *Ageing in Australia*, International Perspectives on Aging 16, DOI 10.1007/978-1-4939-6466-6_9

industrial revolution. Between 1750 and 1850, society was transformed by a series of technological innovations, supported by legislation that allowed the development of large pools of capital for their commercialisation. The population of Europe and North America increased by six times over this period, and per capita income increased tenfold. Life expectancy, measured at birth, started increasing. For example, in England, while life expectancy increased from 37 in 1750 to 40 in 1850, this trend accelerated in the early twentieth century: life expectancy reached 48 by 1900 and 69 by 1950 (Fogel 2004).

These developments meant that for the first time, ordinary people enjoyed a standard of living that enabled them to contemplate retirement and society was becoming rich enough to finance it. At the same time, the organisation of labour changed such that people had to move to enjoy the benefits of the revolution, weakening intergenerational linkages within families. The old were less able than previously to rely on their children to support them through their final years, which provided fertile ground for a society-wide solution.

The second historical imperative came in the form of a series of inter-related global events. The 75-year period between the end of the Franco-Prussian War in the 1870s and the end of the Second World War in 1945 were very tumultuous for the nations of Europe and North America, and for their people. This period witnessed two 'global' wars and the Great Depression. The social fallout from this sequence of events—unemployment, poverty, and family separation among others—precipitated government response in the form of social security programmes that encompassed not only retirement provision, but also unemployment and sickness insurance.

There are many possible policy paradigms for retirement income provision, but all of them rely, to a greater or lesser degree, on the resources of the currently working population to support older cohorts. Two schematic models are widely recognised, although they are by no means the only possibilities. The 'Bismarckian' model links benefits to contributions made while employed, graduated according to income. The 'Beveridge' model covers the entire population, is primarily funded from general revenue, and is linked much more loosely to pre-retirement income. Most countries have elements of both in their social security systems.

The economic rationale for government intervention in retirement provision was first systematically articulated by Diamond (1977).[1] In a remarkably prescient article, Diamond identified the major reasons for a retirement social security programme as redistribution, market failure, paternalism, and administrative efficiency in social insurance delivery.

He points to both the intra- and inter-generational aspects of redistribution, and the importance of a life cycle (or life-course) perspective in thinking about the question of retirement income provision. There are many possible market failures in the retirement income context, but the most important relates to 'adverse selection'[2] in

[1] Recently updated for the Australian context in Diamond (2011).

[2] This market failure stems from information asymmetry—the presumption is that the annuity issuer knows less about the annuitant's life expectancy than he does. Two effects—moral hazard and adverse selection—flow from this. *Moral Hazard* exists whenever the liability of the insurer is affected by actions of the insured party about which the insurer has incomplete information.

the life annuity market. He emphasises risk associated with the length of life as one which private markets would find difficult to insure against.

Subsequently, what might be termed 'cohort risk' was added to this list (Gordon and Varian 1988). Cohorts born in the early part of the twentieth century, for example, faced two world wars and the Great Depression in the prime of their lives. Their human development was impacted in many ways—poor nutrition, lack of work experience, family disintegration. Government intervention in social insurance allowed some of these risks to be shared across generations. Market institutions are ineffective in this kind of risk management and families, which had been the traditional institution for intergenerational risk sharing, have become less effective, for reasons already explained.

Because these social security systems carry with them an ongoing chain of transfers from younger to older generations, through the payment of taxes or social security contributions, they work best when the population is expanding and the economy is growing. In recent decades, of course, a demographic shift has been observed, with a large birth cohort—the baby boomers, born between 1946 and 1965—now approaching retirement, and relatively smaller younger cohorts making up the labour force. The provision of retirement income on this basis at a time of demographic shift has become one of the major challenges in public policy. Around the world, countries are struggling to balance equity, adequacy, and sustainability in their retirement income provision models, and many countries have reformed their retirement income systems, or are planning to do so.

This chapter describes and assesses the Australian paradigm. Like in other countries, this has changed greatly over the last 30 years. Because its design is quite different from that encountered in most other developed countries, the chapter begins by providing an overview of retirement income policy design and describes how the Australian system fits this taxonomy. Having discussed the design of the system it then turns to how well it functions, making use of key metrics and comparisons with other developed countries. The chapter concludes with some thoughts about future challenges and opportunities.

9.2 Retirement Income Policy Design

Governments have responded in different ways to the retirement income provision needs of older people. A much used framework for posits three 'pillars' of support.[3] Different writers have different definitions of what each pillar represents; the following demarcations will be used for the purposes of this chapter. The first pillar

---

Adverse Selection arises if individuals know their own riskiness but the insurer does not; the possibilities offered are then less efficient than if individually tailored policies could be offered. See Sect. 9.4.2.

[3] The idea of 'three pillars' originated in the seminal World Bank publication 'Averting the Old Age Crisis' (World Bank 1994).

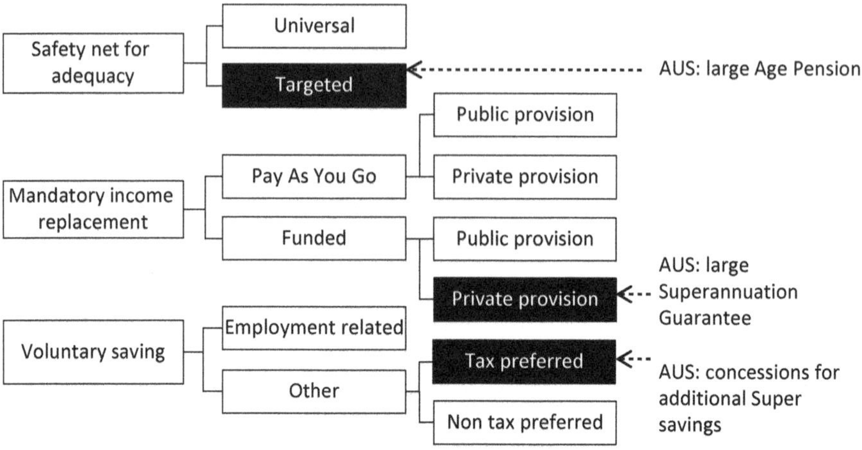

Fig. 9.1 Taxonomy of pension system design

operates as a non-contributory transfer, which means it is available to all.[4] It is possible to think of this as a 'Beveridge' pillar. The second pillar offers payments as a function of pre-retirement labour income—achieved via some form of compulsory contributions by either the employer, employee, or both. This may be thought of as a 'Bismarck' pillar. The third pillar comprises voluntary retirement saving.

At the cost of some oversimplification, most Organisation for Economic Co-operation and Development (OECD) retirement provision policies can be characterised using this schema. Figure 9.1 provides a guide to some of the major alternatives.

Consider the first (Beveridge) pillar. Transfers may be universal or targeted (i.e. subject to a means test). The earnings related compulsory retirement saving contribution component, the second pillar, may be government funded and provided (the most common OECD model) or privately funded and provided. In other countries, they are in the *private* sector in the form of pension or superannuation funds. This has implications for administrative costs and also governance of the system. Another distinction here is that the source of the payments may be past accumulations—a *pre-funded* scheme, or current contributions—an *unfunded* or *pay-as-you-go* plan. This is especially important because an unfunded plan relies on current worker contributions to finance the benefits of the current retired, while a pre-funded scheme transfers resources from an individual's youth to old age. Voluntary saving for retirement (Pillar 3) is generally encouraged through some kind of tax incentive. This is sometimes part of the pension or retirement saving system, or it may be separated, as with Individual Retirement Accounts (IRAs) in the US or Individual Savings Accounts (ISAs) in the UK.

[4]It may only be available subject to a means test, since its function is to ensure the elderly are adequately provided for.

Another area of distinction, not shown in the diagram, which often aligns with pre-funded and unfunded schemes, separates *defined contribution* and *defined benefit* paradigms. Defined contribution schemes rely on a set level of savings which, based on taxes, investment returns and fees, build up to form an accumulation at retirement. Defined benefit, on the other hand, are schemes where the benefit promises are based on some predetermined formula, often relating to salary and years in work. Defined benefit schemes may be pre-funded or unfunded (financed on a pay-as-you-go basis).

As the aged dependency ratio, defined as the ratio of the population over age 65 to the population aged 15–64 years, increases, publicly provided, unfunded, defined benefit schemes place an increasing tax burden on current workers, impacting on their welfare and labour force incentives. The fiscal sustainability of such plans then becomes a major challenge. All over the world, social security reforms have reduced the fiscal burden by reneging on past promises associated with publicly provided pay-as-you-go defined benefit plans. This may take the form of changing the promised indexation of payments (e.g. from wage indexation to price indexation); increasing the access age; changing the method of benefit calculation (e.g. how past wages are indexed to the year of retirement) or changing the level of benefits themselves.

The typical alternative of funded, privately managed defined contribution plans is less vulnerable to political interference but are not immune from tax or regulatory change.

9.3 The Australian Model[5]

Australia's retirement income system, as indicated in Fig. 9.1, comprises a means-tested Age Pension financed from general tax revenues; a mandatory employer financed defined contribution scheme as the second pillar, known as the Superannuation Guarantee; and tax incentives to encourage voluntary superannuation contributions and other private savings. A summary of the three pillars is provided in Table 9.1.

9.3.1 The Age Pension

The Age Pension is financed from general revenue and currently paid at a rate equivalent to 27.7 % of male average full-time earnings for single pensioners and 41.3 % for couples. Net replacement rates (the comparison of net income before and after retirement) are relatively higher because no tax is payable on the Age Pension. The

[5] See Bateman and Piggott (1997, 1998) for a discussion of the historical evolution of the Australian system. The discussion of Australia's retirement income system draws on Bateman and Piggott (2011) and Bateman (2011).

Table 9.1 Features of the Australian retirement income system's three pillars

	Age pension	Superannuation guarantee and voluntary superannuation
Established	1909	1850s (voluntary superannuation), 1992 (Superannuation Guarantee)
Eligibility/coverage	Residency; age (male and female equalised at 65 since 2013; increasing to 67 by 2023); means tested (income and assets)	Superannuation Guarantee: Employees aged 18+ with earnings >$A450/ month; Self-employed not compelled but can contribute on a voluntary basis
		Voluntary superannuation: Tax incentives for contributions, subject to contribution caps
Funding	General tax revenue	Fully funded; individual accounts
		Superannuation Guarantee: 9.5 % mandatory employer contribution (set to increase to 12 %)
		Voluntary superannuation: voluntary employer and employee contributions and limited access to a government co-contribution
Benefit	Single rate is 27.7 % and couple rate is 41.3 % of male average weekly earnings; Indexed to greatest of CPI, PBI and male AWE	Mostly based on defined contributions. No mandatory benefit requirement at retirement. Choice of income stream (account-based pension and/or annuity) and/or lump sum
Other benefits/ features	Pension supplement, rent allowance, pensioner concession card	Vested, portable, preserved to age 56 for both male and female (increasing to 60 by 2025)
Taxation	Pensioner tax rebate fully exempts full Age Pension from tax, partial exemption for part Age Pension	Employer contributions taxed at 15 %, employee contributions generally after-tax; investment earnings taxed at 15 % (reduced by dividend imputation credits and capital gain discounts); benefits mostly tax exempt from age 60
Means tests	*Income test:* Withdrawal of 50c per dollar of private income in excess of free area. Concessional treatment of income from labour	n/a
	Assets test: Withdrawal of $3 per fortnight for every $1000 of assets above thresholds which differ by single/couple home/ non-home owner	
	Age Pension based on lower rate of the two tests; thresholds indexed to CPI	

benefit level is indexed to the greater growth of male average earnings, the consumer price index (CPI), and a pension and beneficiary living cost index (PBI). As a result, the Age Pension keeps up with wages in the rest of the economy and by implication with standards of living. Recipients of the Age Pension also gain access to other benefits such as a pensioner concession card and assistance with rent. Eligibility age is currently 65 but will increase to age 67 between 2017 and 2023 (Australian Government 2009).

The Age Pension is available on the basis of residency, irrespective of contributions or work history, but is means tested. Tests apply to both assets and income and have the effect of excluding the best-off quarter of the eligible population from receiving the Age Pension. As such, rather than a safety net mechanism that targets the poor, the Age Pension functions as a poverty alleviation instrument that excludes the rich. The means tests are comprehensively defined, although the asset test excludes owner-occupied housing. Prior to the Simplified Super reforms (implemented in 2007), differential application of these tests was used as a policy lever to encourage purchase of specific types of retirement income products (Australian Government 2006).

9.3.2 Compulsory Saving: The Superannuation Guarantee

Traditionally, Australia relied mainly on its Age Pension for retirement income provision. Tax breaks for voluntary superannuation (the third pillar) were introduced in 1915 and subsequently strengthened in 1936. However, the level of preservation (the saving of superannuation assets until later ages) was poor and coverage (the proportion of workers or working population with formal savings arrangements in place) was low. By the mid-1980s only about a third of workers in the private sector and less than half of all workers were covered by superannuation. Unlike many other OECD countries, a government provided earnings- or employment-related retirement income scheme has never been introduced in Australia, despite several broadly supported attempts to do so.

The introduction of the Superannuation Guarantee was rooted in key events that took place in the 1980s. A key plank of the Hawke Labor government's economic strategy was an ongoing agreement with the unions that included the idea of building superannuation arrangements into a national centralised wage decision. This came to fruition in 1986, when the poor macroeconomic environment, particularly high inflation relative to trading partners, required wage restraint and provided the opportunity for the introduction of broad coverage superannuation. A central element in the 1986 National Wage Case was what became known as the productivity award superannuation: that an increase in employee compensation should be 6 % in order to keep pace with inflation, but that half of that increase would take the form of a 3 % employer superannuation contribution in order to control further inflationary pressures and by the same token to formalise retirement saving arrangements.[6]

Following the introduction of the broad productivity award superannuation, the negotiation of individual industrial agreements increased superannuation coverage markedly. This was particularly true for workers in private sector industries dominated by women, casual and part-time workers, such as the retail industry, where coverage increased from 24 % in 1986 to 82 % in 1993. Overall coverage doubled to 79 % of all workers.

However, the award-based system was costly and difficult to enforce. An attempt, supported by government and the unions, to increase contributions by a further 3 % was rejected by Australian industrial court in 1991. The government response was to introduce legislation that would enshrine what is now known as the Superannuation Guarantee. Starting in 1992, it required employers to make contributions on behalf of their employees into an approved superannuation fund. The mandatory contribution rate gradually increased to become 9 % by 2002. In 2012 the government initiated an increase in the mandatory contribution rate to 12 % by 2019, which has now been paused—it is now set to increase from the current 9.5–12 % between 2021 and 2025.

9.3.3 Voluntary Retirement Saving

The first and second pillars of Australia's retirement income system are supplemented by voluntary long-term savings that include superannuation, property, shares, managed investments, and home ownership. The latter is the most important non-superannuation asset for most Australians. Net equity in home ownership in 2009/10 was worth 40 % of household wealth and over 80 % of retirees are owner occupiers (mostly with no mortgage) (Australian Bureau of Statistics (ABS) 2011a).

Voluntary contributions are encouraged by income tax incentives (for all except those on the lowest personal marginal tax rate), and the government co-contribution scheme matches personal contributions from low- and middle-income earners and the self-employed (who are excluded from compulsory superannuation contributions) up to a maximum of $500 a year. Some employees contribute to their superannuation accounts through 'salary sacrifice' arrangements. Since these are treated as employer contributions they attract a tax rate of 15 %, which is often lower than the contributor's marginal tax rate. An offset exists at the bottom of the earnings distribution to ensure those on the lowest tax brackets are not paying the 15% contribution tax.Voluntary contributions from employees or employers mean that about one-third of superannuation fund members enjoy contribution rates of over 9 % of earnings (ABS 2009).

---

[6]The agreement was ratified by Australia's industrial court and survived a constitutional challenge tabled by industry representatives (Dabscheck 1989: 99).

Some 90 % of Australian employees are now covered by superannuation, about double the level of coverage seen at the time mandatory arrangements were introduced in the late 1980s (ABS 1999, 2011b).

9.3.4 *Retirement Benefits*

Benefits from superannuation savings can currently be accessed at age 56, the statutory preservation age, which is increasing to 60 for those born after June 1964. The form of retirement benefits is not mandated. These are often taken as lump sums or income streams (account-based pensions or annuities) and are predominantly tax exempt after age 60. As the Superannuation Guarantee matures, the proportion of benefits being paid out of Superannuation funds in the form of account-based pensions (a type of phased withdrawal product) rather than as lump-sum payments is increasing. At present, very few lifetime annuities[7] are purchased (Bateman and Piggott 2011).

The Australian Treasury estimates that a fully mature superannuation guarantee can be expected to deliver a net retirement replacement rate of around 90 % including the means-tested Age Pension for a worker on median male earnings and 78 % for a worker on average weekly earnings (Gallagher 2012).

9.4 What Makes a Retirement Policy Good or Bad?

The introduction to this chapter touched on the rationale for government intervention in retirement income provision. But having discussed the general design of Australia's retirement income system in previous sections we now return to how well it serves its purpose. In doing so, it is helpful to think about the retirement income system from two perspectives: how it affects a retiring individual and its impact on the allocation of resources in the overall economy.

9.4.1 *An Individual's Perspective*

From the point of view of the individual the retirement income system is a means of income insurance. As highlighted by Bodie (1989), and extended in the Australian context in Bateman and Piggott (1997), there are a number of uncertainties that a risk-averse individual may wish to insure against. The most important of these include replacement, investment, longevity, and inflation risks.

---

[7] A lifetime annuity is a financial product that can be purchased to provide a regular payment for the lifetime of the annuitant regardless of length of life or market conditions.

Replacement rate risk refers to the possibility that income after retirement will provide a significantly lower standard of living than that enjoyed during working age. In fact, a common way of comparing retirement outcomes is by use of the replacement rate: the ratio of income in retirement to income prior to retirement.

Investment risk refers to the risk that retirement income suffers as a result of poor investment returns on the underlying assets. Where retirement income depends on the accumulation of contributions in a defined contribution scheme, such as the Superannuation Guarantee operating in Australia, this risk is borne by the individual. Such risk can be mediated by the presence of safety nets, such as Australia's Age pension. In defined benefit schemes, the individual is insured by their pension provider.

Longevity risk relates to the possibility that individuals will outlive their savings. For those with account-based savings (such as a defined contribution retirement saving account), such risk can be transferred to a life insurer if the individual is able to purchase a life annuity. The individual can also be protected if they have a defined benefit pension scheme or, in part, if there is a government-provided safety net.

Inflation risk refers to the risk that future price increases will erode the purchasing power of the individual's savings. The ability to insure against this risk depends in part on the presence of inflation-linked annuities in the market.

9.4.2 Economy-Wide Perspective

Key criteria for evaluating public policy relate to economic efficiency and equity. The rationale is that efficiency can be enhanced by administratively effective public intervention in retirement income provision where there is price distortion, market failure, behavioural barriers, or inadequate saving present.

Depending on arrangements, capital income tax, necessary for governments though it may be, makes consumption in retirement more costly by taxing the return on saving. Taxing work also means that there is a price distortion between working and leisure in retirement. Such price distortions could mean that households will under-provide for retirement and retire early unless government offers tax incentives for saving and later retirement.

Adverse selection can lead to market failure for lifetime annuities. This is because annuities are a type of insurance that requires the pooling of longevity risk across a number of individuals. An insurer is able to pay those that live longer from the proceeds of those that don't. The issue arises when annuity purchase is optional and insurers are unable to distinguish between annuity purchasers with respect to their life expectancy. The insurer will set a premium at a level that is too high for those who expect to live shorter lives to want to purchase the insurance; in turn, since the remaining pool of the insured will live longer on average, the price will need to increase, eroding the market further. This process is familiar from health insurance, which healthy people are unlikely to take up unless there is some form of policy intervention. The surest way of eliminating adverse selection is to make insurance purchase compulsory for all members of the relevant population.

Informational, institutional, and behavioural barriers are sometimes cited as justification for government intervention. For example, myopia or a lack of financial understanding may result in households acting in a short-sighted manner and thereby, even by their own judgement, under-providing for their retirement. The retirement income system may thus seek to compel, 'nudge', or provide strong incentives to overcome these barriers.

A low level of individual saving may also impact the macro-economy. A country's savings rate is important since it translates to investment in future productive capacity. Since retirees sell off their assets to finance retirement consumption, population ageing may lead to a decline in the saving rate.

Government intervention in the retirement income system must also be considered with reference to equity. A safety net to provide income to those with inadequate resources is a crucial element of any welfare system. But policy design ought to also safeguard this vertical equity by preventing retirement saving arrangements being used as a tax shelter for those with high incomes. It is on similar grounds that a policy of supplementary saving and income streams is justified: to minimise the exploitation of the retirement social safety net by those with adequate resources during their working lives.

Equity is also relevant from an intergenerational point of view. Policy design can affect how the burden of population ageing and the retirement of the baby boomers is shared between successive cohorts.

Finally, the means of implementing policy can be as important as its ends. That is, the above criteria need to be addressed in a system that is as simple as possible, to maintain transparency to the individual and tractability to the policy maker. If executed well, policy can improve administrative efficiency and reduce costs.

9.5 How Does the Australian Policy Stack Up?

There are various metrics under which the key elements earlier can be assessed. Here, we look at how the Australian system fares under a selection of the criteria discussed earlier. Coverage against financial risks in retirement is assessed by the ability of the retirement income system to insure against the risk of running out of retirement savings due to one or a combination of investment, longevity, and inflation risks. Replacement risk for the individual and the level of vertical equity in the economy is measured by income replacement and poverty outcomes for pensioners. Economic efficiency of the retirement income system and its success in overcoming behavioural barriers are assessed by looking at incentives or effect on saving and retirement decisions. Intergenerational equity and the danger of future distortionary tax hikes are considered by looking at the sustainability of government spending on age-related pensions.

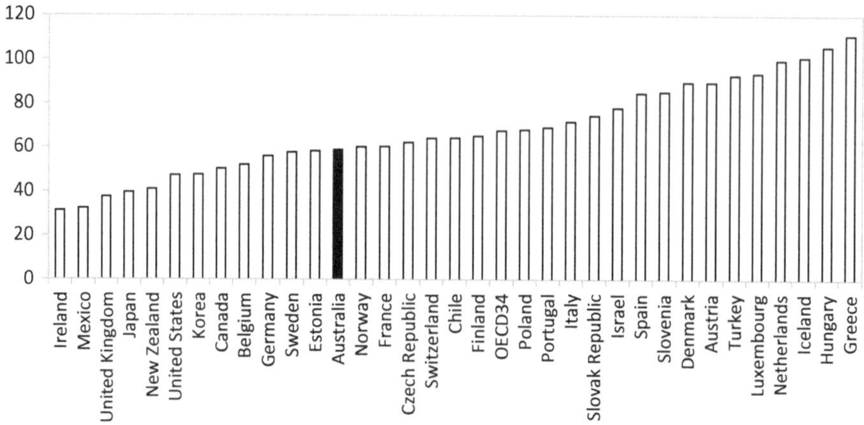

Fig 9.2 Net replacement rates of public and private mandatory schemes for average earners, 2008 (*Source*: OECD (2011))

9.5.1 Retirement Risks

The Superannuation Guarantee, in common with all Defined Contribution schemes, does not guarantee a final benefit. Further, while the Superannuation Guarantee mandates a minimum contribution rate into a superannuation account, there is no requirement that the retirement accumulation be taken in a form that would provide an indexed lifetime income stream. At the aggregate level, take-up of retirement benefits is split evenly between lump sums and a type of phased withdrawal product called an account-based pension. Only 111 lifetime annuities were purchased in 2011, though recent figures suggest a small resurgence. While the public Age Pension does provide some protection against these risks, it is paid at a low rate (27.7 % average earnings) and, due to means test, only about 40 % of those of eligible age receive the full rate of payment (Australian Government 2012).

9.5.2 Adequacy: Income Replacement and Poverty Outcomes

Figure 9.2 compares the net retirement income replacement rates from public and mandatory retirement income schemes across the OECD. Here it is defined as the net pension entitlement divided by career-average net pre-retirement earnings of a hypothetical individual, who enters the labour market at age 20, earns economy-wide average earnings, and retires after 40 years. It is based on calculations conducted by the OECD and some assumptions may differ from those used by the Australian Treasury.

It shows that Australia's replacement rate for an average earner of 59 % is below the OECD average of 68 % and ranks it 13th lowest of the 34 countries. The comparison, however, is based on pension system parameters from 2008, predating considerable increases to the single and couple rate of Australia's Age Pension, which took place as part of a major review in 2009 (Australian Government 2009). The analysis also excludes more recently announced increases to the Superannuation Guarantee mandatory contribution rate, from 9 to 12 % of earnings. Both these changes will improve Australia's replacement rates and ranking.[8]

International comparisons of current mature-age poverty rates also appear to paint a negative picture of Australia's retirement income system. Older Australians have some of the lowest incomes and highest poverty rates in the OECD (2011). However, such figures can be qualified by the data on the average poverty gap—the average difference between the poverty line and the incomes of poor households—which was one of the lowest among developed countries. This means many Australian pensioners have incomes only just below the poverty line (OECD 2008). Furthermore, unlike most OECD countries, older Australians have very high levels of owner-occupied housing, including households with low levels of consumption expenditure. Once housing costs are taken into account, old-age poverty rates are approximately 14 % (at the household level), which is low compared to many other countries (Rodgers and Rodgers 2010; Yates and Bradbury 2009).

9.5.3 Economic Efficiency: Retirement and Saving

There are different ways of introducing incentives to retire later (or rather, blunting the incentives to retire early that are inherent in most pensions where a benefit commences at a specified age). Defined contribution pension schemes, such as the Superannuation Guarantee in Australia, implicitly include such incentives because later retirement translates to more years of contributions and fewer years over which to spread the resulting income. This is illustrated in Fig. 9.3, where the gross superannuation replacement rate is based on the first year income of a simple life annuity for an Australian female and her final year of salary. It shows that later retirement leads to a higher replacement rate under reasonable assumptions. Of course, Age Pension means tests will erode this advantage for some levels of earnings and superannuation asset and income, while the availability of both superannuation and Age Pension income at a given age may encourage retirement. Attempts have been made to address this with a concession in the Age Pension income test for income from employment, a non-trivial income-test free amount and a 50 % rather than 100 % withdrawal rate applying to Age Pension payments. However, aside from higher eligibility ages, which appear to have a strong effect of delaying retirement, there is mixed evidence about the success of these policies in Australia (Headey et al. 2011; Chomik and Whitehouse 2010).

A key outcome of Australia's mandatory superannuation system has been a large growth in private savings. Figure 9.4 shows that superannuation assets have grown

[8] In fact, the OECD (2011: 177) shows that an additional contribution rate of 2.5 % for an Australian average earner would result in their gross replacement rate reaching the average for the OECD.

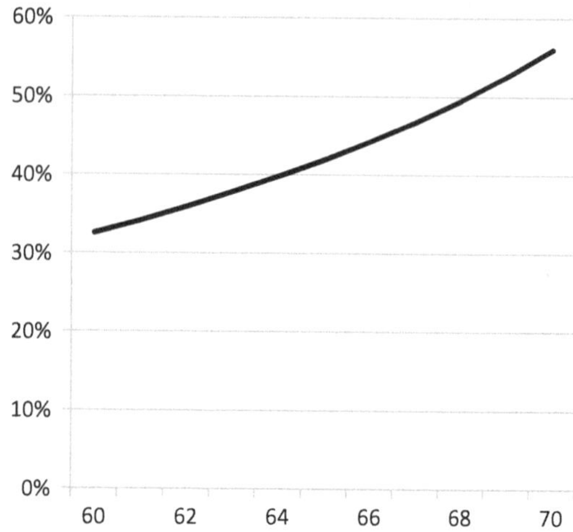

Fig. 9.3 Gross superannuation replacement rate by age of retirement (per cent of final year wages) (*Source*: Author's calculations)

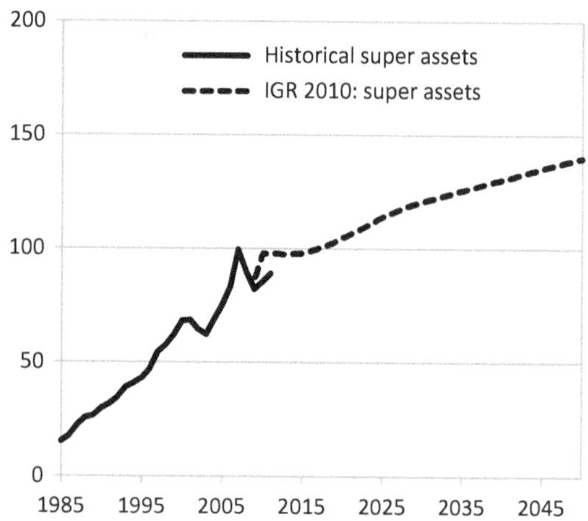

Fig. 9.4 Historical and projected superannuation fund assets, 1985–2050 (per cent of GDP) (*Source*: Australian Government (2010b) and APRA (2012). *Note*: IGR denotes projection from Treasury's Intergenerational Report)

to around 100 % of GDP. At $A1.8 trillion, Australia's superannuation asset fund is one of the largest in the world in spite of our relatively small population Australian Prudential Regulatory Authority (APRA 2014). This growth is projected to continue despite the retirement of baby boomers in the next couple of decades. Once the increases in mandatory contributions are incorporated, the level will be higher still.

9.5.4 *Sustainability*

Australia's retirement income system is more sustainable than those of most comparable countries. Figure 9.5 compares current and projected spending on age-related pensions as a proportion of GDP across the OECD. It shows that Australia's

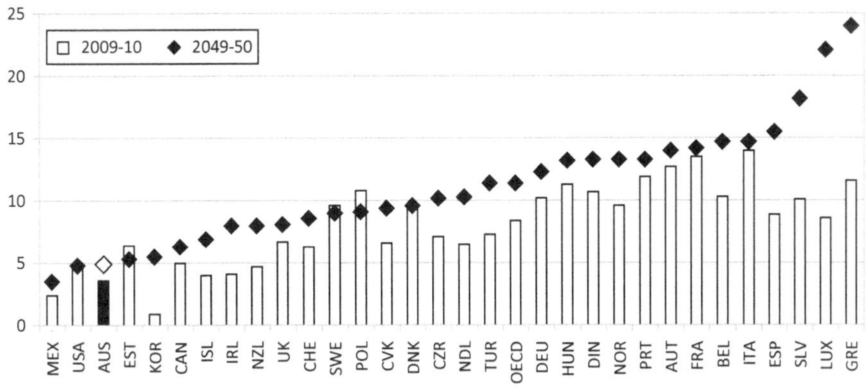

Fig. 9.5 Current and projected public expenditure on age-related pensions, OECD (per cent of GDP), 2010–2050 (*Source*: OECD (2011))

comprehensive but conditional retirement income support provided through the Age Pension and a relatively slower ageing of the population results in some of the lowest current and forecast levels of public spending on pensions in the OECD. The projections suggest that demographic pressures will increase spending on age-related payments from 3.6 % of GDP in 2009–2010 to 4.9 % of GDP in 2049–2050. For age and service pensions alone, the Australian Treasury projects an increase in costs from 2.7 % of GDP in 2012 to 3.8 % of GDP by 2049–2050 (Rothman 2012). This is a non-trivial increase, but is much more sustainable than corresponding projections for other OECD countries (recent implemented and announced policy in Australia, such as a more aggressive asset means test and proposed increases to the pension eligibility age are expected to see future public spending on the program to be lower still).

9.5.5 Administrative Efficiency and Costs

Defined narrowly as the costs passed on to superannuation members through charges, administrative efficiency is poor in Australia's retirement income system. The average annual fees are 0.97 % of asset value and range between 0.7 and 2.53 %: a high level compared to other countries with similar schemes (Cooper 2010; OECD 2011). Superannuation industry efficiency issues were addressed in the Super System (Cooper) Review conducted in 2009–2010. The government has accepted a large number of the review recommendations including 'SuperStream', which will streamline administrative practices, and MySuper, which is designed to introduce competition in default options.[9] Analysis of the

[9] Default options refer to a framework where members are enrolled in a product that does not require decisions on their part.

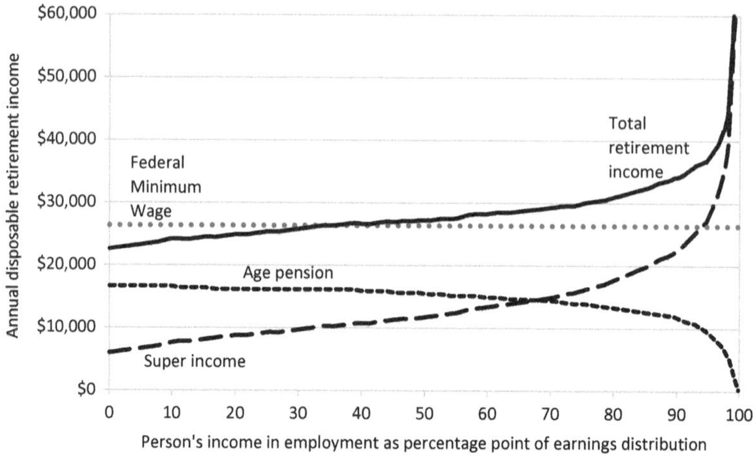

Fig. 9.6 The contribution of superannuation and age pension to total retirement income (*Source*: (Harmer 2009)). *Note*: Modelling based on 9 % contribution rate and 40 % withdrawal rate; position in distribution may change over retirement

proposed reforms suggest that the estimated total annual percentage costs for MySuper products in funds of varying sizes should result in fees ranging between 0.32 and 1.04 % (Cooper 2010).

9.6 The Future of Retirement Income Policy in Australia

9.6.1 Current Reform Agenda

Introduced less than 25 years ago, the Superannuation Guarantee has yet to mature. It would have been 2032 before the mandated 9 % had been in force for 40 years, a full working life. But the recent policy change, which increases the contribution rate to 12 % by 2025, will mean that the 'steady state' will not be in place until 2065. The Age Pension will remain an important source of retirement income for most people for some time to come. Figure 9.6 shows the relative importance of the Age Pension and the superannuation guarantee for a person who retires after a full working lifetime of contributing 9 % to superannuation. The story should not change dramatically despite the increase to the compulsory contribution rate to 12 %.

Perhaps more importantly, Australia's retirement income system has been in a constant state of reform over the past two decades, as outstanding issues have been gradually addressed, and it remains a 'work in progress'. The main changes over the past decade include the introduction of a government co-contribution to encourage voluntary contributions by low- to middle-income earners (2003); the Better (or Simplified) Super reforms of 2007 which eliminated taxes on superannuation benefits for most people aged 60 and over and simplified the means testing of superan-

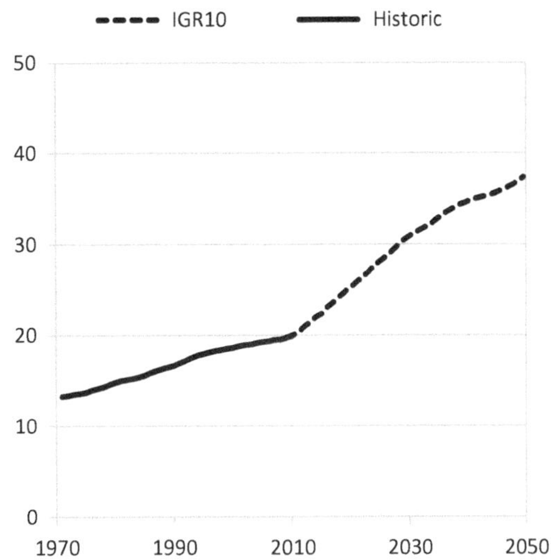

Fig. 9.7 Historic and projected old-age dependency (pop aged ≥65 as per cent of pop 15–64 years), 197–2050 (*Source*: Chomik and Piggott (2012b)). *Note*: IGR relates to Treasury's Intergenerational Report projections

nuation benefits; an increase in the Age Pension age to 67 to be implemented between 2017 and 2023; an increase in the mandatory contribution rate from 9.5 to 12 % to be implemented over 2021–2025 and initiatives to improve the operation and efficiency of the superannuation industry (Cooper 2010).

9.6.2 *Future Challenges*

Preparing Australia's retirement income system for the challenges of population ageing will require further reforms. The old-age dependency ratio, measured as the proportion of the population aged 65 years and over to the population aged 15–64 years, is set to increase from 20 to 36 % over the next 40 years (Fig. 9.7). The concomitant increases in government expenditures on Age Pensions (see spending trajectory in Fig. 9.8), health and aged care, will have an impact on public finances. In the absence of policy change, a fiscal gap would form by the 2030s and widen to 2.75 % of GDP by 2049–2050 and net government debt is expected to, in turn, grow to around 20 % of GDP. Admittedly, such projections are sensitive to assumptions, as has been seen with some more recent projections released by the Australian Treasury.

In this context, a number of areas of retirement income policy are candidates for future attention, several of which are discussed as follows.

Tax is one such area. Despite numerous reforms over the past 30 years, successive governments have left unresolved some inherent flaws in the taxation of Australia's retirement income regime. The taxation of superannuation differs from international practice and departs significantly from the standard paradigms that are more consistent with an economically efficient system.[10] Since employer contributions

[10] See Chomik and Piggott (2012a), Bateman et al. (2001), and Bateman and Kingston (2007).

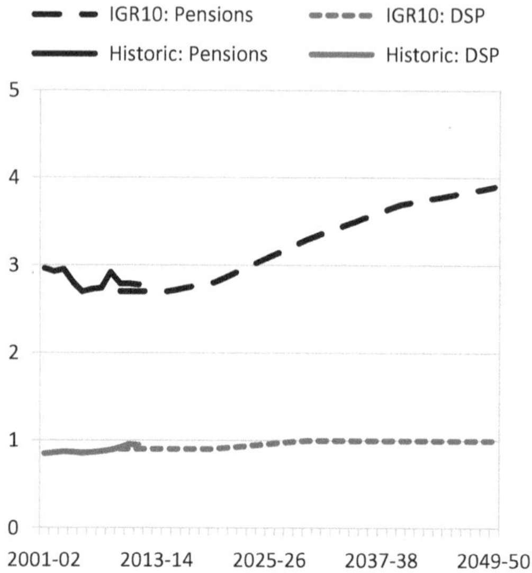

Fig. 9.8 Historic and projected public expenditure on age-related pensions (per cent of GDP), 2001–2050 (*Source*: Chomik and Piggott (2012b)). *Note*: IGR relates to Treasury's Intergenerational Report projections

and superannuation fund earnings are taxed at a flat rate under a specific superannuation tax regime, they are also separated from income tax and its progressivity. Separating superannuation taxes from personal income tax makes tinkering with superannuation taxes tempting for governments because changes don't show up in one's pay packet. The Henry Review recommended that the taxation of contributions be linked to the progressive personal rate schedule, as well as a reduced tax on superannuation fund earnings to help the power of compound interest generate more accumulations and reduce price distortion between assets and over time (Henry 2010). However, the government has largely ignored these recommendations and instead tinkered with taxation of contributions at the top and bottom end of the income distribution in an attempt to address perceived inequities.

As a system based on defined contributions, the superannuation arrangements place considerable responsibility on individuals to make often complex financial decisions, including choosing their superannuation fund, whether to make additional voluntary contributions, the investment option for these contributions and benefit type at retirement. These decisions are further complicated through the interaction of the superannuation system with the Age Pension means tests. It is not clear that Australian retirement savers are yet ready to bear these responsibilities. For example, according to one Australian survey only 37 % of people have the minimum literacy required "to meet the complex demands of everyday life and work in the emerging knowledge-based economy" (ABS 2006: 8).

It is also unclear whether Australian retirement savers are sufficiently 'engaged' in a retirement savings system in which the mandate to make contributions is on the employers (rather than employees). An indication of (dis)engagement is the broad ranging reliance on defaults. Despite choice of superannuation fund, less than 5 % of members actually choose their retirement fund; around 50 % of assets relating to around 60 % of members are in default investment options and only around

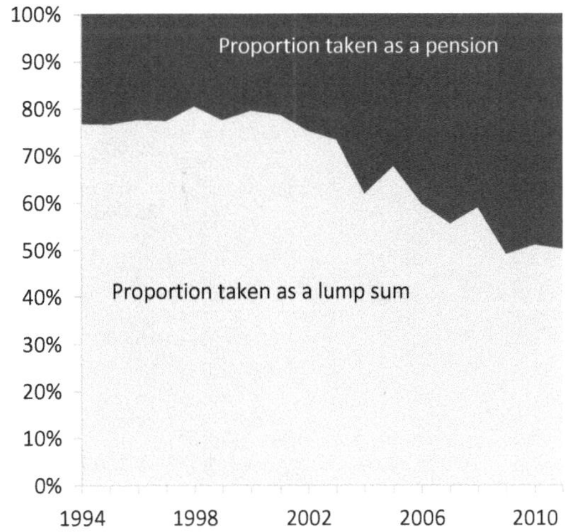

Fig. 9.9 Structure of superannuation benefits, 1994–2011 (per cent share) (*Source*: APRA (2005, 2012))

one-third of members receive (or make) contributions in excess of the mandatory contribution rate. The superannuation industry is understood to be aware of these issues and is trying to better inform and engage fund members.

The drawdown or payout phase is another area of concern. In Australia, the principle of compulsory savings does not appear to extend to either mandating or encouraging that the retirement accumulation be taken as an income stream rather than as a lump sum. The result is a vulnerability of Australian retirees to longevity risk if they fail to purchase a lifetime income stream.

Data show that half of all superannuation fund withdrawals are in the form of a lump sum and that there is a trend towards benefits being taken as income streams (Fig. 9.9).[11] Of those benefits taken as income streams, most are phased withdrawal type products (known as account-based pensions) rather than annuities (Fig. 9.10), which leaves Australian retirees exposed to longevity, investment, and inflation risk. The very low take-up of life annuities can be attributed to demand-side issues (Bateman and Piggott 2011), supply side factors (Evans and Sherris 2011), and regulatory barriers (Henry 2010). However, these factors are not insurmountable and can be addressed by using incentives and mandating or creating default decumulation products, while also supporting industry.

A final issue worth highlighting is mature-age labour force participation, which benefits both government finances and the individual's retirement income. Policy has generally moved towards reducing early retirement incentives and barriers on the labour supply side as well as introducing various anti-discrimination measures on the demand side (Department of Education and Employment and Workplace Relations [DEEWR] (2011)). Mature-age participation is climbing, particularly among men. Overall participation levels among 55–64 year olds stood at 60.6 % in 2010, the sixth highest rate in the OECD. This was 10 percentage points higher than

[11] However, this trend should be interpreted with caution. The annual data includes lump sums and newly started income streams as well as income streams which commenced in previous years.

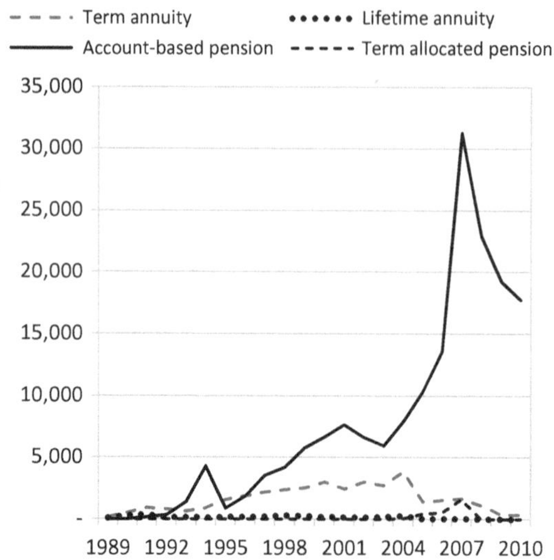

Fig. 9.10 Value of private retirement income streams ($A million) (*Source*: Plan for Life (2012)). *Note*: Only includes products purchased in private market

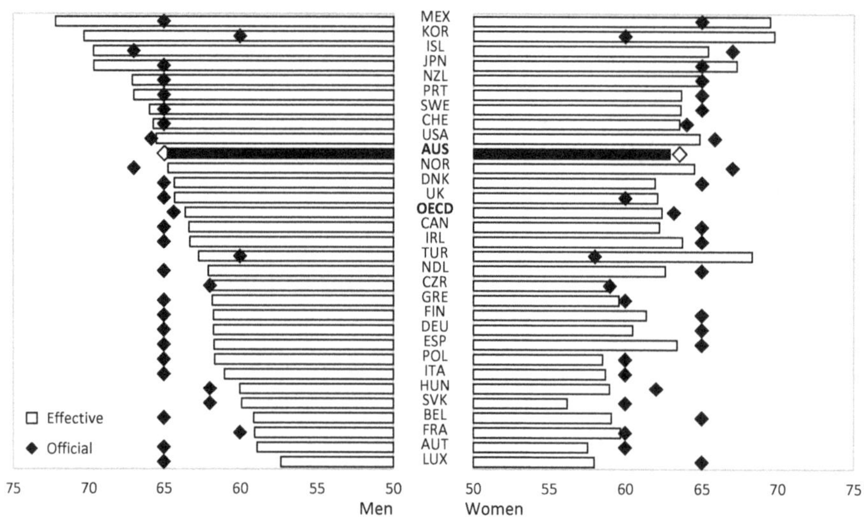

Fig. 9.11 Average effective retirement age and normal pensionable age (*Source*: OECD (2011))

in 2003, when Australia's rate was close to the OECD average (OECD 2013). Compared to other countries[12] and to the participation rates of older men in the past, Australia's mature-age participation and retirement ages have scope to increase further. As shown in Fig. 9.11, average effective age of labour-market exit in Australia in 2004–2009 was 64.8 years for men and 62.9 years for women, below the Age Pension age and that seen in nine other countries.

[12] Greater growth in Australia's mature-age labour force participation may be in part due to its relatively higher economic growth and life expectancy.

Both the financial incentives created by tax and benefit systems and the employment barriers present in the labour market may require considerable attention. For instance, the incentives to retire are strong for those facing the means tests of the Age Pension (Kudrna and Woodland 2009). The lower access age to superannuation savings compared to the Age Pension age is another area which may encourage early labour market exits and act as a signal for when someone should retire.[13]

9.7 Conclusion

This chapter describes and assesses the Australian paradigm for retirement income system provision. Like other countries, this has changed greatly in past decades and is expected to continue to evolve.

The unique design of Australia's retirement income provision system, comprising the 'three pillars' of a means-tested Age Pension, mandatory superannuation under the Superannuation Guarantee, and other, voluntary long term savings, compares well internationally.

Replacement rates are below the OECD average–they are expected to improve with foreshadowed increases in the mandatory superannuation contribution rate, though delays mean this will be far in the future. Old-age poverty after housing costs is low by international standards. Superannuation assets are among the highest in the world, and while the administrative costs are high, there are policies in place to help resolve the administrative inefficiency. Spending on age-related pensions is approximately 3.6 % of GDP, among the lowest rates in the OECD, suggesting that the system is far more sustainable than many others.

Future challenges and opportunities have also been highlighted particularly those relating to tax, decumulation, the challenges associated with choice, and mature-age labour force participation.

9.8 Case Study

Karen and Andrew are a married couple. Karen is 60 and Andrew is 62. Andrew works in the public service and Karen works in retail. Andrew currently has $350,000 in his superannuation account. Karen has been working part time for the past 10 years and has $80,000 in her superannuation account. Their family home is worth $980,000 and they have an outstanding mortgage of $200,000. They have about $30,000 in other savings. They have a daughter (age 28) who works as an accountant and lives in a rented apartment. Their son (age 30) is living overseas with a young family but hopes to move back to Australia and buy a house in Sydney

[13] See Spoehr et al. (2009) and Chomik and Piggott (2012b).

within the next 5 years. Both Karen and Andrew are in excellent health but do not have private health insurance. Karen's parents are still alive in their 90s and are still living in their own home.

Karen and Andrew would like to retire soon. They expect to be able to maintain their current standard of living throughout their retirement. They have been speaking with a financial adviser who has informed them that they would be eligible for at least a part Age Pension from age 67. The financial adviser has spoken to them about options for their superannuation at retirement. That is, they could take their superannuation as a lump sum, and/or take an account-based pension, and/or buy a life annuity or a term annuity. They were also told that if they purchased an account-based pension, they would be able to choose the asset allocation from a range of options with different proportions of growth and defensive assets. The growth assets would be mainly domestic and international shares.

After speaking with friends from work, Karen and Andrew think that they will most likely both take account-based pensions invested in a growth option with all of their superannuation saving.

9.8.1 Discussion Points

- The key risks facing retirees are replacement risk, investment risk, longevity risk, and inflation risk. To what extent would Karen and Andrew be covered for each of these risks?
- How could the couple arrange their retirement finances differently to better address these risks?
- What other factors could affect the ability of retiree couples such as Karen and Andrew to make appropriate financial decisions for their retirement?

Acknowledgement Support from the ARC Centre of Excellence in Population Ageing Research (CEPAR) (project number CE110001029) is gratefully acknowledged.

References

Australian Bureau of Statistics. (1999, August). *Employee earnings, benefits and trade union membership* (Cat. No. 6310.0). Canberra, Australia: ABS.

Australian Bureau of Statistics. (2006). *Microdata: Adult literacy and life skills survey, basic CURF* (Cat. No. 4228.0.30.001). Canberra, Australia: ABS.

Australian Bureau of Statistics (2009). *Employment arrangements, retirement and superannuation, Australia* (April to July 2007, Re-issue) (Cat No. 6361.0). Canberra, Australia: ABS.

Australian Bureau of Statistics. (2011a). *Australian social trends, December 2011* (Cat. No. 4102.0). Canberra, Australia: ABS.

Australian Bureau of Statistics. (2011b, August). *Employee earnings, benefits and trade union membership* (Cat. No. 6310.0), Canberra, Australia: ABS.

Australian Government. (2006, May 9). *A plan to simplify and streamline superannuation*. Canberra: Commonwealth of Australia.
Australian Government. (2009, May 12). *Secure and sustainable pensions*. Canberra: Commonwealth of Australia.
Australian Government. (2010a, May 2). *Stronger, fairer, simpler: A tax plan for our future*. Canberra: Commonwealth of Australia.
Australian Government. (2010b). *Australia to 2050: Future challenges. The 2010 intergenerational report overview*. Canberra: Commonwealth of Australia.
Australian Government. (2012, October). *FaHCSIA annual report 2011-2012*. Canberra, Australia.
Australian Prudential Regulation Authority. (2005, January 11). *Statistics: Superannuation trends* (September 2004). Sydney, Australia: APRA.
Australian Prudential Regulation Authority. (2012, August). *Quarterly superannuation statistics* (June 2012). Sydney, Australia: APRA.
Australian Prudential Regulation Authority. (2014, February). *Quarterly superannuation statistics* (December 2013). Sydney, Australia: APRA.
Bateman, H. (2011). *Structuring the payout phase in a defined contribution scheme in high income countries: Experiences of Australia and New Zealand*. Presented at Conference: Selected Issues in Strengthening Social Protection in East Asia, Singapore, March 2011.
Bateman, H., & Kingston, G. (2007). Superannuation and personal income tax reform. *Australian Tax Forum, 22*(3), 137–163.
Bateman, H., Kingston, G., & Piggott, J. (2001). *Forced savings: Mandating private retirement incomes*. Cambridge, UK: Cambridge University Press.
Bateman, H., & Piggott, J. (1997). *Private pensions in OECD countries—Australia* (OECD Labour Market and Social Policy Occasional Papers No. 23). Paris: OECD.
Bateman, H., & Piggott, J. (1998). Mandatory retirement saving in Australia. *Annals of Public and Cooperative Economics, 69*(4), 547–569.
Bateman, H., & Piggott, J. (2011). Too much risk to insure? The Australian (non-) market for annuities. In O. S. Mitchell, J. Piggott, & N. Takayama (Eds.), *Securing lifelong retirement income: Global annuity markets and policy* (pp. 81–105). New York: Oxford University Press.
Bodie, Z. (1989). *Pensions as retirement income insurance* (NBER Working Paper No. 2917). Cambridge, MA: National Bureau of Economic Research.
Chomik, R., & Piggott, J. (2012a). Pensions, ageing and retirement in Australia: Long-term projections and policies. *The Australian Economic Review, 45*(3), 350–361.
Chomik, R., & Piggott, J. (2012b). *Mature-age labour force participation: Trends, barriers, incentives, and future potential* (CEPAR Briefing Paper 2012/01). Sydney, Australia: ARC Centre of Excellence in Population Ageing Research.
Chomik, R., & Whitehouse, E. (2010). *Trends in pension eligibility ages and life expectancy, 1950-2050* (OECD Social, Employment and Migration Working Papers, No. 105). Paris: OECD.
Cooper, J. (2010). *Super system review: final report*. Canberra: Commonwealth of Australia.
Dabscheck, B. (1989). *Australian industrial relations in the 1980s*. Melbourne, Australia: Oxford University Press.
Department of Education, Employment and Workplace Relations [DEEWR]. (2011). *Ageing and the barriers to labour force participation in Australia*. A rcport by National Seniors Productive Ageing Centre, Commonwealth of Australia on behalf of the Consultative Forum on Mature Age Participation, Canberra.
Diamond, P. (1977). A framework for social security analysis. *Journal of Public Economics, 8*(3), 275–298.
Diamond, P. (2011). Economic theory and tax and pension policies. *Economic Record, 87*, 2–22.
Evans, J., & Sherris, M. (2011). The development of a life annuity market in Australia: An analysis of supplier risks and their mitigation. *JASSA, 2*(2011), 11–15.
Fogel, R. (2004). *The escape from hunger and premature death, 1700–2100: Europe, America and the Third World*. Cambridge, UK: Cambridge University Press.
Gallagher, P. (2012). *Treasury measurement of retirement income adequacy*. Presentation to FEAL Pre-Forum Discussion, February 9, 2011.

Gordon, R., & Varian, H. (1988). Intergenerational risk sharing. *Journal of Public Economics, 37*(2), 185–202.
Harmer, J. (2009). *Pension review report*. Canberra: Department of Families, Housing, Community Services and Indigenous Affairs, Commonwealth of Australia.
Headey, B., Freebairn, J., & Warren, N. (2011). *Dynamics of mature age workforce participation: Policy effects and continuing trends*. Melbourne: Melbourne Institute of Applied Economic and Social Research.
Henry, K. (2010). *Australia's future tax system. Report to the treasurer, December 2009. Part one: Overview*. Canberra: Commonwealth of Australia.
Kudrna, G., & Woodland, A. (2009). *An inter-temporal general equilibrium analysis of the Australian age pension means test*. Sydney: School of Economics, Australian School of Business, University of New South Wales.
Organisation for Economic Co-operation and Development. (2008). *Growing unequal: Income distribution and poverty in OECD countries*. Paris: OECD.
Organisation for Economic Co-operation and Development. (2011). *Pensions at a glance*. Paris: OECD.
Organisation for Economic Co-operation and Development. (2013). *Employment and labour markets: Key tables from OECD*. ISSN 2075-2342.
Plan for Life. (2012). *Detailed annuity statistics*. Plan For Life, Asset International Australia Pty Ltd. Retrieved from http://www.pflresearch.com/
Rodgers, J. R., & Rodgers, J. L. (2010). Chronic and transitory poverty over the life cycle. *Australian Journal of Labour Economics, 13*(2), 117–136.
Rothman, G. (2012). *Modelling the sustainability of Australia's retirement incomes*. Presentation to the 20th Annual Colloquium of Superannuation Researchers, July 2012, UNSW. Sydney, Australia: University of New South Wales.
Spoehr, J., Barnett, K., & Parnis, E. (2009). *Experience works: The mature age employment challenge*. A discussion paper prepared for National Seniors Australia. Adelaide, South Australia: Australian Institute for Social Research, University of Adelaide.
The Treasury. (2014, September 2). *Media release—Failed mining tax finally gone*. Canberra: Commonwealth of Australia.
World Bank. (1994). *Averting the old age crisis: Policies to protect the old and promote growth*. Washington: World Bank.
Yates, J., & Bradbury, B. (2009). *Asset rich, but income poor: Australian housing wealth and retirement* (Luxemburg Income Study Working Paper Series, Working Paper No. 8).

Chapter 10
Housing and the Environments of Ageing

Debbie Faulkner

10.1 Introduction

The relationship between housing and the life course has in recent years been significantly reshaped for many in society (Beer et al. 2011). This trend is particularly true for the older population whose housing status and needs for a long time were seen as stable and therefore predictable (Kendig and Bridge 2007). With increasing life expectancy, an extended period in retirement and the capabilities and desirability to make choices, new opportunities now exist for older people to have the freedom and flexibility to decide where they want to live, the type of housing they want to live in, with whom they want to live and the lifestyle they wish to lead. These opportunities are unlike any that have existed for previous generations of older people. They also have, and will, with the passage of the baby boomers into the older ages, impact on the Australian housing market as well as changing and challenging society's views of older age and the housing, physical and social environments necessary for a decent quality of life. Housing and the communities in which we live are now valued more than ever as important to people's well-being and this is influencing policy agendas with relation to housing affordability, housing design, access to care and developing inclusive and accessible environments.

The ageing of the population will have a significant impact on Australia's housing sector. By 2028 over 28 % of all households will comprise people aged 65 years and over and around half of these households will be lone person households (National Housing Supply Council 2010). The dwelling and tenure choices of these older households will have an impact across the housing market, presenting key challenges such as ensuring there are sufficient options available that enable older people, within reason, to live where they want and in the housing of their choosing.

D. Faulkner (✉)
Centre for Housing, Urban and Regional Planning, Faculty of Arts, University of Adelaide, Room G40, North Terrace, Adelaide, SA 5005, Australia
e-mail: debbie.faulkner@adelaide.edu.au

K. O'Loughlin et al. (eds.), *Ageing in Australia*, International Perspectives on Aging 16, DOI 10.1007/978-1-4939-6466-6_10

This chapter explores the changing role of housing and the environment in older people's lives. The chapter begins with a review of the current housing status of the older population before exploring future housing and lifestyle directions and the need to develop age-friendly communities.

10.2 Current Housing and Tenure Status

10.2.1 Home Ownership

As outlined in Chap. 9, home ownership remains a strong aspiration and preference among Australians. Australia's older population has one of the highest rates of home ownership in the world, with around 80 % of households headed by someone aged 65 and over being owner occupied: a trend that has been stable across many decades (Ong et al. 2013b). High rates of home ownership nationally are unsurprising in many ways as the tenure offers a range of positive benefits including security of tenure and maintenance of living standards for older Australians (as it protects older people from unpredictable and generally rising housing costs over time); it provides control over one's life including safety and privacy; the ability to alter the home environment as needs and preferences change; and most importantly for many older people it is a place of significant personal attachment and meaning.

As a result of high levels of home ownership, older people have gained considerably from the capital growth in house prices over the last decade or so (Ong et al. 2013a) and with superannuation, savings and investments included in the aggregate wealth for the older population, they are clearly the wealthiest cohorts nationally (Kendig et al. 2012). Recent statistics confirm these trends. In 2011–2012, households with a person aged 65 years and over owned 30 % of the country's wealth (Daley et al. 2014). Much of this wealth is tied up in the family home, especially among the oldest cohorts (i.e. for those aged 75 and over) meaning that many among these groups are asset rich but income poor. Consequently, many are heavily dependent on the age pension and this affects their ability to maintain or modify their homes, or make other choices within the housing market. According to the Australian Bureau of Statistics (ABS), in 2011–2012, people living in older households (households where the reference person was aged 65 and over) had the lowest average weekly equivalised disposable household income. For nearly two-thirds of this group (64 %) government pensions and allowances were their main source of income (ABS 2013).

The situation for the baby boom cohorts (born 1946–1965) is already, and will likely continue to be, more variable. Many baby boomers are home owners and may also have had the opportunity to diversify their investments in residential property. With rising incomes and superannuation assets they have a higher standard of living than has been possible for previous generations. Importantly, they also have higher expectations and aspirations about their future housing and lifestyle preferences

(Beer and Faulkner et al. 2011; Olsberg and Winters 2005). Some have been financing these aspirations by accessing their housing equity to fund general, everyday consumption needs, rather than conserving this as an asset to be accessed in extreme circumstances only, as has generally been the case for previous generations of older people.

An upshot of the changing use of housing equity among baby boomers has been considerably higher levels of debt in the primary home than has been the case in the past. The proportion of home owners with a mortgage at ages 50–54 years, 55–59 years and 60–64 years for example, increased by 14.0 %, 15.1 % and 14.3 %, respectively, between 2002 and 2010 (Kelly 2012). In 2012, 31 % of people aged 60–64 years still had a mortgage compared to just 16 % in 2002. Over this period baby boomers increased their level of property debt by nearly 50 %, with recent research suggesting that superannuation is being used to reduce such mortgage debt (Ong et al. 2013a). These changing trends have significant implications both for baby boomers and the retirement income support policies necessary to maintain standards of living for older people (Ong et al. 2013a). In short, for baby boomers their aspirations into older ages will likely need to change, as they may not be able to enjoy the same high living standards they have in their working lives. Additionally, for those with few assets other than superannuation, reliance on the age pension while paying off mortgage debt may place them in housing stress, that is, paying more than 30 % of their income in housing costs (Lovering 2014).

In discussing home ownership among older Australians, it is important to also note that over the last two decades home ownership rates have been falling for all age groups except those aged 65 years and over (Daley et al. 2014; Yates et al. 2008). These now well-established trends are not only due to the difficulty of entering the home ownership market for the first time at any age, but because of the increasing difficulty sustaining home ownership over time. Relinquishment of this tenure has been largely precipitated by such life course events as separation, divorce, bereavement and unemployment (Wood et al. 2010). This means that older people need to be able to access other forms of housing tenure.

10.2.2 Rental Housing

Around 13 % of the population aged 65 years and over currently occupy rental accommodation—whether publicly or privately provided. Demand for this tenure type is projected to continue to increase over time, largely because of declining levels of home ownership evident at the younger age groups. Older people will also feel this pressure, with lower income households in particular bidding for limited affordable rental market options. By 2028 it is projected the underlying demand for rental accommodation among the older population (households with a reference person aged 65 years and over) will increase by 120 % (National Housing Supply Council 2010, p. 143) placing pressures on both the social housing sector (public and community housing) and the private rental market. For the public or social housing sector, underlying demand is highly unlikely to ever be met, regardless of need, including age-related needs. This is because, for over four decades now, government policy has turned away from supply side interventions, such as the provision of housing, in favour of demand-driven housing assistance measures—mainly in the form of Rent Assistance and some other smaller scale programmes to assist people in obtaining and maintaining tenancies in the private rental market (Australian Institute of Health and Welfare (AIHW) 2014). Limited supply of public housing has also seen tighter targeting of eligibility criteria for such housing based on specific needs. While in the past older people were prioritised in public housing wait lists, age is no longer a criterion for tenant selection (AIHW 2013). The upshot of these trends for older people is that this tenure is now an option for a much more limited number of older people.

Public rental housing has been a valued and affordable housing option for older people in the post-war period. It is an option that has provided good quality housing with rents capped at 25 % of income, as well as providing security of tenure and dwellings that are relatively easily adapted to suit the needs of occupants. Surveys of the satisfaction of residents within social housing indicate that the majority of people are satisfied or very satisfied overall with their housing, particularly its condition, location and specific amenities such as size of the dwelling, modifications for special needs, ease of access and entry, privacy and security of the home and neighbourhood (AIHW 2012, 2014). On this issue however, it is noteworthy that many of the older people in public housing today were allocated housing earlier in life and are now ageing in place.

Increasingly, older people who are not home owners and cannot gain access to social housing are having to seek affordable accommodation in the private rental sector. This reality poses a number of challenges and concerns for older people. Fundamental among these is the absolute lack of affordable private rental options for low-income households (Hulse et al. 2014); a supply issue that has persisted for some time now despite attractive rental yields across the sector for investors and the provision of government incentives to increase the number of affordable private rental options. A further challenge raised repeatedly in the research for this group concerns the inappropriateness of private rental accommodation for older low-income households (Faulkner 2009; Fiedler 2014) because of limited security of tenure; the costs associated with moving repeatedly; the inability of such housing to be adapted to changing needs and the economic burden of paying (generally increasing) market rents. Recent statistics confirm this latter concern for older people living in private rental accommodation. In 2011–2012, housing costs accounted for 30 % of gross income for couple only households aged 65 years and over and 40 % for lone person households aged 65 years and over (ABS 2013). With rental costs being at least 30 % of income, these households are considered to be in housing stress and their ability to spend on essentials like food, social outings and health care is compromised (Yates and Milligan 2007). The fact that more older person households are now long-term private renters than ever before (that is renting continuously for 10 or more years) (Stone et al. 2013) compounds these concerns.

10.2.3 Marginalised Housing and Homelessness

When the cost or the security of housing becomes an issue, older people can end up in marginal housing, or worse still end up homeless, the ultimate undesirable housing outcome for vulnerable households. Marginal housing is a particular concern for older households, for as Goodman et al. (2013, p. 2) note it is "highly managed or controlled housing, with fewer occupancy rights and some degree of shared facilities and spaces". It includes precarious housing options such as boarding houses and manufactured home villages. While some of this accommodation may be adequately meeting the specific needs of residents, the high level of control in the hands of marginal housing operators leads to higher levels of insecurity and a sense of disempowerment for residents.

Sadly, for a small, but growing group of older people, the exhaustion of appropriate and affordable housing options has seen homelessness become a reality. At the time of the 2011 Census 14% of the homeless population (14,851 people) were aged 55 years and over. Unlike in the past, where homelessness in the older ages was generally the continuation of a trend of homelessness started much earlier in life, we are increasingly seeing older people who are homeless for the first time later in life, many of them after having had a conventional housing career but having then fallen out of stable housing options for life course, lifestyle and other reasons (Petersen et al. 2014). Specific programmes are in place to assist older people who are homeless or at risk of homelessness. The Assistance with Care and Housing for the Aged (ACHA) programme now a component of the Commonwealth Home Support Programme, for example, funded by the Department of Social Services (and operated through a range of community and government agencies) assists in finding affordable housing and connecting older people with the care and support needs required.

10.3 Future Housing Directions

Despite the difficulties facing some older people in terms of their housing, for most, and particularly the baby boomer generation, their ability to make (and fund) their housing choices will be (and already is) greater than has ever been in the past. Past research by Olsberg and Winters (2005) on the housing intentions of 7000 Australians aged 50 years and over strongly noted this, indicating the changing opinions of older people, especially young older people such as the baby boomers, with regard to preserving wealth for inheritance. For this group there is more of a focus on enjoying and living independent, flexible, consumer-driven lifestyles in older age rather than saving accumulated wealth to be passed onto children. These changes mean it is difficult to predict the choices older people and the baby boomers will make with regard to their housing in the coming years.

As a starting point in understanding the factors shaping the possible housing directions of older people over the coming decades, Pinnegar et al. (2012) devised a preliminary typology based on existing tenure, aspirations for future housing and location, and the constraints on housing and locational choices. This typology comprises six groups of people, three of which are constrained in their choices and three which have the capacity to choose the housing and location that suits their lifestyle:

- 'Older renters', where the availability of rental accommodation dictates housing choice and location (as discussed earlier).
- 'Increased dependency', a group of people who may wish to stay in their current home but health and dependency issues may compromise these choices.
- 'Constrained retreat', people wanting to remain in their current home and area but who, due to financial constraints, need to make some compromises.
- 'Age in place', the people who want to, and are able to keep living in the family home.
- 'Local adaptors', the group who want to and are able to move out of their current home but to stay in the local community and neighbourhood, or who have recently done so.
- 'Scene changers', people who move for greater amenity, often into coastal areas or other areas of locational or aesthetic amenity.

While choice and adjustments in older age have occurred in the past, what is different now is the scale and pace at which aspirations and circumstances may change. Additionally, such change may occur in different ways and at different times during older age, meaning that belonging to one of the above groups is not set in stone, and households may move between groups. These movements across and between housing and locational 'choices' are a challenge for planners and property developers, prompting a greater need for them to understand the changing needs as well as desires of the older population.

10.3.1 Ageing in Place

The predominant choice amongst older Australians at present, particularly for home owners, is to age in place, that is, to remain in their current home, and where living in the current home is not appropriate for whatever reason, to move to housing in the same general location. This desire of older people to 'stay put' in their homes is reflected in Census data which indicates that historically older people are among the least residentially mobile groups in Australia (Hugo 2014).

There are a number of factors influencing older people's desire to age in place. Prominent among these is the attachment to home, community and place forged through home ownership. Home ownership has long been a strong aspirational goal in Australia such that it is the expected outcome of participation in the housing market. Home ownership is generally the major investment in life both financially and emotionally. Striving to achieve home ownership and personalising the space and surroundings of home inevitably leads to a physical, social and emotional attachment to it—home represents a person's life course, their memories and is a personal expression, a place of privacy, refuge, independence and autonomy. Home is thus a determinant of well-being (Kendig et al. 2012).

Familiarity and connection to the neighbourhoods and communities in which homes are located reinforces the desire to age in place (Judd et al. 2010). Consequently, for many decades now the vast majority of older people have lived in detached dwellings of three or more bedrooms in the suburbs of Australia's major cities. Conversely few older people live in flats and apartments and the proportion of older people living in these types of dwellings has been decreasing since 2001 (Judd et al. 2014, p. 2). Whether these trends in types of dwellings occupied and attachment to place and home will continue among older age groups into the future however is somewhat unpredictable.

The extent to which the baby boomer generation will make 'residential adjustments' as they approach and enter retirement, for example, is as yet largely unknown. Research by Olsberg and Winters (2005) found baby boomers appeared less attached to the home and more open to the idea of residential mobility; however, among the first cohorts of baby boomers, ageing in place appears to be the primary aspiration (Pinnegar et al. 2012). Hugo (2014) proffers a possible explanation for this finding: that the mobility of the baby boomers may be delayed as the age at which people leave the work force has increased over the last decade or so (see Chap. 8).

Reflecting older people's preference to stay in their familiar (and often familial) home and surroundings, government policy is now strongly focussed on encouraging and supporting people to 'remain at home rather than in a home'. The current aged care reform package *Living Longer, Living Better* (Government of Australia 2012) centres on providing older people with increased consumer choice and control (so termed Consumer Directed Care) facilitated by more affordable and easier to access age-related support services, and improved and expanded in-home care and support (see also Chap. 15). The provision of such personal support for people to remain in the home of their choosing is important for positive ageing in place outcomes.

Alongside these personal support services, however, is a clear need for well developed and funded home maintenance and modification services to assist older people to make their current homes more suitable as they age. A review of the provision of these types of services across the country in 2008 found such services existed in a haphazard manner, with many older people simply unaware of their existence (Jones et al. 2008). Helpfully, *The Living Better. Living Longer* policy reforms (Government of Australia 2012) included a review of basic home maintenance and modification services, including eligibility, prioritisation, access and current service delivery models and identified the barriers to consumer access (Walker n.d.), offering some hope these reforms would both reflect the need for, and meet some of the demand for these services for older people to allow and extend ageing in place (see Chap. 14).

The development and proliferation of assistive living technologies will also help in supporting older people to remain independent at home. This is particularly the case in terms of dementia care where there is currently a significant technological focus. A wide range of products exist to help with safety and security (e.g. fall detection devices, mobility aids, smoke monitors, door locks), treatment (e.g. monitoring, telemedicine, health medication compliance) and social connectedness (e.g. mobile phones, video and email). However, uptake has been minimal. According to Tegart (2010) this is due partly to a lack of willingness of the current generation of older people to embrace technological innovation, but also because many of the products are incompatible with each other in the home environment and have not incorporated 'human-centred' design. Additionally, a small number of smart-homes have been developed and built showcasing what is currently possible to fully support ageing in place. While all of these technologies offer potential for supporting older people within their homes there remains a lot to learn about how to retrofit such innovation to the existing housing stock—where the majority of older people live. The cost of implementing these innovations is also of concern, with lower income and more vulnerable households at real risk of missing out on these critical supports without government assistance to ensure they can fully embrace them.

10.3.2 Housing and Locational Mobility

From a purely housing market perspective the desire of older people to age in place has, since the early 1990s, raised concern amongst policy makers, about the under-utilisation of housing that results from the ageing of the population in family homes (Millane 2015; Judd et al. 2014). This concern has been one of the driving factors in policy directives for greater diversity of housing forms to accommodate an ageing Australia. To date however, current research indicates a lack of demand for smaller housing options (Wulff et al. 2004; Judd et al. 2010, 2014). Research undertaken into the housing needs of older South Australians strongly indicated older people want housing that is appropriately designed, providing a sense of internal space while providing both sufficient bedrooms and scope for hobbies and other activities (Beer et al. 2009). This is corroborated by Judd et al. (2014) in examining how older Australians utilise their dwellings and properties. In essence, a high proportion of participants in the study by Judd et al. found their homes to be very well suited to their needs and they utilised the space available to them to accommodate their changing interests and requirements with age. In addition to being satisfied and happy with their homes, a number of other factors have been identified to explain the reluctance of older people to downsize. These included that it takes too much effort, both practically and psychologically (Adair et al. 2014; Judd et al. 2014); the financial burden of stamp duty incurred on home sales and the impact the sale may have on their age pension due to the assets test (Adair et al. 2014); the lack of appropriate housing options in familiar surroundings (Judd et al. 2014); and the limited information and advice available to assist people with the decision to move (Millane 2015). Over time however, a number of strategies have been recommended to encourage older people to move, including programmes to overcome the barriers and to promote information on the benefits of moving, assistance with decision-making, managing the move, promoting greater housing diversity and removing financial disincentives. Even with the implementation of such policies and programmes though, Judd et al. (2014) conclude most older people are likely to continue to live in their current homes for as long as they are able and, moreover, if people do choose to downsize they are more likely to move into retirement villages than any other form of accommodation.

Retirement villages have been an important housing option for older Australians for many decades now. They are arguably the most recognisable housing option for older people. At the 2011 Census only 6 % of the population aged 65 years and over was recorded as living in a retirement village. Towart (2013), however, suggests this may be a considerable undercount as her research indicates approximately 30 % of all operational retirement villages were not recorded as retirement villages by the Census. Retirement villages are generally operated by the non-government sector, through both profit and non-profit organisations. As noted by Kendig et al. (2014a) retirement community models have evolved and adapted over time to meet the needs of older people. Importantly, they will need to continue to do this to remain relevant and attractive to the baby boom generation.

While retirement villages are available for all people aged 55 years and over, most people do not make the move until they are 70 years and over (Beer et al. 2009). Australian research indicates people move to this type of accommodation for a range of reasons, but predominantly because of the need for greater social interaction and concerns about both household and property management and future health needs (Stimson and McCrea 2004; Beer et al. 2009; Crisp et al. 2013a, b). Additionally, with the increasing difficulty older people face in the private rental market, non-profit operators of independent living units are seeing increasing demand for the limited number of affordable rental properties they hold (personal communication (n.d.), ACH Group and ECH Inc., South Australia).

Overall there is a high degree of satisfaction with retirement village living. Surveys of residents have consistently found people to be very happy with their move to these housing options and for most their move has been a positive life experience (Kennedy and Coates 2008; Beer et al. 2009; Kendig et al. 2014a). International (Lum et al. 2005; Kneale 2011) and national (Productivity Commission 2011; Kendig et al. 2014a) research indicates that living in retirement communities, where there is an emphasis on the integration of housing with care, results in improved well-being overall and delays in hospital and residential care admissions. How coming generations of older people will interact with the retirement village industry—including the now retiring and soon to be retired baby boomers—remains largely unknown and an area of some conjecture.

In trying to understand what the baby boomers will do, Pinnegar et al. (2012) point out that households are less likely to move by choice if the options are less favourable than staying where they are or they face constraints curtailing their choices or preferences. While the overwhelming trend is currently to age in place (and there are seemingly few indications this will change among baby boomers), some older people will choose to move. For the older age groups generally though, mobility is greatest at two key points in their older age. First at the 'young old' stage of life, around retirement when voluntary moves are often made to environmentally amenable locations (so-called sea change or tree change moves) or to what is considered more appropriate accommodation in anticipation of ageing. Second, older people generally are also mobile at the 'old old' stage of life, where the onset of disability or the loss of a spouse precipitates a move (Hugo 2014).

Notably, among those that may move, or wish to move to more suitable or appropriate accommodation, the vast majority move only within the same local area. Consequently, Pinnegar et al. (2012) label this group 'local adaptors'—people who wish to maintain their connections with the local community but may move because of circumstances like changing household structure (e.g. adult children leaving home or partnership changes) or because of aspirational reasons, such as a desire to travel. While it may be desirable to move, at present older people are limited in their choice of housing options particularly if they are seeking accommodation that will suit their changing needs in terms of adaptability and accessibility over time. Over the last few years there has been increasing discussion about, and a push for the adoption of universal design in housing (Aged and Community Services Australia (ACSA) 2015). There appears to be strong support for adaptable universal design approaches and such a design approach will become increasingly important in terms of the home environment as a place of care. Quinn et al. (2009, p. 13) strongly emphasise this in their research on dwelling, land and neighbourhood use by older home owners, pointing out that "with increasingly higher levels of care being delivered in the home environment, the capacity of housing to accommodate safely the more demanding needs of residents and their carers is an important issue, in regard to both new and existing stock". A significant move forward in this area has been the introduction of the Livable Housing Design (LHD) Guidelines in 2010 (Livable Housing Australia 2012). These guidelines are the product of collaboration between all three levels of government, the aged, disability, community and building and construction sectors. At present, the implementation of livable or universal design features in Australia remains one of choice rather than a mandatory obligation through regulations in the building code. Recently, the Australian Network for Universal Housing Design (ANUHD) called for minimum access features in the National Construction Code for all new and extensively modified housing (ANUHD 2015) and ambitious targets have been set for all new dwellings to meet livable housing design standards by 2020 (ACSA 2015).

A small proportion of older Australians, labelled 'scene changers' by Pinnegar et al. (2012) seek out pleasant environmental areas in coastal, riverine and other non-metropolitan areas (Hugo 2014; Gurran et al. 2005; Burnley and Murphy 2004). Pinnegar et al. (2012) characterise these people as being financially secure, well travelled and from more culturally, ethnically and geographically diverse backgrounds. Concentrations of older people are particularly noticeable along the northern and southern coasts of NSW, in south east Queensland (Hugo 2014) on the Fleurieu and Yorke Peninsulas in South Australia (City of Victor Harbor 2013; Harvey et al. 2008), along much of the length of the River Murray (Hugo 2014), and in the south west of Western Australia (South West Development Commission 2010) for example. Growth in second home ownership over the last 30–40 years is often identified as a determinant of the locational choices of retirees (Burnley and Murphy 2004; Hugo 2014). A detailed study of second home ownership and changing populations in South Australia, however (Paris et al. 2014, p. 11) found "there is no evidence of widespread retirement migration by baby boomers to second homes though some coastal areas are affected as retirement migration adds to already-ageing populations." Additionally, movement of people from more populated areas to regional areas and centres for lifestyle reasons has also been countered by some movement of rural residents, in particular farmers, to nearby regional centres to take advantage of the generally more readily available services. These trends have seen further ageing of regional Australia, which already has a proportionally older population and one that is growing at a faster rate than in the capital cities (Hugo 2014).

While around one-third of older Australians live in places with a resident population of less than 100,000 (Hugo 2014), the majority of older people in Australia continue to reside in Australia's primary cities. Within these cities there is ongoing change in the distribution of the older population. In essence, concentrations of older people have moved from the inner sectors of cities to the middle and now outer suburbs of metropolitan areas—areas of lower density housing with nucleated shopping centres, less intense public transport networks and often limited services. This changing distribution of the older population, accompanied by the changing role and participation of older people in society (particularly as life expectancy increases), has profound implications for the planning, design and provision of housing, infrastructure, services and facilities. It also has clear implications for government at all levels.

10.4 Age-Friendly Communities

It is now well recognised the design of the built environment is a crucial determinant in optimising the health, participation and safety of older people and their overall quality of life. Being physically and socially active in a safe environment is essential to health and well-being (Kendig and Phillipson 2014). A large body of literature now exists on the benefits of physical activity (C3 Collaborating for Health 2011) and the need for social networks and social integration in daily life (The PLoS Medicine Editors 2010) and this is reflected in policy goals (see for example Government of South Australia 2013; Department of Family and Community Services NSW 2012). Walking is considered to be one of the best forms of physical activity and the form of exercise most appropriate for older people. It is low impact (reduced stress on joints), weight bearing (can improve bone density) and can assist with weight loss. Importantly it can reduce the risk of chronic diseases, improve sleep patterns, reduce anger and depression, alleviate anxiety and slow cognitive decline (C3 Collaborating for Health 2011; Udell et al. 2014). Walking also ensures older people 'get out of the house', increasing opportunities for interaction with the broader community. The physical structure of our cities and towns can inhibit or facilitate active lifestyles that include planned and incidental physical activity and social interaction.

Developing age-friendly communities has become a policy imperative in many countries, having been initiated by the United Nations' (2002) focus on enabling and supportive environments, and endorsed by the World Health Organisation's emphasis on active ageing and age-friendly cities (WHO 2007). Over 250 cities and communities worldwide have partnered with the WHO to become age friendly, including eight communities in Australia (Kendig et al. 2014b).

The Australian, as well as State and Territory governments, have introduced initiatives and strategies to support an ageing population and to develop age-friendly communities. However, no national guidelines have been developed around this to date (Productivity Commission 2011). In many areas though, local governments, as the agency with responsibility for the development and maintenance of services, programmes and infrastructure (Productivity Commission 2011) within communities, have taken centre stage in promoting active ageing and age-friendly communities. Their actions in this regard have included promoting age-friendly built environments; creating safe and secure pedestrian environments; fostering age-friendly community planning and design; improving mobility options; supporting recreational facilities, parks and trails; and encouraging housing choices (ALGA 2006).

The ability of local governments to develop age-friendly environments varies however across the country with rural, regional and fringe locations facing greater challenges than their metropolitan counterparts. A review of the adequacy of neighbourhood design for older people in Australian cities and towns indicates a lack of, or poor quality of footpaths; difficult access to buildings; lack of public transport; poor proximity to open spaces; insufficient street lighting and street furniture (Quinn et al. 2009; Judd et al. 2010). These barriers restrict or inhibit older people's full participation in the community.

Recent research by O'Brien (2014) raises further concerns over the development of age-friendly communities. Based on an examination of 20 local government areas across New South Wales O'Brien notes the considerable difficulties local governments face in developing age-friendly environments because of the inadequate maintenance and renewal of government infrastructure, and the need for additional facilities in the face of insufficient resources. Developing age-friendly communities has remained of low priority, a stark contrast to cities worldwide which are now assessing their achievements in meeting the goals of an age-friendly community or city (Tinker and Ginn 2015).

Moving the age-friendly environments agenda forward in Australia will require it to become a more formalised imperative of governments and for it to be central in the long-term strategic, asset and financial planning of state and local governments (Kendig et al. 2014b; O'Brien 2014). Moreover, as a number of organisations have noted (Udell et al. 2014) and Judd et al. (2010, p. 19) comment:

> [B]ringing about the necessary changes to achieve more age-friendly housing and neighbourhoods also presents some challenges for consumers that will require trade-offs. It is unlikely, for example, that current forms of low density suburban development can deliver the mixed-use neighbourhood outcomes and better quality transport systems that best support an ageing society.

In addition, creating age-friendly communities has to be about more than the physical design of environments; there is a need to reconsider what community means. Many of the factors associated with reduced social participation in the community reflect the processes and structures of modern society, including the functioning of communities, prejudices such as ageism, racism and the confluence of rising individual self-sufficiency and declining localised support within society (Peel 2000). Modern society's collective responsibility to care for others, particularly the most vulnerable, has been reduced in our pursuit of privacy, self-sufficiency and independence (Beck 1992). With increasing numbers of older people living on their own and fewer having family on which to rely, social isolation amongst the older population will, potentially, be an increasing reality. In moving forward there is a need to reorientate our way of thinking, to consider more fully what can collectively be done as a community, what people can do for each other, to create supportive age-friendly environments. Governments have a greater role to play here in establishing and supporting partnerships with families and communities to invest in, and strengthen community networks in providing necessary support. As noted by McNeil and Hunter (2014) for Britain, but equally applicable here in Australia, improving the well-being of the older population has resulted in policies and practices that narrowly focus on physical and health needs. Community projects are often viewed as pilot programmes, given one-off funding, and essentially fail to become part of the mainstream spending on service provision and care for the older population. This needs to change.

10.5 Conclusion

Older people's current housing situation and future housing options are the result of a range of factors including: cumulative lifetime opportunities and experiences; present economic, social and personal characteristics; the ability of current policy settings and market forces to address the increasingly diverse needs of the older population; and the way older people and society view older age. These factors can influence older people's housing and impact on their ability to make choices that best suit their needs to age as they wish. Overall however, older people today, and in the coming years, have much greater opportunities and therefore aspirations about how they would like to live the later parts of their lives. Their needs, desires and aspirations are no different to those of other age groups within the community. While an increasing and significant proportion will be reliant on rental accommodation in the coming years, and in particular the private rental market (with all the well-documented challenges this poses for vulnerable groups), the majority of older people will actually have significant property and financial assets to draw on as they age. However with increased longevity, delays in disabling health conditions and substantial aspirations for the post-working part of their life, older people will put increasing pressure on a range of local services, facilities and infrastructures. As the discussion in this chapter shows, impacts will be felt across housing markets, driving supply and demand, as well as requiring that the whole gamut of community and social services are responsive to the specific needs of older Australians as they try to fulfil their aspirations for a decent and desired lifestyle. How much, and to what extent, the housing and residential choices of the baby boomer generation will deviate or differ from preceding older cohorts is yet to be established, but they will have considerable political power to bring about change. This change will require choice and diversity of housing options within the market place including the need for additional appropriate and affordable housing and ways and means to ensure existing properties meet age-related needs. Creating age-friendly communities will require giving greater consideration to the shape and form of neighbourhoods and the meaning and purpose of community, but also require an understanding of the influence of a range of other policy arenas including aged care, health, employment and retirement income policies.

Housing and the residential locations in which it is situated are central to people's well-being and quality of life. This is a time of great opportunity and if we can more fully meet the needs of our ageing population then we will create the housing, environments, communities and opportunities for choice that will better meet and support the aspirations and needs of all in society, regardless of age.

10.6 Case Studies of Housing Paths and Trajectories

Jan and David

Jan and David are in their 60s, in good health and retired now. They own their home in a regional coastal centre. They share this home with their daughter and her family but for 6 months of the year they travel the country in their caravan. They are so-called grey nomads. They love getting away, off the beaten track, exploring places that most people do not even know exist. They wish to continue this lifestyle for as long as they can.

Mary

Mary, aged 75 years, has lived in a regional centre about 1 hour's drive from the capital city. She has lived in her public housing dwelling for 45 years. Mary is reliant on her age pension and grateful for her public rental housing. She has a three bedroom house and while very satisfied, a smaller backyard would be nice. Mary still drives and this is very important as public transport is limited. She has good neighbours and is active in the broader community. Mary sees her future as staying in her current home and requesting modifications when they are needed.

Judith and Ken

Judith and Ken, self-funded retirees, are aged 75 years and consider themselves to be fit and healthy. They retired to the south coast, attracted by the environmental and aesthetics of the area (weather, sea/river, local shops, community atmosphere). Rather than downsizing they have moved from a three bedroom house into a four bedroom, two storey home because they felt they needed more space and to take advantage of the views. Judith and Ken plan to stay in their larger house until their health fails.

George

George, aged 60, is a single man reliant on government support, living in a low socio-economic area of a city. He lives in private rental accommodation and has moved five or six times in the last 3 years. Life is a daily struggle especially as 50 % of George's income goes towards the rent. George feels very isolated, as his financial circumstances do not provide him with spare money to get to social activities. George hopes one day to be offered public housing.

Peter

Peter is 65 years old. Due to a challenging marriage breakdown and deteriorating mental health, Peter decided to take off in his van. The van became Peter's home for 20 years—he slept in his van, ate in the van, changed his clothes in the van. He picked up work where he could. Peter's health deteriorated while living such a transient existence. Peter was put in touch with the Aged Care and Housing Assistance programme and his life changed. For the first time he had a unit to live in—stability and security in his life, his health improved and he had his dignity back. Of importance he had a place for his grandchildren to visit and this made life worthwhile and meaningful.

10.6.1 Discussion Points

- Identify and discuss the factors which have influenced and impacted on the choice of housing in each of the above scenarios and how these shaped their ability to make housing decisions which best suit their needs as they age.

References

Adair, T., Williams, R., & Menyen, T. (2014). *Downsizing decisions of senior Australians: What are the motivating and discouraging factors*? Melbourne, Australia: National Seniors Productive Ageing Centre.

Aged and Community Services Australia. (2015). *The future of housing for older Australians position paper*. Deakin, Australia: ACSA.

Australian Bureau of Statistics. (2013). *Household income and income distribution, Australia 2011–12* (Cat. No. 6523.0.). Canberra, Australia: ABS.

Australian Institute of Health and Welfare. (2012). *National social housing survey: State and territory results 2010* (AIHW Cat. No. HOU 264). Canberra, Australia: AIHW.

Australian Institute of Health and Welfare. (2013). *Australia's welfare 2013*. Australia's Welfare Series No. 11 (AIHW Cat. No. AUS 174). Canberra, Australia: AIHW.

Australian Institute of Health and Welfare. (2014). *Housing assistance in Australia 2014* (AIHW Cat. No. HOU 275). Canberra, Australia: AIHW.

Australian Local Government Association. (2006). *Age-friendly built environments: Opportunities for local government*. Deakin, Australia: ALGA.

Australian Network for Universal Housing Design. (2015). Position statement and declaration of support. ANUHD. Retrieved April 2015, from http://www.anuhd.org.

Beck, U. (1992). *Risk society: Towards a new modernity*. London: Sage.

Beer, A., Faulkner, D., Baker, E., Tually, S., Rafter, P., & Cutler, C. (2009). *Our homes, our communities: The aspirations and expectations of older people in south Australia*. Adelaide, Australia: Flinders Institute for Housing, Urban and Regional Research and ECH.

Beer, A., Faulkner, D., Paris, C., & Clower, T. (2011). *Housing transitions through the life course: Aspirations, needs and policy*. Bristol: Policy Press, University of Bristol.

Burnley, I. H., & Murphy, P. (2004). *Sea change: Movement from metropolitan to Arcadian Australia*. Sydney, Australia: UNSW Press.

C3 Collaborating for Health. (2011). *Review: The benefits of physical activity for health and well-being*. London: C3 Collaborating for Health.

City of Victor Harbor. (2013). *Urban growth management strategy*. South Australia: City of Victor Harbor. Retrieved April 2015, from http://www.victor.sa.gov.au/page.aspx?u=840.

Crisp, D. A., Windsor, T. D., Anstey, K. J., & Butterworth, P. (2013a). What are older adults seeking? Factors encouraging or discouraging retirement village living. *Australasian Journal on Ageing, 32*(3), 163–170.

Crisp, D. A., Windsor, T. D., Anstey, K. J., & Butterworth, P. (2013b). Considering relocation to a retirement village: Predictors from a community sample. *Australasian Journal on Ageing, 32*(2), 97–102.

Daley, J., Wood, D., Weidmann, B., & Harrison, C. (2014). The wealth of generations. Melbourne, Australia: Grattan Institute. Retrieved April 2015, from http://grattan.edu.au/wp-content/uploads/2014/12/820-wealth-of-generations3.pdf.

Department of Family and Community Services. (2012). *NSW ageing strategy*. Sydney, Australia: Department of Family and Community Services Office, NSW Government. http://www.adhc.nsw.gov.au/__data/assets/file/0011/257276/1282_ADHC_NSW-AgeingStrategy_WEB.pdf. Accessed Apr 2015.

Faulkner, D. (2009). The fragile links between private rental accommodation, quality of life and the threat of homelessness for older people. *Parity, 22*(3), 14–15.

Fiedler, J. (2014). *Australian senate economics references committee inquiry into affordable housing submission*. Melbourne: Education & Housing Advice, Housing for the Aged Action Group. http://www.aph.gov.au/Parliamentary_Business/Committees/Senate/Economics/Affordable_housing_2013/Submissions. Accessed Jan 2015.

Goodman, R., Nelson, A., Dalton, T., Cigdem, M., Gabriel, M., & Jacobs, K. (2013). *The experience of marginal rental housing in Australia* (AHURI Final Report No. 210). Melbourne, Australia: Australian Housing and Urban Research Institute.

Government of Australia. (2012). *Living longer. Living better*. Australia: Department of Health. Retrieved 10 July, 2015, from http://apo.org.au/node/29086.

Government of South Australia. (2013). *Prosperity through longevity: South Australia's ageing plan, our vision 2014–2019*. Adelaide, Australia: South Australia Office for the Ageing.

Gurran, N., Squires, C., & Blakely, E. (2005). *Meeting the sea change challenge: Sea change communities in coastal Australia*. Sydney, Australia: Planning Research Centre, University of Sydney.

Harvey, N., Rudd, D., & Clarke, B. (2008). The 'Sea Change' phenomenon in South Australia. *South Australian Geographical Journal, 107*, 69–85.

Hugo, G. (2014). *The demographic facts of ageing in Australia*. Adelaide, Australia: University of Adelaide.

Hulse, K., Reynolds, M., & Yates, J. (2014). *Changes in the supply of affordable housing in the private rental sector for lower income households, 2006–11* (AHURI Final Report No. 235). Melbourne, Australia: Australian Housing and Urban Research Institute. Retrieved January 2015, from http://www.ahuri.edu.au/publications/download/ahuri_51018_fr1.

Jones, A., de Jonge, D., & Phillips, R. (2008). *The role of home maintenance and modification services in achieving health, community care and housing outcomes in later life* (AHURI Final

Rep. No. 123). Melbourne, Australia: Australian Housing and Urban Research Institute. Retrieved March 2015, from http://www.ahuri.edu.au/publications/download/ahuri_20335_fr.

Judd, B., Liu, E., Easthope, H., Davy, L., & Bridge, C. (2014). *Downsizing amongst older Australians* (AHURI Final Rep. No. 214). Melbourne, Australia: Australian Housing and Urban Research Institute. Retrieved April 2015, from http://www.ahuri.edu.au/downloads/publications/EvRevReports/AHURI_Final_Report_No214_Downsizing_amongst_older_Australians.pdf.

Judd, B., Olsberg, D., Quinn, J., Groenhart, L., & Demirbilek, O. (2010). *Dwelling, land and neighbourhood use by older home owners* (AHURI Final Report No. 144). Melbourne, Australia: Australian Housing and Urban Research Institute, UNSW-UWS Research Centre. Retrieved March 2015, from https://www.be.unsw.edu.au/sites/default/files/upload/pdf/cf/research/ahuriprojects/theefficienuseofdwellings/FR144.pdf.

Kelly, S. (2012). Household savings and retirement. Where has all my super gone? A report on superannuation and retirement for CPA Australia. VIC, Australia: KELLY research and CPA Australia. Retrieved April 2015, from https://www.cpaaustralia.com.au/~/media/corporate/allfiles/document/professional-resources/superannuation/household-savings-retirement.pdf.

Kendig, H., & Bridge, C. (2007). Housing policy for a long-lived society. In A. Borowski, S. Encel, & E. Ozanne (Eds.), *Longevity and social change in Australia* (pp. 219–238). Sydney, Australia: UNSW Press.

Kendig, H., Clemson, L., & Mackenzie, L. (2012). Older people: Well-being, housing and neighbourhoods. In *International encyclopedia of housing and home* (vol. 5, pp. 150–155). Amsterdam: Elsevier.

Kendig, H., Crisp, D., Gong, C., Conway, E., & Squires, B. (2014). *An investigation of the IRT retirement community landscape: A pilot study*. Wollongong, Australia: IRT.

Kendig, H., Elias, A.-M., Matwijiw, P., & Anstey, K. (2014). Developing age-friendly cities and communities in Australia. *Journal of Aging and Health, 26*(8), 1390–1414.

Kendig, H., & Phillipson, C. (2014). Building age-friendly communities: New approaches to challenging health and social inequalities. In N. Denison & L. Newby (Eds.), *"If you could do one thing..." nine local actions to reduce health inequalities* (pp. 102–111). London: The British Academy. Retrieved from http://www.britac.ac.uk/policy/Health_Inequalities.cfm.

Kennedy, D. J., & Coates, D. (2008). Retirement village resident satisfaction in Australia: A qualitative enquiry. *Journal of Housing for the Elderly, 22*(4), 311–334.

Kneale, D. (2011). *Establishing the extra in extra care*. London: International Longevity Centre UK.

Livable Housing Australia. (2012). *Livable housing design guidelines* (2nd ed.). Sydney, Australia: Livable housing Australia. Retrieved April 2015, from http://livablehousingaustralia.org.au/library/help/Livable_Housing_Design_Guidelines_Web1.pdf.

Lovering, M. (2014). *More retire with larger housing debt* (AHURI Evidence Review 055). Melbourne, Australia: Australian Housing and Urban Research Institute. Retrieved April 2015, from http://www.ahuri.edu.au/housing_information/review/evrev055.

Lum, J. M., Ruff, S., & Williams, A. P. (2005). *When home is community. Community support services and the well-being of seniors in supportive and social housing*. Greater Toronto: Ryerson University, Neighbourhood Link/Senior Link, the University of Toronto and United Way of Greater Toronto. Retrieved March 2015, from http://www.teamgrant.ca/M-THAC%20Greatest%20Hits/M-THAC%20Projects/All%20info/When%20Home%20is%20Community/Publications/p400534.pdf.

McNeil, C., & Hunter, J. (2014). *The generation strain: Collective solutions to care in an ageing society*. London: Institute for Public Policy Research.

Millane, E. (2015). *The head, the heart & the house: Health, care and quality of life*. Australia: Percapita. Retrieved March 2015, from http://percapita.org.au/wp-content/uploads/2015/01/Head-Heart-House_Final.pdf.

National Housing Supply Council. (2010). *National housing supply council 2nd state of supply report*. Canberra, Australia: Australian Government.

O'Brien, E. (2014). Planning for population ageing: Ensuring enabling and supportive physical-social environments—Local infrastructure challenges. *Planning Theory & Practice, 15*(2), 220–234.

Olsberg, D., & Winters, M. (2005). *Ageing in place: Intergenerational and intrafamilial housing transfers and shifts in later life* (AHURI Final Rep. No. 88). Melbourne, Australia: Australian Housing and Urban Research Institute. Retrieved January 2015, from http://www.ahuri.edu.au/downloads/publications/EvRevReports/AHURI_Final_Report_No88_Ageing_in_place_intergenerational_and_intrafamilial_housing_transfers_and_shifts_in_later_life.pdf.

Ong, R., Haffner, M., Wood, G., Jefferson, T., & Austen, S. (2013a). *Assets, debt and the drawdown of housing equity by an ageing population* (AHURI Positioning Paper No. 153). Melbourne, Australia: Australian Housing and Urban Research Institute. Retrieved April 2015, from https://www.melbourneinstitute.com/downloads/hilda/Bibliography/Other_Publications/2013/Ong_etal_Assets_debt_and_the_drawdown_of_housing_equity_by_an_ageing_population.pdf.

Ong, R., Jefferson, T., Wood, G., Haffner, M., & Austen, S. (2013b). *Housing equity withdrawal: uses, risks, and barriers to alternative mechanisms in later life* (AHURI Final Rep. No. 217). Melbourne, Australia: Australian Housing and Urban Research Institute.

Paris, C., Thredgold, C., Jorgensen, B., & Martin, J. (2014). *Second homes and changing populations: Impacts and implications for local government in south Australia*. Adelaide, Australia: Centre for Housing, Urban and Regional Planning (CHURP), University of Adelaide. http://www.council.robe.sa.gov.au/webdata/resources/files/Second%20Homes%20&%20Changing%20Populations%20%20Impacts%20for%20Local%20Government%20in%20SA.pdf. Accessed Feb 2015.

Peel, M. (2000). Between the houses: Neighbouring and privacy. In P. Troy (Ed.), *A history of European housing in Australia* (pp. 269–286). Cambridge: Cambridge University Press.

Petersen, M., Parsell, C., Phillips, R., & White, G. (2014). *Preventing first time homelessness amongst older Australians* (AHURI Final Rep. No. 222). Melbourne, Australia: Australian Housing and Urban Research Institute. Retrieved February 2015, from http://www.ahuri.edu.au/publications/download/ahuri_21005_fr.

Pinnegar, S., van den Nouwelant, R., Judd, B., & Randolph, B. (2012). Understanding housing and location choices of retiring Australians in the 'baby boom' generation. Sydney: City Futures Research Centre, University of New South Wales. Retrieved from https://www.be.unsw.edu.au/research-centres-and-clusters/city-futures/other-research-reports#sthash.kaymXiK2.dpuf.

PLoS Medicine Editors. (2010). Social relationships are the key to health, and to health policy. *PLoS Medicine, 7*(8), 1–2.

Productivity Commission. (2011). *Caring for older Australians, Report no 53, final inquiry report*. Canberra, Australia: Commonwealth of Australia. Retrieved January 2015, from http://www.pc.gov.au/inquiries/completed/aged-care/report.

Quinn, J., Judd, B., Olsberg, D., & Demirbilek, O. (2009). *Dwelling, land and neighbourhood use by older home owners* (*AHURI Positioning Paper No. 111*). Melbourne, Australia: Australian Housing and Urban Research Institute. Retrieved January 2015, from http://www.ahuri.edu.au/publications/download/ahuri_70392_pp.

South West Development Commission. (2010). South West Active Ageing Research Project. Western Australia: South West Development Commission. Retrieved April 2015, from http://www.swdc.wa.gov.au/media/68943/active%20ageing_web_2.pdf.

Stimson, R. J., & McCrea, R. (2004). A push–pull framework for modelling the relocation of retirees to a retirement village: The Australian experience. *Environment and Planning A, 36*(8), 1451–1470.

Stone, W., Burke, T., Hulse, K., & Ralston, L. (2013). *Long-term private rental in a changing Australian private rental sector* (AHURI Final Rep. No. 209). Melbourne, Australia: Australian Housing and Urban Research Institute. Retrieved February 2015, from http://www.ahuri.edu.au/downloads/publications/EvRevReports/AHURI_Final_Report_No209_Longterm_private_rental_in_a_changing_Australian_private_rental_sector.pdf.

Tegart, W. J. McG. (2010). Smart technologies for healthy longevity. Melbourne, Australia: Australian Academy of Technological Sciences and Engineering (ATSE). Retrieved February 2015, from http://www.atse.org.au/Documents/Publications/Reports/Health%20&%20Tech/Smart%20Tech%20for%20Healthy%20Longevity%20-%20Report.pdf.

Tinker, A., & Ginn, J. (2015). *An age friendly-city—How far has London come?* London: Kings College. Retrieved April 2015, from http://www.ifa-fiv.org/wp-content/uploads/2015/03/Age-Friendly-London-Report-final.pdf.

Towart, L. (2013). *Who lives in retirement villages; Are they wealthy enclaves, ghettos or connected communities?* Retrieved March 2015, from http://www.stantonsresearch.com.au/pdffiles/nov13.pdf.

Udell, T., Daley, M., Johnson, B., & Tolley, R. (2014). *Does density matter? The role of density in creating walkable neighbourhoods*. Melbourne, Australia: National Heart Foundation of Australia.

United Nations. (2002). *Political declaration and Madrid international plan of action on ageing*. New York: United Nations. Retrieved March 2015, from http://undesadspd.org/Ageing/Resources/MadridInternationalPlanofActiononAgeing.aspx.

Walker, B. (n.d.). Living longer. Living better aged care reform overview. Canberra, Australia: Department of Health and Ageing. Retrieved from http://www.nswhmms.org.au/sites/default/files/bernadette_walker_living_longer._living_better.pdf.

Wood, G., Colic-Peisker, V., Berry, M., & Ong, R. (2010). *Asset poverty and older Australians' transitions onto housing assistance programs* (AHURI Final Rep. No. 156). Retrieved March 2015, from Melbourne, Australia: Australian Housing and Urban Research Institute. http://www.ahuri.edu.au/publications/download/ahuri_30563_fr.

World Health Organisation. (2007). *Global age-friendly cities: A guide*. Geneva: WHO. Retrieved March 2015, from http://www.who.int/ageing/publications/Global_age_friendly_cities_Guide_English.pdf.

Wulff, M., Healy, E., & Reynolds, M. (2004). Why don't small households live in small dwellings?—Disentangling a planning dilemma. *People and Place, 12*(1), 57–70.

Yates, J., Kendig, H., Phillips, B., Milligan, V., & Tanton, R. (2008). *Sustaining fair shares: The Australian housing system and intergenerational sustainability* (AHURI Final Rep. No. 111). Melbourne, Australia: Australian Housing and Urban Research Institute. Retrieved February 2015, from http://www.ahuri.edu.au/publications/download/ahuri_60314_fr.

Yates, J., & Milligan, V. (2007). *Housing affordability: A 21st century problem* (AHURI Final Rep. No. 105). Melbourne, Australia: Australian Housing and Urban Research Institute. Retrieved February 2015, from http://www.ahuri.edu.au/publications/download/ahuri_nrv3_final_report.

Chapter 11
Transport and the Older Australian: Freedom, Challenges and Opportunities

Graham Currie and Alexa Delbosc

11.1 Introduction

Access to affordable mobility has generated significant economic and social benefits for most people. In particular, the car has provided enormous mobility benefits, particularly for the baby boomer generation which was the first to experience these benefits throughout their lifetime (Browning and Sims 2007). However, the baby boomers' love affair with the car has also set them up for considerable challenges into the future. Unlike previous generations, their lives have been built around access to cars, but inevitably age will force many out from behind the wheel. Mobility and access are essential for healthy and happy ageing and significant challenges are faced when these are impeded.

This chapter examines how cars have revolutionised the mobility of baby boomers and changed the shape of Australia's cities. It describes how the mobility of older people will change going forward and discusses how a lack of mobility can have serious negative impacts on the quality of life of older people. A case study, based on research from Melbourne and regional Victoria, illustrates these impacts. A range of other mobility and access challenges are then identified, and options to address these challenges into the future are discussed.

G. Currie (✉) • A. Delbosc
Public Transport Research Group, Institute of Transport Studies, Monash University, Clayton Campus, Wellington Road, Clayton, VIC 3800, Australia
e-mail: Graham.Currie@monash.edu; Alexa.Delbosc@monash.edu

K. O'Loughlin et al. (eds.), *Ageing in Australia*, International Perspectives on Aging 16, DOI 10.1007/978-1-4939-6466-6_11

11.2 The Mobility Revolution

Australia's baby boomer generation includes over 5.4 million people born between 1946 and 1965, with the oldest baby boomers now entering their 70s (Australian Bureau of Statistics 2011). This generation was the first to grow up with increasing access to car-based mobility. The boomers are healthier, wealthier, better educated and more reliant on cars than any previous generation (Goulias et al. 2007; Coughlin 2009; D'Ambrosio et al. 2012).

Cars were not just a means of travel for the baby boomers, they were a 'dream machine' bringing the family together on weekend outings and making it possible to enhance social as well as economic activities (Davison 2004). The car also provided for new freedoms within the family and facilitated social and economic change that shaped cities and society.

Women felt some of the most profound changes of this generation. Baby boomer women were more likely than their mothers to work outside the home, feminism and equality were growing concerns and divorce became more common (D'Ambrosio et al. 2012). The traditional role of women as homebound carers (often without a driver's licence) and men as travelling workers began to shift with this generation. This is reflected in a more rapid increase in licensing and driving among older women than older men (Rosenbloom 2001, 2012; Coughlin 2009; Li et al. 2012). Table 11.1 demonstrates this trend using licensing data from Melbourne: over 90 % of baby boomer women have a driving licence, whereas in earlier generations men were almost twice as likely as women to have a licence.

The mobility provided by cars was also the spark that set off the rapid suburbanisation of Australian cities after World War II (Dodson 2007). In the past, cities were shaped by what could be reached on foot or by public transport. The generation before the baby boomers grew up in an era when not everyone had a car and driving was one of several transportation options (Rosenbloom 2012). In contrast, the baby boomer generation grew up in an era of rapid motorisation and made their housing, work and lifestyle decisions based on the mobility provided by the private car. This car-centred mobility created the sprawling suburban form of Australia's cities that we know today.

Because the baby boomers grew up with cars, they are entering retirement with significantly higher mobility expectations than previous generations. Already, older people are more mobile than ever before in history and there is every indication that they will not be giving up the keys to their cars as previous generations did (Rosenbloom 2001, 2012). As shown in Table 11.1, in 1994 over 90 % of baby boomers (then aged 29–48 years) held a driver's licence; by 2009 this cohort was aged 44–63 years but showed no signs of relinquishing their driver licence. This will have implications not just for the travel of baby boomers, but the environmental sustainability of Australia's cities (Rosenbloom 2001).

Table 11.1 Car licensing rates by age cohort, 1994 and 2009

		Age group (%)		
Year	Gender	29–48	44–63	65+
1994	Male	**95**	94	82
	Female	**91**	81	46
2009	Male	94	**95**	90
	Female	90	**92**	65

Note: cells highlighted in bold represent the baby boomer generation
Source: Author's analysis of the Victorian Activity and Travel Survey 1994 (VATS) (Victorian Integrated Survey of Travel and Activity 2001) and Victorian Integrated Survey of Travel and Activity 2009 (VISTA) (Department of Transport 2009)

11.3 New 'Drivers' of Change

The baby boomer generation is just beginning to transition from working age to retirement age. This demographic shift will change the nature and extent of their personal travel needs and raise new opportunities and challenges.

It is important to note that by the time people enter their 50s, most adults have established work, social and leisure lives centred around the place they live. For the baby boomers, this means that they are likely to age in middle and outer suburbs, with a high concentration in regional and rural areas (Frith et al. 2012). Those who do move in later life are likely to make a 'tree change' or 'sea change' to rural and coastal areas for lifestyle reasons (Olsberg and Winters 2005; (see Chap. 10)). Many of these locations require boomers to maintain their dependence on the car for personal travel.

There are some suggestions that boomers are more likely than previous generations to retire later in life and to mix retirement with part-time work, with long-term projections suggesting that workforce participation rates for older people are increasing (see Chaps. 8 and 9). Older workers tend to transition into part-time work, often combined with a move towards less physically demanding, service-oriented jobs (Srinivasan et al. 2006). This can result in the timing of some work trips changing to outside of peak periods, which may reduce congestion during these periods. However, this is also likely to make these trips less suitable for public transport, which concentrates its services into peak periods (Srinivasan et al. 2006).

Retirement does not herald an end to the need for mobility and travel. It is likely that boomers will continue to expect a vibrant social life, including travel, family time, hobbies, volunteering and physical activity (D'Ambrosio et al. 2012). A Canadian study found that baby boomers were the only cohort to *increase* activities over time rather than decrease as they aged (Miranda-Moreno and Lee-Gosselin 2008). Baby boomers are also considered a 'sandwiched' generation, needing to travel both in order to care for elderly parents and to provide childcare to grandchildren (Fingerman et al. 2012; Millward 1999).

11.4 Lack of Mobility and Access

Section 11.3 illustrated the importance of mobility and travel, even as older people transition from work into retirement. Research has now established strong links between mobility, access to activities and a high quality of life (Metz 2003). In addition, mobility can provide feelings of independence and increased self-esteem (Whelan et al. 2006). What this evidence also suggests is that without mobility these benefits are at risk.

Although there is every indication that baby boomers will be more mobile than previous generations as they age, ageing will nevertheless eventually bring physical limitations to mobility. Many older people do not plan for the inevitable loss of car-based mobility. In a survey of elderly residents of Arizona, USA, between 10 and 40 % said they simply did not know how they would shop for basics such as groceries or make social visits if they could not drive (Rosenbloom 1999). It is important, therefore, to understand how limitations to mobility impact the quality of life of older people.

Early work on the influence of mobility on quality of life found that driving cessation is a strong predictor of depression, even when controlling for age, education, marital status, physical health and cognitive function (Marottoli et al. 1997; Ragland et al. 2005). This indirectly implies that a sudden loss of mobility (through driving cessation) can significantly reduce quality of life.

Other studies have explored this relationship by directly measuring mobility and quality of life. An early exploration into this relationship found that in a sample of 1000 older people in Great Britain, those who had difficulty walking, as well as those with long-standing illness had a lower quality of life (Banister and Bowling 2004). A similar study of older people in five European countries examined the relationship between out-of-home mobility and quality of life. The analysis found that elderly persons who experienced more out-of-home activities and had more transport options had a higher quality of life (Mollenkopf et al. 2005).

More recently, a study of time-use data among elderly Canadians categorised out-of-home activities based on how they might provide psychological, health and community benefits. It found small but significant relationships between psychological well-being and time spent in exercise and ‘psychologically beneficial’ activities (Spinney et al. 2009).

To explore lack of mobility and access in the Australian context in greater depth, a case study is presented. This research is part of a larger study conducted in Victoria (Currie et al. 2009).[1]

---

[1] Australian Research Council Industry Linkage Program Project LP0669046 ‘Investigating Transport Disadvantage, Social Exclusion and Well-being in Metropolitan, Regional and Rural Victoria’, Monash University, in association with the University of Oxford (UK), University of Ulster (UK), Department of Transport, Victoria, the Bus Association of Victoria and the Brotherhood of St. Laurence. The principal chief investigator is Prof. G. Currie and the project Research Fellow is Ms. Alexa Delbosc. The chief investigators are Prof. T. Richardson, Prof. P. Smyth and Dr. D. Vella-Brodrick. The partner investigators are Prof. J. Hine, Dr. K. Lucas, Mr. J. Stanley, Dr. J. Morris, Mr. R. Kinnear and Dr. J. Stanley.

11.5 Research in Urban and Regional Victoria: A Case Study

A large, multi-disciplinary study was conducted in Victoria in 2008–2009 to explore the impact of 'transport disadvantage' on social exclusion and psychological well-being. People who experience transport disadvantage have difficulty accessing locations important to maintain daily life. This can be due to problems with the transport system, physical access or psychological characteristics of the individual. The study was one of the first to demonstrate the impacts that transport disadvantage can have on other aspects of people's lives.

This section describes two key findings from this study: that transport disadvantage can have particularly negative impacts on older people and that some locations are more vulnerable to transport disadvantage than others.

11.5.1 Study Approach

The aim was to explore mobility and lifestyle aspects in depth, as a means of establishing links between the two. Surveys targeted individuals with both good and poor access to mobility, covering Melbourne and the rural and regional parts of the Latrobe Region of Victoria (Currie 2011). The research covered all age cohorts and a range of social groups but this analysis will focus on the situation of older Australians within the community.

11.5.2 Transport Problems That Matter

Transport disadvantage is a multi-dimensional concept. In this survey it was measured using 18 separate facets (e.g. 'getting to places quickly' or 'buses/trains/trams operating frequently') and a statistical analysis (Delbosc and Currie 2011b) isolated four specific forms of transport disadvantage (see Table 11.2). People could belong to none, one or more of these categories.

Analyses established that these four types of transport disadvantage did not all have the same negative impacts on people's lives. The 'Vulnerable/Impaired' group, in particular, stood out as suffering the most negative consequences from transport disadvantage. This group identified difficulty with physical access, fear for their own safety and trouble understanding where to go in the transport system. Members of this group were more likely to be older women and on a low income than the rest of the survey and they were less likely to walk or use public transport. They were less likely to feel safe while travelling or in their own home, they were more likely to be disabled and they were unsatisfied with their general health.

Table 11.2 Four categories of transport disadvantage

Category of transport disadvantage	Sub-scale components
General transport disadvantage	Being able to travel when you want to
General difficulties in travelling by any mode	Finding transport so you can travel
18 % of survey sample	Being able to get around reliably
	Getting to places quickly
	Finding the time to travel when you need to
Transit disadvantage	Buses/trains/trams being available at night
Issues related to public transport	Buses/trains/trams being available at weekends
38 % of survey sample	Buses/trains/trams operating frequently
	Being able to make bus/train/tram connections
Vulnerable/impaired	Being able to physically get onto/off buses/trains/trams
Physical access and sense of vulnerability	Needing help to get around on your own
10 % of survey sample	Being able to understand where to go
	Feeling safe from theft/attack when travelling on your own
Rely on others	Having to rely on others for transport
Require assistance to travel	Finding someone to provide assistance when transport is available
25 % of survey sample	Covering the costs of your transport

Only 10 % of the survey sample fell into this group but the impacts on these people were notable. They were more likely to be socially excluded than any other transport disadvantage group, especially through low income and a lack of participation in social activities. They were also more likely to suffer from lower psychological well-being.

These findings represent an important warning for ageing baby boomers. Although they will enter old age healthier than previous generations, some may find themselves in a position where their travel is limited by physical disability or fear for personal safety. If this is the case, they are likely to suffer negative consequences such as social exclusion and poorer well-being.

11.5.3 Locational Disadvantage

The study also explored how geographic location can make people more vulnerable to transport disadvantage. Although these findings do not focus on older people specifically, the issues raised are important for all age groups.

Transport disadvantage was measured using two indicators: how often people had difficulty accessing activities because of transport issues, and whether there were activities people could not do because of transport problems. The study explored whether these factors influenced people's psychological well-being measured on four scales:

1. Satisfaction with Life Scale (SWLS): Participants indicate how much they agree with five statements about their life conditions and how close their life is to their ideal (Diener et al. 1985).
2. Personal Well-being Index (PWI): Participants indicate how satisfied they are with nine different aspects of their life (International Wellbeing Group 2006).
3. Positive affect schedule (PA): Participants rated how much they generally felt a range of positive emotions (Watson et al. 1988).
4. Negative affect schedule (NA): Participants rated how much they generally felt a range of negative emotions (Watson et al. 1988).

Analyses explored these issues for inner Melbourne (within 20 km of the city centre), outer Melbourne, 'fringe' Melbourne (living in the outer ring of suburbs in areas of little or no public transport) and the rural/regional setting of the Latrobe Valley.

People living in fringe and regional areas were further from shops/activity areas and public transport, they travelled farther each day and they were more likely to rely on cars. They were also more likely to report transport difficulties—as many as 20 % in fringe areas and 24 % in regional areas said there were activities they could not do because of transport problems.

The most important differences were revealed when the correlations between transport difficulties and well-being were compared between the geographic areas. Table 11.3 shows the correlation between the four different measures of psychological well-being and the two measures of transport difficulties. The larger the absolute value of the correlations, the stronger the relationship between the two variables.

In inner and outer Melbourne the correlations between the two were small and inconsistent; this suggests that in these areas, experiencing transport difficulties does not necessarily have a negative impact on psychological well-being. However in fringe areas, and especially in regional areas, the correlations were higher. This suggests that if you live in fringe or regional areas with high car dependency and little public transport, experiencing transport difficulties can have a serious negative impact on your well-being.

11.5.4 Outcomes and Implications

Overall this case study presents some important findings for mobility in ageing. Some of the strongest negative impacts of lack of mobility found in the study were associated with older Australians. In addition, problems were found to be particularly important in fringe urban and regional locations. The locational aspects of these issues are particularly important because, as noted earlier, baby boomers are concentrated in middle and outer suburbs, as well as making up a considerable portion of Australia's rural population (Frith et al. 2012). These areas are characterised by low-density development and are often far from quality public transport, continuing the boomers' lifelong dependence on the car for mobility.

Table 11.3 Correlations between transport difficulties and well-being by geographic areas

	Metro overall	Inner Melbourne	Outer Melbourne	Fringe Melbourne	Regional
Correlation between "Frequency of difficulties accessing activities due to lack of transport" and well-being					
Satisfaction with Life Scale	–0.19*	–0.24*	–0.16*	–0.20	–0.41*
Personal Well-being Index	–0.21*	–0.27*	–0.17*	–0.33*	–0.44*
Positive affect	–0.02	–0.11	0.02	–0.10	–0.08
Negative affect	0.21*	0.15**	0.24*	0.18	0.34*
Correlation between "Number of activities cannot do because of transport problems" and well-being					
Satisfaction with Life Scale	–0.14*	–0.09	–0.13*	–0.32*	–0.30*
Personal Well-being Index	–0.07**	–0.07	–0.05	–0.24**	–0.33*
Positive affect	0.05	–0.02	0.08	–0.08	0.06
Negative affect	0.07	0.00	0.07	0.19	0.22*

*$p<0.01$, **$p<0.05$

Source: Delbosc and Currie (2011a)

This has important implications for future mobility of this cohort as research has shown that urban density can have a large impact on mode choice, particularly walking. An American study found that areas with very high urban density encourage fewer car trips and more bus, rail and walking trips regardless of age (Giuliano et al. 2003). Similarly, a study in Virginia, USA, found that in walkable, mixed-use towns, seniors took 2.3 trips on foot per week, compared to 0.7 trips in suburban areas and 0.4 trips in rural areas (Lynott 2006).

11.6 Other Mobility/Access Challenges

An increasing population of older drivers has been seen as a possible risk in terms of road safety to the community. However, evidence suggests that, due to experience, older drivers have many safe driving advantages compared to younger people (Rosenbloom 2011). However, due to population ageing the increase in the sheer number of older drivers will increase the absolute number of accidents within this cohort. Evidence also shows that when accidents do occur, older cohorts have more adverse outcomes, including higher risk of death (Browning and Sims 2007).

A range of disabilities of many kinds are also associated with ageing and can have significant implications for access and mobility (Currie and Allen 2007). An emerging trend has been growth in the use of 'powered mobility aids', notably mobility scooters; between 1998 and 2003 there was a 78 % increase in the number of motorised scooters being used in Australia. Due to poor safety regulation this trend has been associated with an increase in accidents (Rose and Richardson 2010; Gibson et al. 2011).

In cohorts where disabilities are less severe, the car remains a dominant mode of travel in Australia for all age groups: 57 % of people with mild physical activity limitations drive a car. However, even when disabilities prevent individuals from driving themselves, the car still dominates as a means of mobility: 52 % of trips to work/school made by people with profound/severe activity limitations were made as a passenger in a car (Currie and Allen 2007). Disability in ageing has been strongly associated with a need for more door-to-door travel due to limitations on walk access distance. This makes car and taxi travel attractive, yet these options can be expensive for cohorts with a relatively lower income. Having a family member or friend assist by providing car transport shifts the expense of car travel to these carers as well as requiring additional time commitments from them.

Indeed the affordability of private vehicle travel is a major concern for the mobility of older people into the future. Although demand for fossil fuels is likely to increase in the near term, much evidence points to constraints on the future supply of carbon-based fuels. The likely outcome, of higher prices for fuel and lower affordability of mobility for older Australians, has been identified as a significant concern (Kalata 2005). The low-density geographical distribution of Australian cities and rural and regional centres makes them particularly difficult to service by public transport, walking and cycling (Currie et al. 2007). For this reason, fuel price rise vulnerability is an emerging concern particularly associated with Australian cities (Dodson and Sipe 2006).

There is also evidence that older Australians are increasing their use of urban public transport systems (Currie and Delbosc 2010), encouraged by concessionary fare schemes and increasing implementation of accessible infrastructure. However, public transport use is not without its own challenges for an ageing and disabled population. Barriers to services and information, concerns about personal safety, poor walking environments, worries about falls on moving vehicles and lack of facilities for older people are common complaints noted by older Australians about urban public transport (Currie 2010). However, the major challenge for older Australians seeking to use public transport is the lack of reasonable service quantity in locations other than central cities—a problem faced by travellers of any age.

11.7 Options and Opportunities

Government policy approaches to the transport challenges presented by an ageing population have struggled with a limited funding base, potentially enormous funding needs and complex coordination issues across agencies and levels of government. In practice all governments have limited capacity to address these complex issues. In Australia most mobility is self-funded and self-managed; it is not within the means of government to fill all mobility gaps for the population as a whole into the future.

Nevertheless, targeted interventions have occurred with some success in the past. As early as the 1990s the Federal Labor Government recognised that lack of transport was a major component in barriers to social justice (Travers Morgan 1992). Around a decade later the UK government included transport and its impacts on social exclusion as a major component of its national strategy. A significant outcome of this was the mandating of 'accessibility planning' tools in the development of all transport plans (Social Exclusion Unit 2003). In effect, planners were required to recognise transport disadvantage as a part of the development of plans to address transport futures.

The UK has more of a planning culture than Australia. In Australia transport planning is not mandated; rather plans are developed when authorities choose to make them. A review contrasting the approach in the UK with strategies in Victoria noted a considerable difference in transport funding outcomes between the two governments in the late 2000s. At the time, the Victorian Labor government implemented considerable investments to provide transport solutions, while the UK approach involved changes in processes, without significant investment (Lucas and Currie 2012). Another point about these approaches is that they address all forms of mobility challenges, of which only some concern older Australians.

From a regulatory perspective the Federal government's Disability Discrimination Act (Australian Government 1992) has acted to address accessibility barriers to transport, notably within cities with available public transport systems. At a state level, several policy-based solutions have sought to mandate consideration of future mobility challenges into policy development. In Victoria the Transport Integration Act now mandates that social implications of policy development be included in planning (Morris and Kinnear 2011).

At an operational level, all levels of government provide some form of transport for people without access to services; these can range from taxi subsidy schemes to concessionary fare schemes on public transport. Community transport, including local government and health vehicles, is another operational solution to fill gaps in transport. The provision of new public transport services to fill social gaps in access was a strategy developed in Victoria during the late 2000 Labor government era. Termed 'social transit' (Betts 2007), the concept was to ensure a more equitable spread of bus service provision in outer suburban and regional areas to provide a minimum level of service for access by the community.

All Australian States are also seeking to better inform older Australians of the need to plan ahead for mobility when driving is no longer an option (e.g. Department of Infrastructure 2007).

From a research perspective there have also been recent innovations in how improved mobility can be valued in transport planning processes such as cost–benefit analysis (Currie et al. 2009; Stanley et al. 2011). This means agreed national processes for the evaluation of transport project proposals can include some quantification of what was previously anecdote-based evidence of mobility benefits.

Overall however, these solutions may well be considered relatively small in relation to the scale of the mobility problems emerging for Australia's older generations. Baby boomers will continue to 'age in place', often in rural, regional and outer urban locations, which will require a continued dependence on private cars for mobility. Australia has never had planning laws which effectively curtail development in these areas and it is now common for residential homes for the aged to be built in cheaper fringe/remote locations away from activities and alternative transport options. This has led some to suggest that Australia is going to face 'mobility crises' (Currie 2009) into the future, with serious negative implications for the quality of life of older generations.

11.8 Conclusion

This chapter has shown that mobility and access are essential to healthy and happy ageing. Australians have always faced challenges in maintaining mobility as they age; however, the ageing of the baby boomers, the single largest generation in history, presents unique challenges going forward. In the short term the mobility that cars provide to baby boomers will enable them to maintain their mobile lifestyles. But as ageing, disability and affordability begin to weigh against the car, Australia must be willing to pursue personal, policy and planning alternatives.

References

Australian Bureau of Statistics. (2011). *Australian demographic statistics* (Cat No. 3101.0). Canberra, Australia: ABS.

Australian Government. (1992). *Disability discrimination Act 1992*, (Australia) Act No. 135 of 1992. Australia.

Banister, D., & Bowling, A. (2004). Quality of life for the elderly: The transport dimension. *Transport Policy, 11*, 105–115.

Betts, J. (2007). Transport and social disadvantage in Victoria: A government perspective. In G. Currie, J. Stanley, & J. Stanley (Eds.), *No way to go: Transport and social disadvantage in Australian communities*. Melbourne, Australia: Monash University ePress.

Browning, C., & Sims, J. (2007). Ageing without driving: Keeping older people connected. In G. Currie, J. Stanley, & J. Stanley (Eds.), *No way to go: Transport and social disadvantage in Australian communities*. Melbourne, Australia: Monash University ePress.

Coughlin, J. F. (2009). Longevity, lifestyle and anticipating the new demands of aging on the transportation system. *Public Works Management & Policy, 13*(4), 301–311.

Currie, G. (2009). Australian urban transport and social disadvantage. *Australian Economic Review, 42*(2), 201–208.

Currie, G. (2010). *Transport perspectives on ageing, social exclusion and well being*. Clayton, Australia: Monash Research for an Ageing Society Forum, Monash Club.

Currie, G. (Ed.). (2011). *New perspectives and methods in transport and social exclusion research*. Bingley, England: Emerald.

Currie, G., & Allen, J. (2007). Australians with disabilities: Transport disadvantage and disability. In G. Currie, J. Stanley, & J. Stanley (Eds.), *No way to go: Transport and social disadvantage in Australian communities*. Melbourne, Australia: Monash University ePress.

Currie, G., & Delbosc, A. (2010). Exploring public transport usage trends in an ageing population. *Transportation, 37*, 151–164.

Currie, G., Richardson, T., Smyth, P., Vella-Brodrick, D., Hine, J., Lucas, K., et al. (2009). Investigating links between transport disadvantage, social exclusion and well-being in Melbourne: Preliminary results. *Transport Policy, 16*(3), 67–105.

Currie, G., Stanley, J., & Stanley, J. (Eds.). (2007). *No way to go—Transport and social disadvantage in Australian communities*. Melbourne, Australia: Monash University ePress.

D'Ambrosio, L. A., Coughlin, J. F., Pratt, M., & Mohyde, M. (2012). The continuing and growing importance of mobility. In J. F. Coughlin & L. A. D'Ambrosio (Eds.), *Aging America and transportation: Personal choices and public policy*. New York: Springer.

Davison, G. (2004). *Car wars: How the car won our hearts and conquered our cities*. Sydney, Australia: Allen & Unwin.

Delbosc, A., & Currie, G. (2011a). The spatial context of transport disadvantage, social exclusion and well-being. *Journal of Transport Geography, 19*(6), 1130–1137.

Delbosc, A., & Currie, G. (2011b). Transport problems that matter—Social and psychological links to transport disadvantage. *Journal of Transport Geography, 16*(1), 170–178.

Department of Infrastructure. (2007). *Maintaining mobility—The transition from driver to non-driver*. Melbourne, Australia: Policy and Intergovernmental relations Division.

Department of Transport. (2009). Victorian integrated survey of travel and activity (VISTA) 2007. State Government of Victoria. https://www.data.vic.gov.au/data/dataset/victorian-integrated-survey-of-travel-and-activity.

Diener, E., Emmons, R. A., Larsen, R. J., & Griffin, S. (1985). The satisfaction with life scale. *Journal of Personality Assessment, 49*(1), 71–75.

Dodson, J. (2007). Transport disadvantage and Australian urban planning in historical perspective. In G. Currie, J. Stanley, & J. Stanley (Eds.), *No way to go: Transport and social disadvantage in Australian communities*. Melbourne, Australia: Monash University ePress.

Dodson, J., & Sipe, N. (2006). Shocking the suburbs: Urban location, housing debt and oil vulnerability in the Australian city. Research Paper 8, Urban Research Program. Brisbane, Australia: Griffith University.

Fingerman, K., Pillemer, K., Silverstein, M., & Suitor, J. (2012). The baby Boomers' intergenerational relationships. *The Gerontologist, 52*(2), 199–209.

Frith, W., Mara, M. K., & Langford, J. (2012). *Demand for transport services: Impact on networks of older persons' travel as the population of New Zealand ages* (NZ Transport Agency Research Rep. 481). Wellington, New Zealand: NZ Transport Agency.

Gibson, K., Ozanne-Smith, J., Clapperton, A., Kitching, F., & Cassell, E. (2011). Targeted study of injury data involving motorised mobility scooters. Melbourne, Australia: Monash University, Australian Competition and Consumer Commission, Monash University Accident Research Centre.

Giuliano, G., Hu, H.-H., & Lee, K. (2003). *Travel patterns of the elderly: The role of land use* (Metrans Project 00-8). Los Angeles, CA: University of Southern California, School of Policy, Planning and Development.

Goulias, K. G., Blain, L., Kilgren, N., Michalowski, T., & Murakami, E. (2007). Catching the next big wave: Do observed behavioral dynamics of baby boomers force rethinking of regional travel demand models? *Transportation Research Record: Journal of the Transportation Research Board No. 2014* (pp. 67–75). Washington, DC: National Academies of Science, Engineering and Medicine

International Wellbeing Group. (2006). *Personal wellbeing index*. Waurn Ponds, Australia: Australian Centre on Quality of Life, Deakin University. http://www.acqol.com.au/reports/auwbi.php.

Kalata, J. (2005). *The effects of gasoline costs on U.S. Residents 50+*. Washington, DC: AARP. http://www.aarp.org/research/housing-mobility/transportation/gas_prices.html.

Li, H., Raeside, R., Chen, T., & McQuaid, R. W. (2012). Population ageing, gender and the transportation system. *Research in Transportation Economics, 34*, 39–47.

Lucas, K., & Currie, G. (2012). Developing socially inclusive transportation policy– transferring the United Kingdom policy approach to the state of Victoria? *Transportation, 39*(1), 151–173.

Lynott, J. (2006). *Senior travel: Understanding differences by the types of communities in which our aging population lives.* Transportation Research Board 85th Annual Meeting (Paper #06-2956). Washington, DC.

Marottoli, R. A., Mendes de Leon, C. F., Glass, T. A., Williams, C. S., Conney, L. M., Jr., Berkman, L. F., et al. (1997). Driving cessation and increased depressive symptoms: Prospective evidence from the new haven EPESE. *Journal of the American Geriatrics Society, 45*(2), 202–206.

Metz, D. (2003). Transport policy for an ageing population. *Transport Reviews, 23*(4), 375–386.

Millward, C. (1999). *Understanding links between family experience, obligations and expectations in later life* (Working Paper No. 19). Melbourne, Australia: Australian Institute of Family Studies.

Miranda-Moreno, L. F., & Lee-Gosselin, M. (2008). A week in the life of baby boomers: how do they see the spatial-temporal organization of their activities and travel? *Transportation, 35*, 629–653.

Mollenkopf, H., Baas, S., Marcellini, F., Oswald, F., Ruoppila, I., Szeman, Z., et al. (2005). Mobility and quality of life. In H. Mollenkopf, F. Marcellini, I. Ruoppila, Z. Szeman, & M. Tacken (Eds.), *Enhancing mobility in later life: Personal coping, environmental resources and technical support.* Amsterdam: IOS Press.

Morris, M., & Kinnear, R. (2011). Transport planning and policy perspectives. In G. Currie (Ed.), *New perspectives and methods in transport and social exclusion research.* Bingley, England: Emerald.

Olsberg, D., & Winters, M. (2005). *Ageing in place: Intergenerational and intrafamilial housing transfers and shifts in later life* (AHURI Final Rep. No. 88). Melbourne, Australia: Australian Housing and Urban Research Institute.

Ragland, D. R., Satariano, W. A., & MacLeod, K. E. (2005). Driving cessation and increased depressive symptoms. *The Journals of Gerontology Series A: Biological Sciences and Medical Sciences, 60*(3), 399–403.

Rose, G., & Richardson, M. (2010). *Operational impacts of alternative private passenger vehicles, for Austroads* (Austroads Project No. NS1205, Austroads Publication No. AP–R351/10). ISBN: 978-971-921709-921704-921703.

Rosenbloom, S. (1999). Mobility of the elderly: Good news and bad news. In Conference Proceedings 27: Transportation in an aging society: A decade of experience. Bethesda, MD: Transportation Research Board.

Rosenbloom, S. (2001). Sustainability and automobility among the elderly: An international assessment. *Transportation, 28*, 375–408.

Rosenbloom, S. (2011). Driving off into the sunset. In K. Lucas, E. Blumenberg, & R. Weinberger (Eds.), *Auto motives—Understanding car use behaviors.* Bingley, England: Emerald.

Rosenbloom, S. (2012). The travel and mobility needs of older people now and in the future. In J. F. Coughlin & L. A. D'Ambrosio (Eds.), *Aging America and transportation: Personal choices and public policy.* New York: Springer.

Social Exclusion Unit. (2003). *Making the connections: Final report on transport and social exclusion* (p. 145). London: Social Exclusion Unit, Office of the Deputy Prime Minister.

Spinney, J. E. L., Scott, D. M., & Newbold, K. B. (2009). Transport mobility benefits and quality of life: A time-use perspective of elderly Canadians. *Transport Policy, 16*(1), 1–11.

Srinivasan, N., McGuckin, N., & Murakami, E. (2006). Working retirement: Travel trends of the aging workforce. *Transportation Research Record: Journal of the Transportation Research Board, No., 1985*, 61–70.

Stanley, J., Henscher, D., Stanley, J., Currie, G., Greene, W., & Vella-Brodrick, D. (2011). Social exclusion and the value of mobility. *Journal of Transport Economics and Policy, 45*(2), 197–222.

Travers Morgan Pty Ltd. (1992). *Strategies to overcome transport disadvantage*. Report (Social Justice Research Program into Locational Disadvantage (Australia)); Department of the Prime Minister and Cabinet. Canberra, Australia: Australian Government Publishing Service.

Victorian Integrated Survey of Travel and Activity. (2001). *User manual volume 1 user's guide: A companion document to the VATS94–VATS99 databases* (pp. 1–96). Melbourne, Australia: Transport Research Centre.

Watson, D., Clark, L. A., & Tellegen, A. (1988). Development and validation of brief measures of positive and negative affect: The PANAS scales. *Journal of Personality and Social Psychology, 54*(6), 1063–1070.

Whelan, M., Langford, J., Oxley, J., Koppel, S., & Charlton, J. (2006). *The elderly and mobility: A review of the literature*. Clayton, VIC, Australia: Monash University Accident Research Centre.

Part III
Policy and Practice in Health and Care

Chapter 12
Health Policies for an Ageing Australia

Stephen Duckett

12.1 Introduction

It is now more than 50 years since Kenneth Arrow, Nobel prize winner in economics, demonstrated that the health care system is so riddled with peculiarities that ordinary markets (and market assumptions) don't work in the sector (Arrow 1963). Market failure in health care means that governments need to intervene through regulation, subsidies and/or direct provision, to ensure that health care needs are met.

An important focus of health policy is about redistribution (the 'Robin Hood' function), combating social exclusion so that people who are less well off can get access to necessary care. Another important aim of health policy (effected through government subsidies and provision of health care) is to allow equalisation of payments for health care over the life cycle (a 'piggy bank' function). The latter function is about equalising contributions to public funding (via tax) over a person's life, even though health needs are typically greatest after the person retires and has less income to pay for any care needs (Barr 2001).

Australian Bureau of Statistics' (ABS) data show that average household total spending for people aged over 65 years is about 60 % of that of the whole adult population (ABS 2011). Given that average weekly health care spending by people aged over 65 is about the same as the all household average ($69 versus $66 per week), this means that health care spending actually makes up a larger proportion of weekly spending for the over 65 age group (9.5 % on health care) compared to the whole population (5.3 %). For pharmaceuticals, spending by people aged over 65 is considerably higher than the average ($30 per week versus $18), although the ABS

S. Duckett, PhD, DSc, FASSA, FAHMS (✉)
Health Program, GRATTAN Institute, 8 Malvina Place, Carlton, VIC 3053, Australia
e-mail: stephen.duckett@grattan.edu.au; http://www.grattan.edu.au/home/health

K. O'Loughlin et al. (eds.), *Ageing in Australia*, International Perspectives on Aging 16, DOI 10.1007/978-1-4939-6466-6_12

has highlighted that this data point has a high relative standard error, suggesting that there is significant variability in the spending patterns.

This brief analysis of health spending by households reveals some of the strengths and weaknesses of Australia's health policies:

- The 'piggy bank' function is working: despite older people's significantly greater need for and use of health care, government subsidies to (or direct provision of) key aspects of health care means that average spending is roughly the same for older people as the rest of the population; but
- The 'Robin Hood' function may not be working so well. Spending patterns vary so some subsets of older people may face significant out-of-pocket costs to access health care.

Other chapters in this book focus on components of and specific issues relevant to health policy, for example demography of ageing, health of older people and primary health care. In this chapter, the focus is on policy related to one of the most expensive parts of the health system: hospitals.

12.2 Hospital Utilisation

Older people are hospitalised more frequently than younger people and for different conditions (see Fig. 12.1).

Eight Major Diagnostic Categories account for about two-thirds of all admissions; the relative importance of these varies with age. For example, pregnancy, the fourth most frequent reason for admission for people under 75, obviously does not figure as a reason for admission in older age groups.

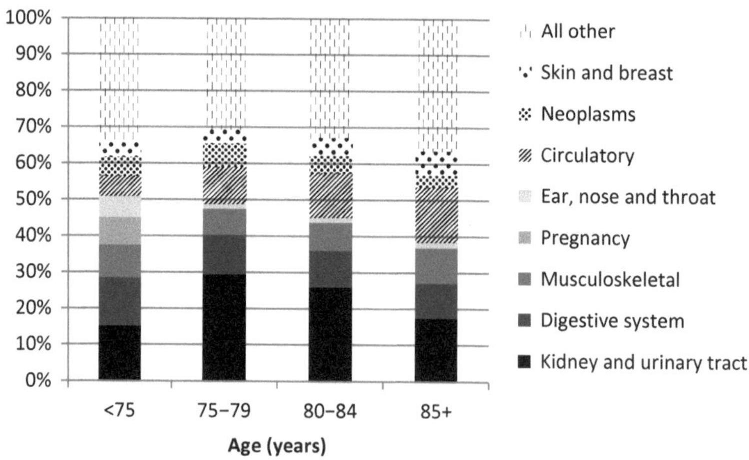

Fig. 12.1 Australia: reasons for hospital admissions (major diagnostic category), 2009–2010, by selected age groups (AIHW 2011)

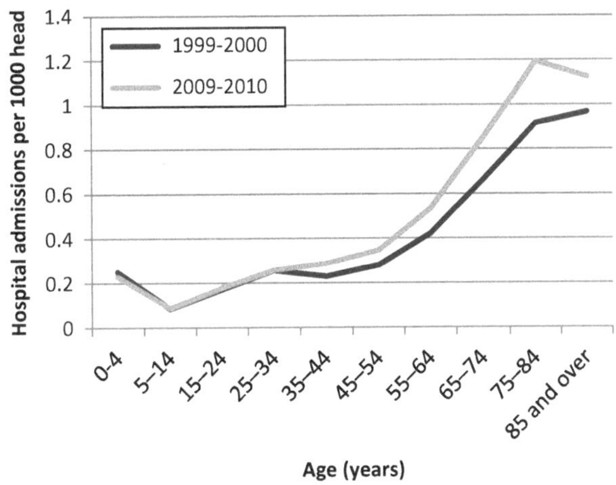

Fig. 12.2 Australia: hospital admissions per thousand population, 1999–2000 and 2009–2010 (AIHW 2011)

For people aged under 75 years, diseases and disorders of the kidney and urinary tract account for about 15 % of admissions (the most frequent condition being admissions for dialysis). However, for people in the 75–79 years age group, kidney and urinary tract conditions account for almost twice this proportion (29 %), with high proportions in this Major Diagnostic Category also at higher age groups.

Circulatory diseases and diagnoses account for 6 % of admissions for people under 75, increasing to 10 % of admissions for people in the 75–84 age group, 12 % for people aged 80–84 and 15 % of admissions for people aged 85 and over.

Health care use increases with age, and hospital utilisation is no exception to that pattern (see Fig. 12.2).

Figure 12.2 shows the typical U-shaped pattern of health care utilisation: people 85 and over are admitted to hospitals at more than three times the Australian average of 0.37 admissions per 1000 head. The increase in hospital use from around age 45 to 54 is a steady one: there is no age cut-off below which everyone is healthy and above which everyone uses hospitals extensively. This has a number of policy implications, for example 'healthy ageing' policies (discussed in Chap. 5), are increasingly relevant from the 45–54 age group upwards, not just to those who have retired.

Figure 12.2 also shows the increase in hospital utilisation at older age groups is a continuation of a long-term trend of increasing admissions (Gray et al. 2004; Travers et al. 2008). The admission rate for people aged 75–84 increased by 30 % over the decade and by 16 % for those aged 85 and over. At first glance it might seem that the health of older people has deteriorated over the decade. However, utilisation statistics are affected by both consumer characteristics such as health status as well as provider factors. The most obvious provider factor is supply: clearly if services are not provided they cannot be used. However, utilisation can also be affected by provider practice patterns and choices. A more detailed examination of the utilisation data shows that a significant proportion of the increase in admissions was for renal dialysis. These admissions are mostly same day and subject to reporting artefacts: some states count renal dialysis treatments as same-day admissions whilst others count them as outpatient attendances. There may also have been changes in these reporting practices over the decade. We don't know how much of the increase in renal dialysis admissions is real and how much is due to a switch in reporting practices. But if we take out the renal dialysis effect, the increase in admission rate for 75–84-year-olds was less than half a per cent and about 3 % for 85 and over. Even if all the renal dialysis effect is real, because renal dialysis admissions are relatively less costly than other admissions, the total cost impact of this growth is not as great as the raw data might suggest.

12.2.1 The 'Ageing Apocalypse'?

This simple analysis of hospital utilisation rates gives some insight into the impact of ageing on the health system. Certainly there is an ageing effect: there will be a larger number of older people in the future, and older people do use health care more than younger people. Unfortunately, it is often this visible reality that is the only basis for apocalyptic projections of the future demand for health care. However, if we look at the Australian data, adjusted for the probable renal dialysis effect, there is an almost imperceptible change in the utilisation rate, so the ageing effect is entirely about growth in the numbers of older people, a slow and steady effect not warranting the cataclysmic language so often used in the media.

The simplistic 'more older people, more problems' analysis is often all there is behind the forewarnings of an 'ageing avalanche' or the 'silver tsunami' that will overwhelm the health system: Schulz (Schulz 1998) disparagingly refers to the basis of this argument as "voodoo demographics" (p82). Reputable demographers now see a major part of their job as being to "remove the preconceptions and inaccuracies" that surround debates about the impact of ageing (Leone 2010, p. 4). Distressingly, the apocalyptic argument implicitly conveys the impression that older people are simply a burden, belying any contributions in their past or their contemporary roles. The apocalyptic scenario also ignores the reality that, over the long term, increased life expectancy leads to increased gross domestic product (GDP) and GDP per capita (Swift 2011), which in turn will make increased health costs more affordable.

The small and gradual impact of ageing, referred to as a 'glacier' in the title of one paper (Barer et al. 1995), leads to the conclusion that any impact of ageing can be absorbed as part of the normal evolution and innovation in the health system. If a 'grey glacier' is the right analogy, then the appropriate policy response to this gradual impact of ageing is incremental change over time, rather than a single tectonic shift.

12.3 Policy Directions

Two main types of strategies have been pursued to reduce demand for acute hospital care amongst the elderly: prevention and diversion.

12.3.1 *Preventable Admissions*

About 1 in every 12 hospital admissions is estimated to be partly or wholly preventable (Page et al. 2007). Health conditions where good primary care or public health might eliminate or reduce the incidence of hospitalisations include those preventable by immunisation, diabetic complications (e.g. foot amputations), asthma and congestive heart failure (Ansari 2007; Ansari et al. 2006). The prevalence of preventable admissions (the technical term is 'ambulatory care sensitive conditions') increases with age (see Fig. 12.3).

More than one quarter (27 %) of all potentially avoidable admissions occur among people aged over 75, compared to the 17 % of total admissions accounted for by this age group. One out of every eight admissions of a person aged over 75 is either wholly or partly preventable with good primary care or public health interventions.

A number of strategies have been developed to facilitate early detection of older people at risk of hospitalisation and to prevent hospital admissions or readmission (Codde, Arendts, et al. 2010; Codde, Frankel, et al. 2010; Hillen et al. 2011; Katterl et al. 2012). Health policies operating at the interface with aged care policies have also been pursued, including developing better health services in residential aged care facilities and strategies to reduce hospital admissions from residential aged care facilities.

12.3.2 *'Diversion' Strategies*

Older people tend to have a longer length of stay in hospital than younger people and are at risk of remaining in hospital when there is no longer a medical reason to do so because their previous support arrangements may be inadequate. 'Diversion' strategies are designed to ensure that people who no longer need acute inpatient care

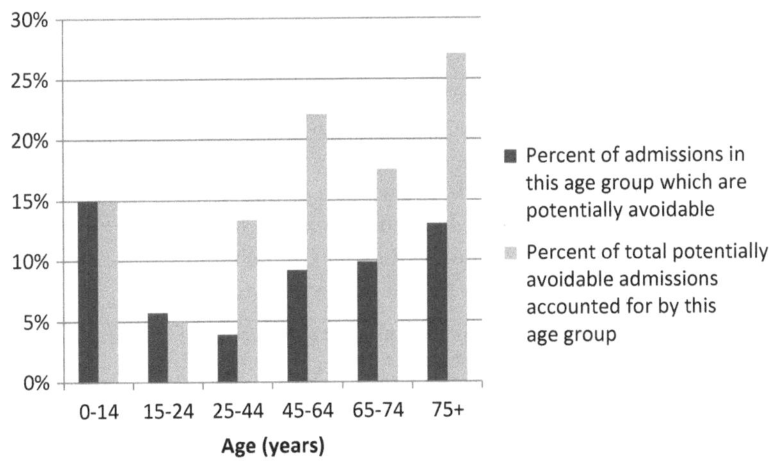

Fig. 12.3 Australia: ambulatory care sensitive conditions by age group, 2009–2010 (AIHW 2011)

are able to go home or to other accommodation (such as a residential aged care facility) as soon as possible.

Access to residential aged care facilities is not evenly distributed across Australia. Nor has residential aged care provision kept pace with demand, resulting in long waiting times for admission (Gray et al. 2006). One policy response has been to invest in 'transition programmes' to facilitate the hospital-aged care service transition. Ensuring that it is a person's care needs which determine their care location (i.e. hospital, residential aged care facility, home) should be a primary objective of both health and aged care policy. However, this is not always the case and development of transition programmes in Australia has had a political dimension mired in the politics of Commonwealth–State relations. Cameron, Crotty et al. (Cameron et al. 2010) have argued that:

> …it is likely that the establishment of the (national transition) program was driven by tensions between the Australian Government which funds residential aged care, and the States who fund public hospital care. Both levels of government believed that cost shifting was occurring. The States' hospitals said that the Australian Government was responsible for the lack of residential aged care beds which was producing long queues of older people in hospital waiting for these beds, resulting in blockages to hospital flows and access block in emergency departments. The Australian Government considered that State hospitals were discharging older people rapidly to residential aged care beds to avoid paying for subacute services (p. 147).

The efficacy (and cost-effectiveness) of transition programmes has been mixed (Cameron et al. 2010; Hall et al. 2012) and there is concern that a focus on specific transition care programmes of unproven effectiveness may detract from investment in inpatient sub-acute services of proven effectiveness (Gray et al. 2008). There is more evidence that structured post-acute care programmes and sub-acute provision are efficacious (Lim et al. 2003). However, these programmes have not been as successful in attracting Commonwealth funding support, possibly because of the political factors described earlier, as they operate in an acute care paradigm and the Commonwealth government has eschewed direct support for such services. Older people may also require longer periods of rehabilitation, so a further diversion strategy is to transfer patients to sub-acute units which specialise in rehabilitation or geriatric care.

Some 20 years after the term sub-acute was formally introduced into Australia, there is still ambiguity as to its meaning (Poulos and Eagar 2007). Sub-acute services such as rehabilitation, geriatric evaluation and management, and psychogeriatric care are the 'Cinderellas' of the hospital system: they have suffered neglect and under-investment by both State and Commonwealth governments.

The National Health and Hospitals Reform Commission (National Health and Hospitals Reform Commission 2009), for example, noted that:

> Historically, our health system has focused on tackling the immediate risk of people dying from acute conditions, but has invested less in the ongoing, and sometimes slow, process of helping people recover and reduce the impact of complications following an acute illness (p. 105).

The Commission concluded:

Table 12.1 Australia: sub-acute and non-acute admissions: total patient days and average length of stay in public and private hospitals, 2010–2011, by type of care (AIHW 2011)

	Public hospitals			Private hospitals		
Care type	Separations	Patient days	Average length of stay	Separations	Patient days	Average length of stay
Rehabilitation	86,426	1,501,869	17.4	200,808	964,215	4.8
Palliative care	28,255	319,659	11.3	5507	67,142	12.2
Geriatric evaluation and management	26,484	507,556	19.2	77	575	7.5
Psychogeriatric care	2445	120,869	49.4	6336	43,758	6.9
Maintenance care	20,889	711,297	34.1	2665	46,101	17.3
Total	164,499	3,161,250	19.2	215,393	1,121,791	5.2

> [M]any parts of Australia have limited or poorly developed sub-acute services. This means, for example, that people may not get adequate rehabilitation following a stroke or a heart attack or a hip replacement to allow them to return to as active a life as possible (p. 105).

Commonwealth government neglect can be seen in the poor development of data and recording systems to describe sub-acute care. The Commonwealth has supported and maintained a classification system for acute care (Diagnosis Related Groups) for the last 25 years (Duckett 2008). With this development it is possible to describe, cost and pay for acute services in a nationally consistent way. There has been no similar support for a classification system for sub-acute services, and there is no nationally consistent government collection of detailed sub-acute data.

Sub-acute and non-acute services are important components of the health system accounting for about 17 % of all public hospital bed days and 13 % of all private hospital bed days. The length of stay in public hospital units is substantially longer than in private facilities, averaging about 19 days (see Table 12.1)

Public and private hospitals have substantially different sub-acute roles:

- More than 90 % of sub-acute admissions and over 85 % of all sub-acute days in private hospitals are in rehabilitation units or programmes. In contrast, sub-acute days in public hospitals are spread over the five care types. Rehabilitation accounts for just over half of admissions and just under half of public hospital sub-acute bed days.
- The length of stay in public and private hospitals is so different that it suggests that the types of patient being admitted are quite different; that is, the differences are so great that they are probably not simply efficiency differences. In rehabilitation, for example, the length of stay in public hospital units is more than three times longer than in private hospital units.

There are also substantial differences between States in sub-acute provision (see Fig. 12.4).

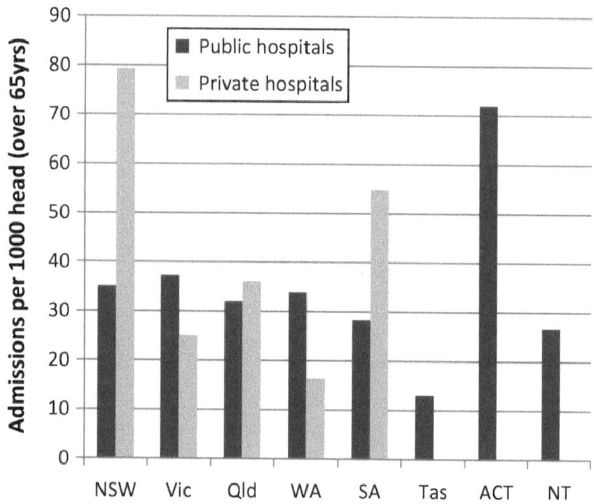

Fig. 12.4 Australia: sub-acute and non-acute services, admissions per thousand population aged over 65, public and private hospitals, 2010–2011 (AIHW 2011)

The rate of admission (public plus private) to sub-acute and non-acute services, for people over age 65 in New South Wales, the State with the highest rate, is more than twice that of Western Australia which has the lowest admission rate (114 versus 50 admissions per thousand population). The differences are even greater when comparing public hospital services alone: there is more than a fivefold difference between the Australian Capital Territory, with an admission rate of 72 per thousand population over age 65, and Tasmania, which has an admission rate of 13 per thousand. There is a thirty per cent difference between larger States: Victoria at 37 admissions per thousand population over age 65, versus South Australia at 28.

As mentioned earlier, the differences between public and private hospital provision are so great that one has to assume there are different practice patterns at play. Here the lack of standardised statistics means that one can only speculate as to the reasons for such a wide variation in admission rates: one hypothesis is that New South Wales' private hospitals are counting same-day rehabilitation programme activity as admissions, whereas private hospitals in other States either do not count in that way or do not provide these programmes.

There are also significant intra-State differences. Giles, Halbert et al (Giles et al. 2009) analysed sub-acute bed provision per thousand older people within each region served by an aged care assessment service. The box plot in Fig. 12.5 shows the median bed ratio as the horizontal line within the box; the outer edges of the box show the upper and lower quartile points, with the vertical lines extending from the box showing the range of most of the observed provision data outside these two bounds, with a handful of extreme regions highlighted separately as 'outliers'. As there is only one Aged Care Assessment Service serving the ACT, there is no box, simply a mark for the single observation (2.6 beds per thousand older population).

Comparing New South Wales and Victoria, for example, it can be seen that although the medians are very similar (2.85 beds per thousand older people in NSW compared to 3.1 in Victoria), there are few regions in New South Wales with sub-

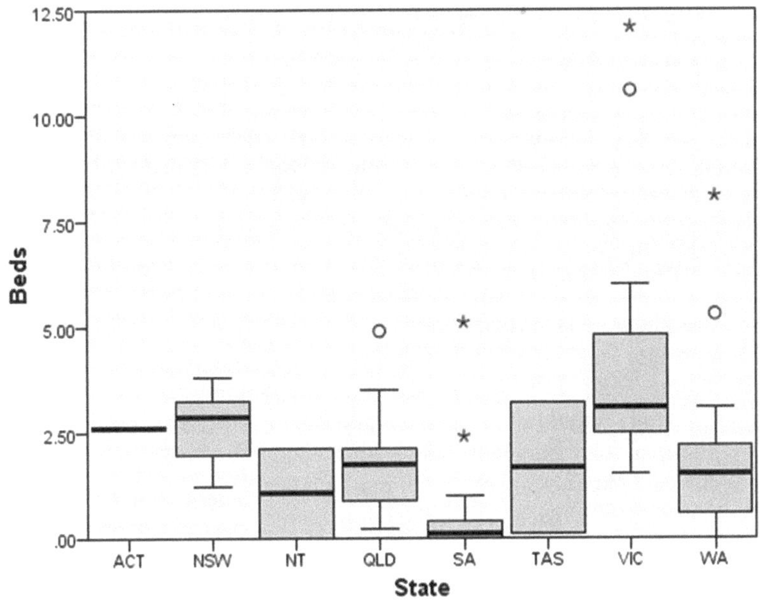

Fig. 12.5 Australia: sub-acute beds per thousand 'older population' (*), by aged care assessment service regions, 2005–2006 (*Data source*: Giles et al. 2009). * 'Older population' is all people over 70 and for Northern Territory includes Aboriginal people over 50

stantially higher or lower provision. The lower and upper quartiles for New South Wales are 2.0 and 3.2, respectively, a variation of just over 50 %. In contrast, one region in Victoria has more than 12 beds per thousand population, and more than 75 % of regions in Victoria have more than 2.5 beds per thousand (the lower quartile), with 25 % of regions having more than 4.8 beds per thousand, which is an almost twofold variation.

There are a large number of regions which have no beds available, and a number of other regions which have very low rates of provision. The significant variation in rates of provision per thousand older people leads to major differences in access across Australia, and to significant difficulties in gaining access to these services in the low provision regions, as was highlighted by the National Health and Hospitals Reform Commission (National Health and Hospitals Reform Commission 2009).

12.4 'The Road Not Travelled'

Although the emphasis on prevention and diversion strategies is laudable, current policies do not cover the full range of potential health policies available to address an ageing Australia. Two gaps can be used to illustrate this claim.

12.4.1 Discharge Planning and Access to Aged Care Assessment

Older people tend to stay longer in hospital. The average length of stay for a person aged over 85 in 2010–2011 was 6 days, compared to the overall average of 3.1 days. Older people are more likely to be discharged to a residential aged care facility (6 % for people aged over 85 versus less than 0.5 % for people aged under 85), or to another hospital (9 % versus 4 %). All of this highlights the importance of planning for older people to transition from acute care to home or another care setting.

Discharge planning is "the development of an individualised discharge plan for the patient prior to leaving hospital, with the aim of containing costs and improving patient outcomes" (Shepperd et al. 2013, p. 2). Contemporary best practice is that discharge planning should start before admission for elective procedures and as early as possible for emergency admissions. Good discharge planning reduces length of stay and readmission rates, but most studies are not well enough designed to make definitive conclusions about the impact of discharge planning on outcomes (Hansen et al. 2011; Shepperd et al. 2010, 2013).

Despite the importance of discharge planning, robust measures of hospital discharge planning performance have not been developed (Jha et al. 2009). Anecdotal evidence, and evidence from the literature, suggests that discharge planning is not always delivered well (Bauer et al. 2009).

Although much of the literature on discharge planning is relatively weak, a United States study (Jack et al. 2009) reported a 'reengineering' of discharge planning based on three principles:

- Clearly delineating the roles and responsibilities of everyone on the health care team.
- Providing patient education throughout the hospitalisation.
- Ensuring easy flow of information from the patient's doctor to the hospital team and back to the doctor, including a written discharge plan.

The revised model showed clear reductions in readmissions. The results have been promoted by the United States' Agency for Healthcare Research and Quality as a model for other US hospitals (Clancy 2009). The key elements of the revised model are shown below in the 'Practice tips' box.

Practice tips: Good practice in discharge planning includes

- educating patients about their diagnoses throughout the hospital stay;
- making appointments for clinician follow-up and post-discharge testing, including making and coordinating appointments, discussing their importance with the patient, and confirming transportation arrangements;
- discussing any tests or studies that have been completed in the hospital and deciding who is responsible for follow-up;

- organising post-discharge services, including making appointments and discussing how to receive each service;
- confirming the medication plan and making sure patients understand changes in routines and which side effects to monitor;
- reconciling the discharge plan with national guidelines and critical pathways;
- reviewing steps to take if a problem arises, such as whom to call and what constitutes an emergency;
- expediting the discharge summary to the physicians and other services responsible for the patient's care after discharge;
- asking patients to explain in their own words the details of the discharge plan;
- giving patients a written discharge plan at the time of discharge explaining the reason for hospitalisation and information about medications and what to do if their condition changes; and
- phoning the patient 2–3 days after discharge to identify and resolve any problems.

Source: Clancy (Clancy 2009); reproduced with permission

There is no evidence that the Australian experience with discharge planning is any better than that in the United States, nor that the directions of good practice are any different (Scott 2010). Although there have been some small-scale Australian projects in this area (Bolch et al. 2005; Wilson et al. 2003), there has been no national, systematic approach to facilitate similar reengineering.

The Australian and New Zealand Society for Geriatric Medicine recommends that, as part of discharge planning, a "[C]omprehensive geriatric assessment should be performed in older patients, particularly those who have cognitive impairment, functional impairment, multiple comorbidities or complex social issues" (Australian and New Zealand Society for Geriatric Medicine 2009, p. 158). Just as sub-acute beds are unevenly distributed, so too with specialist geriatricians. However, this maldistribution can be partly overcome by use of video and online technologies to provide aged care assessment services (Dakin et al. 2011; Gray and Wootton 2008; Gray et al. 2009). Developments such as this have not been part of policy to date.

12.4.2 *Improving Quality of In-Hospital Care for Older People*

Admission to hospital is a risky business with around 10 % of patients experiencing an adverse event while in hospital. In their study of admissions of people aged 40 and over, to Queensland and Victorian public hospitals in 2005–2006, Rowell (2010) found a prevalence of adverse events of 8 %. Older people were more at risk of adverse events than younger people: they found the prevalence in the

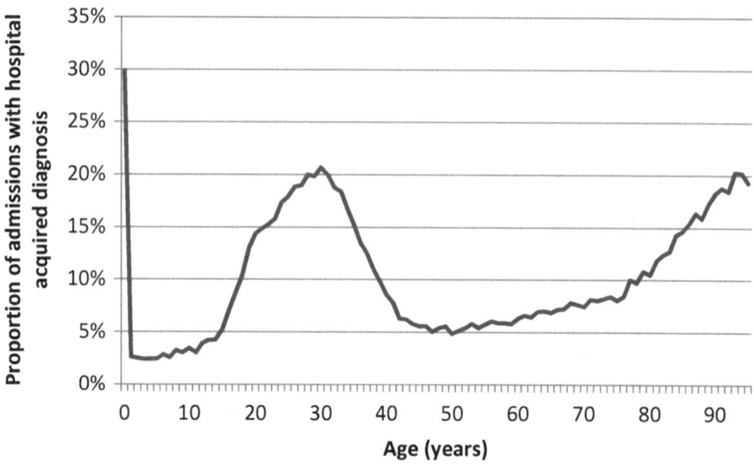

Fig. 12.6 Prevalence of hospital acquired diagnoses, by age, Victorian and Queensland public hospitals, 2005–2006 (*Source*: Rowell et al. 2010; Reproduced with permission)

40–70-year-old aged group was 7 % versus 11 % in the over 70 age group. Rowell et al. measured adverse events by whether an additional diagnosis was recorded during the course of the hospital admission ('hospital acquired' diagnoses). They found more than 200,000 additional diagnoses were recorded, 53 % of which were in patients over age 70; whilst patients over 70 only accounted for 37 % of total admissions. Figure 12.6 (taken from Rowell et al. 2010) shows the proportion of admissions which had an adverse event.

It can be seen that there are a significant number of adverse events in young adulthood, primarily associated with child birth. However, after age 50, there is a steady increase in the risk of adverse events with age. At age 90, for example, 18 % of all admissions have an adverse event. This suggests that an important (but unfortunately neglected) aspect of policy and practice should be to reduce the incidence of adverse events among the hospitalised elderly.

12.5 Conclusion

Health policy for an ageing Australia should be broad based, recognising the importance of the increased incidence of chronic disease and multi-morbidity, and the growing numbers of older people (see Chaps. 4 and 5). The main element of such a policy requires change to the primary health care system (see Chap. 13).

Hospital policies, the focus of this chapter, are also important. What has been shown here is that, contrary to the impression often conveyed in the media, Australia's ageing population is not a tsunami which will overwhelm the acute care system. There will be a glacial impact: large and significant but slow. In turn this means that policies to position the hospital system, and its interface with the aged care system, are required. Some policies relating to prevention and diversion have been implemented, but experience with these suggests they need to go further. Some important areas, such as minimising adverse events in the elderly, have yet to attract significant policy attention.

There are also policy and practice gaps which need to be addressed. The impacts of an ageing Australia on the demand for hospital care can be managed, but only if purposive action is taken on a regular basis. New policies need to be developed, piloted and implemented. Failure to do so may indeed lead to problems in the system, in which case they will be attributable to a failure in policy and foresight, not the consequences of a demographic time bomb.

12.6 Case Study: Ruth's Story

Ruth is an elderly (in her 90s) but feisty woman who has been living at home. She presented to a major teaching hospital with a swollen stomach. Following investigations in the emergency department, Ruth was transferred to a bed. A surgeon appeared the next day and his first words were "About this tumour…". This was the first time Ruth had been given any intimation of cancer. The surgeon then explained that Ruth would need a bowel operation, the result of which was that she would have a colostomy bag attached for the rest of her life for her faecal waste.

Ruth's family knew that her remaining life would be a misery with a bag, and she would probably not be able to manage at home any longer. The family was, through its networks, able to find the name of a surgeon with a better reputation and they arranged for Ruth to be transferred to another teaching hospital. This occurred at midnight, when Ruth was alone. She felt that she was being kidnapped and potentially could be murdered. That paranoia stayed with her for the next few days.

The outcome of the surgery was good, Ruth only had a temporary bag and she eventually went back home. In the meantime she was in a rehabilitation hospital and a nursing home. While in the nursing home Ruth ate little and lost a significant amount of weight, even though she was assessed as being able to feed herself.

The discharge process from the rehabilitation hospital, following the removal of the temporary bag, involved conflict between the family and the home care provider. Ruth's family was given an ultimatum that if they didn't agree to Ruth being discharged on a particular date, her access to a temporary (Commonwealth-funded) home support package would be withdrawn. No amount of pleading that Ruth was not quite ready for discharge changed this rule application. In the end, the family took the risk of poor home follow-up, kept her in hospital the few extra days and got her home for Christmas.

At home, Ruth's family eventually organised for Ruth to have regular visits from a nurse and someone to help her shower. Part of the nurse's role involved bandaging Ruth's lower leg (she has poor circulation and a chronic wound). Despite daily visits, the nurses didn't observe, or didn't report, that Ruth's big toe had become infected. Eventually Ruth started to complain, more doctor visits followed, including an emergency hospital admission. Ruth's toe was amputated because of a deep infection. Ruth returned home walking.

12.6.1 Discussion Points

- Would you describe the process of care Ruth received as good? What elements were not?
- If you assume that all the staff involved were well intentioned, why do you think the elements of care that weren't so good occurred?
- Could any of those elements have been prevented? How?

References

Ansari, Z. (2007). The concept and usefulness of ambulatory care sensitive conditions as indicators of quality and access to primary health care. *Australian Journal of Primary Health, 13*(3), 91–110.

Ansari, Z., Laditka, J. N., et al. (2006). Access to health care and hospitalization for ambulatory care sensitive conditions. *Medical Care Research & Review, 63*(6), 719–741.

Arrow, K. J. (1963). Uncertainty and the welfare economics of medical care. *American Economic Review, 53*, 941–973.

Australian Bureau of Statistics. (2011). *Household expenditure survey, Australia: Detailed expenditure items, 2009-10* (Cat. no. 6530.0). Canberra, Australia: ABS.

Australian Institute of Health and Welfare. (2011). *Australian Hospital Statistics 2009-2010* (AIHW Cat. no. HSE 107). Canberra, Australia: AIHW.

Australian and New Zealand Society for Geriatric Medicine. (2009). Australian and New Zealand Society for Geriatric Medicine Position Statement No. 15: Discharge Planning. *Australasian Journal on Ageing, 28*(3), 158–164.

Barer, M. L., Evans, R. G., et al. (1995). Avalanche or glacier?: Health care and the demographic rhetoric. *Canadian Journal on Aging, 14*(2), 193–224.

Barr, N. (2001). *The welfare state as piggy bank. Information, risk, uncertainty, and the role of the state*. Oxford: Oxford University Press.

Bauer, M., Fitzgerald, L., et al. (2009). Hospital discharge planning for frail older people and their family. Are we delivering best practice? *A review of the evidence. Journal of Clinical Nursing, 18*(18), 2539–2546.

Bolch, D., Johnston, J. B., et al. (2005). Hospital to home: An integrated approach to discharge planning in a rural South Australian town. *Australian Journal of Rural Health, 13*(2), 91–96.

Cameron, I. D., Crotty, M., et al. (2010). Whither transition care. *Australasian Journal on Ageing, 29*(4), 147–149.

Clancy, C. M. (2009). Reengineering hospital discharge: A protocol to improve patient safety, reduce costs, and boost patient satisfaction. *American Journal of Medical Quality, 24*(4), 344–346.

Codde, J., Arendts, G., et al. (2010). Transfers from residential aged care facilities to the emergency department are reduced through improved primary care services: An intervention study. *Australasian Journal on Ageing, 29*(4), 150–154.

Codde, J., Frankel, J., et al. (2010). Quantification of the proportion of transfers from residential aged care facilities to the emergency department that could be avoided through improved primary care services. *Australasian Journal on Ageing, 29*(4), 167–171.

Duckett, Stephen (2008), 'Casemix development and implementation in Australia', in John Kimberly, G De Pouvourville, and T d'Aunno (eds.), The Globalization of Managerial Innovation in Health Care (Cambridge University Press).

Dakin, L., Cutler, A., et al. (2011). Reliability of online geriatric consultation triage decisions: A pilot study. *Australasian Journal on Ageing, 30*(4), 239–240.

Giles, L. C., Halbert, J. A., et al. (2009). The distribution of health services for older people in Australia: where does transition care fit? *Australian Health Review, 33*(4), 572–582.

Gray, L., & Wootton, R. (2008). Innovations in aged care: Comprehensive geriatric assessment online. *Australasian Journal of Ageing, 27*(4), 205–208.

Gray, L. C., Broe, G. A., et al. (2006). Developing a policy simulator at the acute-aged care interface. *Australian Health Review, 30*(4), 450–457.

Gray, L. C., Travers, C. M., et al. (2008). Transition care: Will it deliver? *The Medical Journal of Australia, 188*(4), 251–253.

Gray, L. C., Wright, O. R., et al. (2009). Geriatric ward rounds by video conference: A solution for rural hospitals. *The Medical Journal of Australia, 191*(11-12), 7–21.

Gray, L. C., Yeo, M. A., et al. (2004). Trends in the use of hospital beds by older people in Australia: 1993-2002. *Medical Journal of Australia, 181*, 478–481.

Hall, C. J., Peel, N. M., et al. (2012). Can post-acute care programmes for older people reduce overall costs in the health system? A case study using the Australian Transition Care Programme. *Health & Social Care in the Community, 20*(1), 97–102.

Hansen, L. O., Young, R. S., et al. (2011). Interventions to reduce 30-day rehospitalization: A systematic review. *Annals of Internal Medicine, 155*(8), 520–528.

Hillen, J. B., Reed, R. L., et al. (2011). Hospital admissions from residential aged care facilities to a major public hospital in South Australia (1999–2005). *Australasian Journal on Ageing, 30*(4), 202–207.

Jack, B. W., Chetty, V. K., et al. (2009). A reengineered hospital discharge program to decrease rehospitalization: a randomized trial. *Annals of Internal Medicine, 150*(3), 178–187.

Jha, A. K., Orav, E. J., et al. (2009). Public reporting of discharge planning and rates of readmissions. *New England Journal of Medicine, 361*(27), 2637–2645.

Katterl, R., Anikeeva, O., et al. (2012). *Potentially avoidable hospitalisations in Australia: Causes for hospitalisations and primary health care interventions*. Adelaide, Australian: Primary Health Care Research & Information Service.

Leone, T. (2010). How can demography inform health policy? *Health Economics, Policy, and Law, 5*(1), 1–11.

Lim, W. K., Lambert, S. F., et al. (2003). Effectiveness of case management and post-acute services in older people after hospital discharge. *The Medical Journal of Australia, 178*(6), 262–266.

National Health and Hospitals Reform Commission. (2009). *A Healthier Future For All Australians—Final Report*. Canberra, Australian: National Health and Hospitals Reform Commission.

Page, A., Ambrose, S., et al. (2007). *Atlas of avoidable hospitalisations in Australia: Ambulatory care-sensitive conditions*. Adelaide, Australian: Public Health Information Development Unit, The University of Adelaide.

Poulos, C., & Eagar, K. (2007). Determining appropriateness of rehabilitation or other subacute care: is there a role of utilisation review? *Australian and New Zealand Health Policy, 4*(3), 1–7.

Rowell, D., Nghiem, H. S., et al. (2010). How different are complications that affect the older adult inpatient? *Quality and Safety in Health Care, 19*(6), 1–5.

Schulz, J. A. (1998). The economics and financing of long-term care. *The Australasian Journal on Ageing, 17*(1 Suppl), 82–84.

Scott, I. A. (2010). Preventing the rebound: Improving care transition in hospital discharge processes. *Australian Health Review, 34*(4), 445–451.

Shepperd, S., Lannin, N. A., et al. (2013). Discharge planning from hospital to home. *The Cochrane Database of Systematic Reviews, 1*, CD000313.

Shepperd, S., McClaran, J., et al. (2010). Discharge planning from hospital to home. *Cochrane Database of Systematic Reviews, 1*, CD000313.

Swift, R. (2011). The relationship between health and GDP in OECD countries in the very long run. *Health Economics, 20*(3), 306–322.

Travers, C. M., McDonnell, G. D., et al. (2008). The acute–aged care interface: Exploring the dynamics of 'bed blocking'. *Australasian Journal on Ageing, 27*(3), 116–120.

Wilson, B., Rogowski, D., et al. (2003). Integrated Services Pathways (ISP): a best practice model. *Australian Health Review, 26*(1), 43–51.

Chapter 13
Primary Health Care and Older People

Colette Browning, Jenny Davis, and Shane Thomas

13.1 Introduction

This chapter focuses on the important role that primary health care (PHC) plays in the health and well-being of older people. First, we examine the concept of PHC and how it operates in Australia, with an emphasis on the recent PHC reforms. These reforms have been driven, in part, by increasing multi-morbidity and the ageing of the Australian population (see Chaps. 4 and 5). We discuss how PHC is funded in Australia and provide a brief history of the sector. This discussion provides an important backdrop to the role of PHC in promoting healthy ageing. We next examine PHC service use by older people, issues around accessibility of after-hours PHC services for older people particularly as it relates to the use of emergency department services, and finally the problem of poor service integration in a system where health care and aged care are separate.

Primary health care is "the first level of contact individuals, families and communities have with the health care system" (Primary Health Care Research &

C. Browning (✉)
ARC Centre of Excellence in Population Ageing Research (CEPAR),
University of New South Wales, Kensington, NSW, Australia

Royal District Nursing Service Institute, 31 Alma Road, St Kilda, VIC 3182, Australia

International Institute for Primary Health Care Research, Shenzhen, China

Peking University, Beijing, China
e-mail: cbrowning@rdns.com.au

J. Davis
College of Healthcare Sciences, James Cook University QLD 4811, Australia

S. Thomas
International Institute for Primary Health Care Research, Shenzhen, China

Primary Health Care Research, Monash University, Clayton, VIC 3800, Australia

Peking University, Beijing, China

K. O'Loughlin et al. (eds.), *Ageing in Australia*, International Perspectives on Aging 16, DOI 10.1007/978-1-4939-6466-6_13

Information Service 2015). There are many different definitions of PHC but the definitive one is provided by the World Health Organization (WHO) (1978), in which PHC was defined as:

> Essential health care based on practical, scientifically sound and socially acceptable methods and technology made universally accessible to individuals and families in the community through their full participation and at a cost that the community and country can afford to maintain at every stage of their development in the spirit of self-reliance and self-determination. It forms an integral part both of the country's health system, of which it is the central function and main focus, and of the overall social and economic development of the community. It is the first level of contact of individuals, the family and community with the national health system bringing health care as close as possible to where people live and work, and constitutes the first element of a continuing health care process (p. 2).

Primary Health Care sits at the intersection of society and deeper levels of engagement with the formal health system including specialist care provided in hospitals. The Australian PHC system is a pivotal component of the overall health system.

13.2 Funding of Primary Health Care Services in Australia

In Australia, the majority of health care services are provided by the PHC sector, involving a broad range of services including general practice and allied health services, delivered across public, private and non-government providers (Australian Institute of Health and Welfare 2014b) amidst a complex mix of State and Commonwealth funding and governance arrangements (Australian Institute of Health and Welfare 2012). Government funding for Australian primary health services is mainly achieved through the Medicare system, which itself is undergoing comprehensive review.

In 2012–2013 PHC expenditure in Australia was approximately AUD53billion. In 2011–2012, 70 % of total health expenditure was provided by governments, with the Australian Government contributing 42 % and State and Territory governments providing 27 % of the total national spend. Patients (17 %), private health insurers (8 %) and accident compensation schemes (5 %) provided the remaining 30 % of PHC services. In Australia, approximately equal amounts of money have been spent on PHC and hospital care over the decade to 2013, with PHC expenditure accounting for around 38 % of the total Australian health spend and hospital expenditure for around 40 %. When adjusted for inflation the total spend on health and its constituent components has remained relatively constant over this period with modest overall increases but with some disinvestment by government being taken up by private individuals to achieve the same net spend result (Australian Institute of Health and Welfare 2014a).

13.2.1 The Primary Health Care Workforce

In Australia the PHC workforce consists principally of general practitioners (GPs), nurses and smaller numbers of allied health professionals (physiotherapists, occupational therapists, social workers, osteopaths, psychologists, speech pathologists, audiologists and dietitians), indigenous health workers, pharmacists, dentists, health promotion officers and paramedics (Australian Institute of Health and Welfare 2014b). In 2013, of the 95,000 registered medical practitioners in Australia 35 % were specialists, 33 % were GPs, 18 % were specialists in training and 12 % were hospital non-specialists (Australian Institute of Health and Welfare 2013). GPs are the main providers of PHC in Australia, also acting as gatekeepers to other health and specialist services (Gadzhanova and Reed 2007; Thomson et al. 2013).

The PHC workforce faces a number of challenges as a result of the projected ageing of the population, and resultant changes in the burden of disease, including a significant increase in demand for health services and aged care. Specifically, population ageing related health workforce challenges include: larger numbers of older people requiring services whilst the health care workforce is shrinking, due to both the ageing of its own workforce, and the reduction in both public and private revenue available to support these services, as the relative proportion of working age people falls (Health Workforce Australia 2012).

Existing health workforce shortages in regional, rural and remote areas of Australia are compounded by declining workforce numbers, maldistribution and changing demographics of the PHC workforce, particularly GPs and nurses (Health Workforce Australia 2012; Pain et al. 2014) and other allied health professionals (Health Workforce Australia 2012). It is noteworthy that 66 % of the Australian population live in capital cities, with the remaining 34 % living in regional/rural/remote areas. The sea change and tree change phenomena in retirement (see Chap. 10) is placing more pressure on regional and rural services (Australian Bureau of Statistics 2014).

13.3 Reforms in Primary Health Care Services

Although PHC service expenditure in Australia has been relatively stable, there have been a range of recent structural revisions to the delivery and funding of PHC services in Australia.

The Divisions of General Practice (DGPs) were first piloted by the Federal Government in 1992 to promote a local approach to improve the quality of health care services, particularly for disadvantaged groups, by encouraging GPs to work with other health care professionals (Todd et al. 1998). In response to the 2010 Labor Government's National Health Reforms, by 2011 DGPs had evolved into a national network of Medicare Locals (MLs). A particular objective was to improve the co-ordination and integration of services within PHC, and between PHC and other services, including acute (Local Hospital Networks (LHNs)) and aged care services, in order to improve the patient journey. Medicare Locals provided an expanded role for nurses, allied health practitioners, pharmacists and specialists. Specific initiatives under the National Health Reforms included health promotion and prevention activities, improved access to PHC by older people, improved access to mental health primary care services, expanded after-hours access and better management of patients with chronic illnesses.

Following a review commissioned by the Liberal-National Government Health Minister (Horvath 2014), MLs were replaced by Primary Health Networks (PHNs) in July 2015. The role of PHNs, as outlined by Horvath (2014), is "… to work with GPs, private specialists, LHNs, private hospitals, aged care facilities, Indigenous health services, non-government organisations and other providers to establish clinical pathways of care that arise from the needs of patients (not organisations) that will necessarily cross over sectors to improve patient outcomes" (p. 10). The review found variation in the performance of MLs on the key National Health Reform initiatives and that many patients were still experiencing uncoordinated care contributing to poor outcomes. Horvath (2014) recommended a refocus on the "paramount role" (p. ii) of general practice through the establishment of Clinical Councils with "significant" GP membership (p. 11). Consumer-led Community Advisory Committees were also to be established, reflecting the goal of placing patient needs ahead of organisational needs. Such committees would need to ensure adequate representation by older consumers. As was the case with MLs, PHNs are expected to improve health outcomes for local communities, and also to meet national key performance standards that are to be developed, including preventable hospital admissions, reduced smoking prevalence rates and targets for screening. While the functions of the different PHNs are the same, the organisational structures will vary according to local circumstances.

Thus, the current system of PHC in Australia has evolved over the last two decades to reflect the need for co-ordinated input from a range of PHC practitioners in order to respond to population ageing and increases in the prevalence of people with multiple chronic conditions. The revisions and reforms continue with the government's 2015 PHC review, which is ongoing. In April 2015 the Minister for Health and Sport announced the establishment of the Medicare Benefits Schedule (MBS) Review Taskforce (Taskforce). Its task is to consider how the use of more than 5,500 items on the MBS can improve the health of patients by aligning clinical evidence with practice (Australian Government Department of Health 2015). The review has attracted considerable controversy, with recent trenchant criticism from the Australian Medical Association concerning the perceived focus of the Review upon costs reduction.

Prior to the 2016 Federal election the Coalition government announced the planned implementation of the Health Care Home, a patient-centred approach to chronic illness management similar to models of primary care in Canada and the USA. In July 2016 a roundtable of key stakeholders endorsed the approach arguing that the model should be expanded beyond the management of chronic conditions (Consumer Health Forum, The George Institute for Global Health, RACGP and Menzies Centre, 2016). The model's key elements include informed, active patients and multidisciplinary team based coordinated care but it is unclear how the model will link health and aged care services for older people. The roundtable recommended the addition of social and life issues to the core elements, a key consideration for older people.

Thus, the PHC system in Australia is certainly a work in progress, but one that has, in the main, achieved its major objectives of accessible and equitable services for Australians, including older Australians. However, there are some emerging issues for older people in terms of after-hour access and service integration particularly across the health and aged care systems.

13.4 Older People and Primary Health Care Services in Australia

For the older patient access to the taxpayer subsidised primary health care system is crucial and many older Australians receive the majority of their health care encounters through this system. As in many developed countries the ageing of the population (see Chap. 4) and the increase in the prevalence of long-term chronic conditions (see Chap. 5) have placed pressure on the ability of the system to deliver affordable, accessible care for its older citizens. A number of the reforms in the primary care system discussed earlier have attempted to address some of these pressures including multi-professional case management approaches for complex conditions and health screening to identify opportunities for intervention. The aged care system is essentially separate from the health system but there have been attempts to provide better integration between the two under recent aged care reforms (see Chap. 14).

The inclusion of healthy ageing principles in population ageing has heralded a shift in the role of health services in the health and well-being of older people. Primary health care services for older people, largely delivered by medical practitioners, have traditionally focused on symptom management: how can we manage your blood pressure, what pain relief should we prescribe for your arthritis? However, the evidence is mounting that preventive approaches have the potential to assist older people in managing their health conditions and in delaying the onset of chronic illnesses. The National Health and Hospitals Reform Commission's (2009, p. 171) Final Report recommended that "… primary health care … be embedded as the cornerstone of our health system, reinforcing prevention, early intervention, and connected care." An accessible and responsive PHC system is an important component in maximising health and well-being for older people.

However, more recently there have been momentous changes proposed by the Australian government in the funding and delivery of PHC services. These changes have been controversial (Richardson 2014; Russell 2015) and key elements of them, such as compulsory co-payments, have failed to achieve the required support in the legislature. Such measures are especially salient for older people, particularly for those on fixed incomes.

13.4.1 Primary Care Service Use by Older People

Ageing populations are associated with multiple co-morbidities and an increasing prevalence of chronic disease, particularly diabetes and cardiovascular disease. Older Australians are proportionally heavy users of health services in the Australian health system. Various census studies have shown that older hospital occupants (aged 65 years plus) are consistently 4–5 times the proportion of such people in the community. This service use pattern, combined with demographic ageing in Australia, concentrates the mind on how increased demands upon the PHC system in particular, and the health system in general, are to be managed and funded.

These trends are placing demands on the way we prevent and manage illness and promote healthy ageing (Health Workforce Australia 2012). In Australia, people with chronic conditions and complex care needs have their health care needs managed under the Chronic Disease Management items in the Medicare insurance system. The GP prepares and reviews a management plan and co-ordinates team care with allied health professionals including physiotherapists, dietitians, psychologists and occupational therapists. However, complex funding systems exist, with duplication of assessment and care plan creation, and significant barriers to information sharing and communication between members of the care teams (Lawn et al. 2015). These system issues influence the effectiveness and efficiency of care (Australian Commission in Safety and Quality in Health Care 2011).

While older people can access these Chronic Disease Management services they also receive specialised services such as the 75+ Health Assessment (75+ HA). The 75+ HA is a service provided by GPs under the Enhance Primary Care package, introduced in 1999, which provides a comprehensive assessment of the older person's physical function including falls history, psychological and social well-being, cognitive impairment, blood pressure, medication, continence and immunisation record (Gray and Newbury 2004). It may also include assessment of nutritional and dental requirements. The patient and their carer are provided with a written report with recommendations for further investigations or interventions, including the need for community services. Such assessments are important for older people as they provide an opportunity for the GP to identify health issues and provide appropriate preventive health care and interventions at the early stages of risk. However, it is important that appropriate interventions are available following health assessments and that the patient is monitored regularly by health care professionals (Byles et al. 2004).

A recent 11-year analysis of the 75+ HA found that uptake was in the order of 20 % of the eligible population (Hamirudin et al. 2014). The low rate of 75+ HA is perhaps surprising given the frequency of GP visits by this age group. Those aged 75 and over comprise about 18 % of GP visits (Britt et al. 2014), with this figure having increased over the last 10 years (see Fig. 13.1). The average number of doctor visits, over the period 2005–2006, in those aged 75 years and over, was around nine per person compared with four per person for those aged under 65 years. So there are many opportunities for GPs to conduct the 75+HA. Hamirudin et al. (2014) suggest that the low rate of uptake is due to practices not routinely inviting older people to participate in health assessments, or fear on the part of the older person in confronting negative health conditions, such as cognitive impairment. However, those who do undertake a health assessment find the experience beneficial (Spillman et al 2012).

Interestingly, Hamirudin et al. (2014) found that uptake was higher in rural and regional areas, possibly due to older people having closer and more consistent relationships with health services in non-metropolitan areas.

The most common patient reason for an older person visiting a GP is for a check-up or a prescription (each accounts for around 20 per 100 encounters). GPs report that the most common managed conditions in older people are chronic conditions such as hypertension, diabetes, osteoarthritis and lipid disorders. Under the National

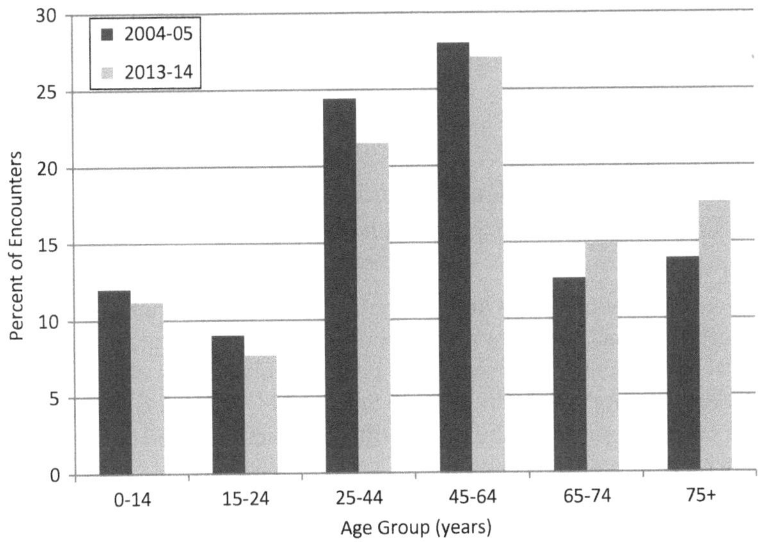

Fig. 13.1 Frequency of GP visits by age group in 2004–2005 and 2013–2014

Chronic Disease Strategy, GPs have a key role in the screening of risk factors for these chronic conditions.

However, concerns about the availability of GP services being adequate to meet the current and future needs of an ageing population are consistently reported in the literature. As is the case for the general community, access to GPs, and to PHC more broadly, by older persons, is often limited by practice capacity; affordability; available after-hours services including home visits (National Aged Care Alliance 2007); willingness to provide care for older people, including those living in aged care facilities (Comino et al. 2012; Taylor et al. 2013; Thomson et al. 2013); and by the number and availability of practitioners in rural and regional areas (Harris et al. 2012). These access issues contribute to the demand for acute services, particularly emergency departments (Harris et al., 2012).

One area of concern is the number of avoidable primary care-type emergency department (ED) attendances by older people. Avoidable presentations are defined as those presentations categorised as triage category 4 or 5 who did not arrive by ambulance, police, or correctional vehicle and, at the end of the episode, were not admitted to hospital or did not die. Across Australia around 15 % of ED attendance is by those aged 70 years and over (Lowthian et al. 2013). A recent study of presentations at ED by older people (REDIRECT) found that 15 % of these were avoidable (Mazza et al. 2015). Reasons for potentially avoidable attendance at ED included lack of availability of primary care services when needed, and lack of awareness of other service options, including locum services. There is little data available in Australia around the effectiveness of primary care interventions to reduce avoidable ED attendance. A recent systematic review (Ismail et al. 2013) provided little evidence concerning the effectiveness of primary care interventions in reducing ED attendance. However, the authors concluded there was limited evidence that the availability of emergency nurse practitioners in the community reduced ED attendance.

13.5 Primary Health Care: Aged Care Interface: The Problem of Poor Service Integration

Integration and coordination within Australian PHC itself, and at the interface with the broader health and aged care system is problematic. Consistent multi-disciplinary care collaboration and communication between sectors and across multiple providers is lacking (Australian Commission in Safety and Quality in Health Care 2011).

Older people living in the community often have multiple and complex points of contact with aged care and health service providers. For example, they may receive aged care services in the form of community care (see Chap. 14) as well as specific community nursing services delivered to their home. At the same time they may consult a pharmacist for medication management and be under the care of a GP who may have initiated a Chronic Care Plan that includes visits to a physiotherapist for a falls intervention. They may have been recently discharged from hospital after receiving treatment for a heart problem or may be a frequent attendee at EDs. Communication between the various providers is often limited due to difficulties in sharing patient records and the logistics of co-ordinating care plans.

The medical care of residents in aged care raises access issues for older people and has historically been provided by independent GPs without formal relationships to individual aged care providers or the broader aged care sector (Australian and New Zealand Society for Geriatric Medicine 2011; Gadzhanova and Reed 2007; O'Halloran et al. 2007). These 'unstructured' service models are increasingly questioned in relation to how well they meet the care needs of residents in aged care facilities (Australian and New Zealand Society for Geriatric Medicine 2011). Existing logistical challenges of primary medical care provision in residential aged care facilities are further complicated by institutional characteristics and systems of care (Arendts et al. 2010; Ching et al. 2014; Tuckett et al. 2014). Residential care facilities report particular difficulty accessing GP services and maintaining their input into routine services such as medication review (Gadzhanova and Reed 2007). The aged care sector more broadly is unattractive to GPs, particularly in relation to reimbursement, lack of equipment and supplies, and time-consuming administrative and clinical processes (Arendts et al. 2010; Gadzhanova and Reed 2007; Tuckett et al. 2014). However, improved access to primary care services for older people, particularly those living in residential care settings has been shown to reduce hospital transfers (Codde et al. 2010).

One potential solution to poor service integration is the use of Integrated Care Models. The National Health and Hospitals Reform Commission (2009) highlighted the need to integrate health and aged care services and recently the WHO (2015) launched the global strategy on people-centred and integrated health services. These models are designed for people with multiple chronic health conditions (often older people) who are at high risk of hospital and ED admissions. Integrated Care Model teams include GPs, specialists, allied health professionals, hospitals and community care services in a team-based approach that aims to treat the person rather than a particular medical condition and relies on electronic communication systems to share patient information to enhance care co-ordination. A core principle of integrated care is incorporating the consumer's perspective in service delivery (Shaw et al. 2011). In evaluating the evidence for integrated care models, the UK-based Kings Fund and Nuffield Trust concluded that there are significant benefits of these approaches, especially for those who experience poor service coordination, such as older people and those with multiple chronic illnesses (Goodwin et al. 2012).

In Australia, the New South Wales Government recently implemented an integrated care strategy that involves Local Health Districts working with primary, community and acute health services to coordinate services (NSW Health 2015). An electronic health record, HealthNet and real-time patient feedback systems will be implemented together with tools to identify people at risk of chronic illness to be targeted for early intervention.

13.6 Conclusion

The provision of accessible, effective and integrated PHC services for older people lies at the heart of the response to the health needs of older people. As has been noted in this chapter, the Medicare funding arrangements in Australia are undergoing major review at this time. This follows on from a series of attempted structural reviews of the PHC system itself. While continuous system improvement is an important expectation to be fulfilled by government, constant change in arrangements is neither helpful for older people attempting to negotiate their way through a complicated system, nor for the practitioners or those attempting to train practitioners to operate within the system.

The interests of older people, who are the major users of health services, have not been explicitly and centrally incorporated in these various reviews. This is concerning and may inadvertently lead to suboptimal results. Primary health care is intended to assist in the prevention of entry of people into more expensive secondary and tertiary care. It is literally the front line response of the health system to the health needs of the population. It is also an important preventive mechanism to assist older people to remain community dwelling and to not enter the very expensive residential aged care service system. As the major system users, the interests of older people must be reflected in reforms to ensure that they are effective and efficient for this key group. Chronic illness health management services, delivered through the PHC system, are pivotal to these efforts. We require a system that can deliver such initiatives with appropriate funding arrangements, clear and well-known access points and a skilled practitioner workforce to deliver them. To meet the anticipated demand for future services for older people and the associated workforce, traditional professional roles and responsibilities will be challenged as community and consumer expectations and new care models are required. A shift towards multi-disciplinary teamwork and integrated care models (Oliver-Baxter et al. 2014), and use of incentives more directly applied to groups of professionals and not individual providers are possible future scenarios. This is a key challenge that confronts the Australian health system. It will be interesting to see what role the proposed Health Care Home plays in addressing this challenge.

13.7 Case Study

Joyce Carter is a 75-year-old woman who lives alone in a small rural town 60 km east of Melbourne. In the last year, Joyce moved into a small unit after she had to sell the dairy farm she ran with her recently deceased husband. Joyce's children live in Melbourne but she has been reluctant to drive to Melbourne to see her children and grand children as she is concerned about her eyesight and worries about her safety on the rural roads. Joyce is active: she walks daily and is a keen cook and has a healthy diet. She manages her osteoarthritis well through exercise and pain medication. Joyce receives some community care services such as assistance with gardening and cleaning. However, her social networks are small and lately she has been feeling a bit lonely. She is still grieving for her husband. She has not seen the doctor in some time as her attitude when she lived on the farm was one of self-reliance and getting on with life. However, she is noticing some worrying symptoms such as shortness of breath and dizziness and her sleep patterns have been disturbed. She is also concerned about her ability to concentrate and sometimes she forgets what she is supposed to be doing. She has made an appointment to see her local doctor to discuss her concerns.

13.7.1 Discussion Points

- How might Joyce's GP approach and assess her concerns?
- Identify and discuss how the changes in Joyce's circumstances over the past year might be impacting on her current well-being.
- What types of services would assist Joyce? Consider medical, psychological and social support services.
- Consider how you would develop a care plan for Joyce that would facilitate the integration of her service needs.

Acknowledgement Britt H, Miller GC, Knox S, Charles J, Valenti L, Pan Y, Henderson J, Bayram C, O'Halloran J, Ng A 2004. General practice activity in Australia 2003–04. AIHW Cat. No. GEP 16. Canberra: Australian Institute of Health and Welfare (General Practice Series No. 16). Britt H, Miller GC, Henderson J, Bayram C, Harrison C, Valenti L, Wong C, Gordon J, Pollack AJ, Pan Y, Charles J. General practice activity in Australia 2013–14. General practice series no. 36. Sydney: Sydney University Press, 2014.

References

Arendts, G., Reibel, T., Codde, J., & Frankel, J. (2010). Can transfers from residential aged care facilities to the Emergency Department be avoided through improved primary care services? Data from qualitative interviews. *Australasian Journal on Ageing, 29*(2), 61–65. doi:10.1111/j.1741-6612.2009.00415.x.

Australian and New Zealand Society for Geriatric Medicine. (2011). Comprehensive Geriatric Assessment and Community Practice, *Position Statement No. 8*.

Australian Bureau of Statistics. (2014). *Population by Age and Sex, Regions of Australia (Cat. no. 3235.0)*. Canberra, Australia: ABS.

Australian Commission in Safety and Quality in Health Care. (2011). *Patient safety in primary health care. Consultation Report July 2011*. Sydney, Australia: Australian Commission in Safety and Quality in Health Care.

Australian Government Department of Health. (2015, 27 September). Medicare Benefits Schedule Review Taskforce. Retrieved 28 September, 2015, from http://www.health.gov.au/internet/main/publishing.nsf/Content/MBSReviewTaskforce

Australian Institute of Health and Welfare. (2012). *Australia's health 2012: In brief*. Canberra, Australia: AIHW.

Australian Institute of Health and Welfare. (2013, March 2015). Workforce. Retrieved 28 September, 2015, from http://www.aihw.gov.au/workforce/

Australian Institute of Health and Welfare. (2014a). *Health expenditure Australia 2012–13: Analysis by sector. Health and Welfare Expenditure Series no. 53. (AIHW Cat. no. HWE 62)*. Canberra, Australia: AIHW.

Australian Institute of Health and Welfare. (2014b). *The measurement of patient experience in Non-GP primary health care settings (AIHW Cat. no. 66)*. Canberra, , Australia: AIHW.

Britt, H., Miller, G., Henderson, J., Bayram, C., Valenti, L., Harrison, C., et al. (2014). *A decade of Australian general practice activity 2004-05 to 2013-14* General practice series no. 37, Retrieved from purl.library.usyd.edu.au/sup/9781743324233.

Byles, J. E., Tavener, M., O'Connell, R. L., Nair, B. R., Higginbotham, N. H., Jackson, C. L., et al. (2004). Randomised controlled trial of health assessments for older Australian veterans and war widows. *Medical Journal of Australia, 181*(4), 186–190.

Ching, J. L., Kwong, M. W.-L., Stuart, R. L., Buising, K. L., Friedman, N. D., Bennett, N. J., et al. (2014). Antibiotic prescribing practice in residential aged care facilities—Health care providers' perspectives. *Medical Journal of Australia, 201*(2), 101–105. doi:10.5694/mja13.00102.

Codde, J., Frankel, J., Arendts, G., & Babich, P. (2010). Quantification of the proportion of transfers from residential aged care facilities to the emergency department that could be avoided through improved primary care services. *Australasian Journal on Ageing, 29*(4), 167–171. doi:10.1111/j.1741-6612.2010.00496.x.

Comino, E. J., Davies, G. P., Krastev, Y., Haas, M., Christl, B., Furler, J., et al. (2012). A systematic review of interventions to enhance access to best practice primary health care for chronic disease management, prevention and episodic care. *BMC Health Services Research, 12*, 415.

Consumer Health Forum, The George Institute for Global Health, RACGP, Menzies Centre (2016) Patient-centred Healthcare Homes in Australia: Towards Successful Implementation, available at chf.org.au/pdfs/chf/Patient-centred-Healthcare-Homes-in-Australia---Towards-Successful-Implementation.pdf.

Gadzhanova, S., & Reed, R. (2007). Medical services provided by general practitioners in residential aged-care facilities in Australia. *Medical Journal of Australia, 187*(2), 92–94.

Goodwin, N., Smith, J., Davies, A., et al. (2012). I*ntegrated care for patients and poulations: Improving outcomes by working together*. Retrieved September 28, 2015 from http://www.kingsfund.org.uk/sites/files/kf/integrated-care-patients-populations-paper-nuffield-trust-kingsfund-january-2012.pdf.

Gray, L. C., & Newbury, J. W. (2004). Health assessment of elderly patients. *Australian Family Physician, 33*(10), 795–797.

Hamirudin, A. H., Ghosh, A., Charlton, K., Bonney, A., & Walton, K. (2014). Trends in uptake of the 75+ health assessment in Australia: a decade of evaluation. *Australian Journal of Primary Health, 21*(4), 423–428. doi: http://dx.doi.org/10.1071/PY14074

Harris, M. F., Davies, P. G. P., Fanaian, M., Zwar, N. A., & Liaw, S.-T. (2012). Access to same day, next day and after-hours appointments: The views of Australian general practitioners. *Australian Health Review, 36*(3), 325–330. doi:10.1071/AH11080.

Health Workforce Australia. (2012). Health Workforce 2025: Doctors, nurses and midwives—Vol. 1. Adelaide, Australia.

Horvath, J. (2014). *Report to the Minister for Health and Minister for Sport, Australia.* Barton, Australia: Commonwealth of Australia.

Ismail, S. A., Gibbons, D. C., & Gnani, S. (2013). Reducing inappropriate accident and Emergency Department Attendances: A systematic review of primary care service interventions. *British Journal of General Practice, 63*(617), e813–e820. doi:10.3399/bjgp13X675395.

Lawn, S., Delany, T., Sweet, L., Battersby, M., & Skinner, T. (2015). Barriers and enablers to good communication and information-sharing practices in care planning for chronic condition management. *Australian Journal of Primary Health, 21*(1), 84-89. doi: http://dx.doi.org/10.1071/PY13087

Lowthian, J., Curtis, A., Stoelwinder, J., McNeil, J., & Cameron, P. (2013). Emergency demand and repeat attendances by older patients. *Internal Medicine Journal, 43*(5), 554–560. doi:10.1111/imj.12061.

Mazza, D., Joe, A., Turner, L., Lowthian, J., Byles, J., Browning, C., et al. (2015). *Avoidable GP-type emergency department presentations by older people: Findings from the REDIRECT Study.* Paper presented at the 2015 Primary Health Care Research Conference, Adelaide, Australia.

National Aged Care Alliance. (2007). *National Aged Care Alliance Position Paper. Transport and access to health care services for older Australians* (p. 18): National Aged Care Alliance.

National Health and Hospitals Reform Commission. (2009). *A healthier future for all Australians—Final report of the National Health and Hospitals Reform Commission.* (P3-5499). Canberra, Australia: Australian Government.

NSW Health. (2015). *Integrated care.* Retrieved September 28, 2015 from: http://www.health.nsw.gov.au/integratedcare/pages/default.aspx.

O'Halloran, J., Britt, H., & Valenti, L. (2007). General practitioner consultations at residential aged-care facilities. *Medical Journal of Australia, 187*(2), 88–91.

Oliver-Baxter, J., Brown, L., Raven, M., & Bywood, P. (2014). *Quality improvement financial incentives for general practitioners. PHCRIS policy issue review.* Adelaide, Australia: Primary Health Care Research & Information Service.

Pain, T., Stainkey, L., & Chapman, S. (2014). AgedCare+GP: Description and evaluation of an in-house model of general practice in a residential aged care facility. *Australian Journal of Primary Health, 20*, 224–227.

Primary Health Care Research & Information Service. (2015). *PHCRIS getting started guides: Introduction to…primary health care.* Retrieved September 28, 2015, from http://www.phcris.org.au/guides/about_phc.php.

Richardson, J. (2014). *Six things you should know about the co-payment scheme.* Retrieved from The Conversation website: http://theconversation.com/six-things-you-should-know-about-the-co-payment-scheme-35329.

Russell, L. (2015). *The AMA and Medicare: A love-hate relationship.* Retrieved from The Conversation website: http://theconversation.com/the-ama-and-medicare-a-love-hate-relationship-36346.

Shaw, S., Rosen, R., & Rumbold, B. (2011). *What is Integrated Care?* London, England: Nuffield Trust. Retrieved from http://www.nuffieldtrust.org.uk/sites/files/nuffield/publication/what_is_integrated_care_research_report_june11.pdf.

Spillman, M., Kimber, D., & Cheffins, T. (2012). Patient views of over 75 years health assessments in general practice. *Australian Journal of Advanced Nursing, 29*(4), 5–10.

Taylor, M. J., Edvardsson, D., Horey, D., Fetherstonhaugh, D., Nay, R., & Swerissen, H. (2013). General practitioner service provision in residential aged care facilities: 1998–2011. *Australasian Journal on Ageing, 32*(1), 56–59. doi:10.1111/j.1741-6612.2012.00647.x.

Thomson, S., Osborn, R., Squires, D., & Jun, M. (Eds.). (2013). *International profiles of health care systems, 2013. Australia, Canada, Denmark, England, France, Germany, Italy, Japan, The Netherlands, New Zealand, Norway, Sweden, Switzerland, and The United States.* The Commonwealth Fund, November 2013.

Todd, R., Sibthorpe, B., & Todd, C. (1998). *Profile of divisions of general practice: 1995/96.* Commonwealth Department of Health and Family Services.

Tuckett, A., Parker, D., Clifton, K., Glaetzer, K., Greeve, K., Israel, F., et al. (2014). What general practitioners said about the palliative care case conference in residential aged care: An Australian perspective. Part 1. *Progress in Palliative Care, 22*(2), 61–68. doi:10.1179/1743291X13Y.0000000066.

World Health Organization. (1978, September 6–12). *Declaration of Alma-Ata: International conference on primary health care*, Alma-Ata, USSR, Geneva.

World Health Organization. (2015). *WHO global strategy on people-centred and integrated health services: Interim report.* Geneva, Switzerland: WHO.

Chapter 14
Care and Support for Older People

Yun-Hee Jeon and Hal Kendig

14.1 Introduction

'Aged care' is becoming a central concern in Australian society at large as well as for governments that aim to sustain the well-being of frail, older Australians. Timely and adequate health and aged care services are required to meet the challenges of the steady increase in people at advanced ages and the rising prevalence of chronic diseases. This chapter constructively examines 'support and care', inclusive of consumers and caregivers, provider organisations and care professionals, policymakers and governments. It presents a forward-looking view and identifies some complex issues and challenges in developing new directions for more effective systems of care and support.

The next section of the chapter describes the evolution of aged care services and legacies that continue to shape them. It reviews the influential Productivity Commission Inquiry Report *Caring for Older Australians* (2011), followed by an overview of existing and newly introduced programmes. The chapter then turns to the viewpoints, needs, and support services available to older people and their caregivers as well as issues concerning the care workforce. It highlights key philosophies underpinning new approaches such as person-centred care, reablement, and

Y.-H. Jeon (✉)
Faculty of Nursing and Midwifery, The University of Sydney,
88 Mallett Street, Camperdown, NSW 2050, Australia
e-mail: yun-hee.jeon@sydney.edu.au

H. Kendig
Centre for Research on Ageing, Health and Wellbeing (CRAHW), Research School of Population Health, Australian National University (ANU), Canberra, ACT 0200, Australia

ARC Centre of Excellence in Population Ageing Research (CEPAR),
University of New South Wales, Kensington, NSW 2033, Australia
e-mail: hal.kendig@anu.edu.au

K. O'Loughlin et al. (eds.), *Ageing in Australia*, International Perspectives on Aging 16, DOI 10.1007/978-1-4939-6466-6_14

consumer directed care, as well as efforts to improve quality assurance and better balance provision across care settings. In the conclusion, we outline issues that we believe will need to be addressed over the 10-year period from 2012 set by the government for implementation of its aged care reforms.

14.2 Policy and Programmes

14.2.1 Policy Legacies and Issues for Building Responsive and Integrated Care Systems

The scope for new policy development and implementation is heavily shaped by legacies from the past. Existing programmes and services reflect and reinforce political and community expectations of what ‘should’ be provided for older people. In policy debates over new initiatives, it is crucial to take into account what can be termed the ‘political economy’ of organisational and professional groups that depend on government support for their ongoing viability and employment of their workers. Influential organisations include professional groups (e.g. Australian Nursing and Midwifery Federation), peak bodies for the aged care industry (e.g. Aged and Community Services Australia and Leading Aged Services Australia), and peak consumer organisations (e.g. COTA Australia, Alzheimer’s Australia, and Carers Australia).

It is important to consider the ways in which policy advocates and government decision makers reflect the interests of their constituencies and appeal to political and community interests, policies, and funding. The complexity of decision-making reflects tensions between levels of government, ‘health’ versus ‘welfare’ conceptions of ageing issues, and tensions between ‘spending departments’, such as Social Services and Health, and the central expenditure control agencies such as Treasury and Finance.

Aged care provisions reflect the continuing influence of earlier historical eras. Kendig and Duckett (2001), and others such as Howe (1997) and Fine and Pross (2009), have provided interpretation of the earlier turning points and developments through to the 1990s. In many respects aged care, as it is now known, began with the establishment, in the mid-1980s, of the Residential Care Program (funded by the Commonwealth Government) and the Home and Community Care (HACC) programme funded jointly by the Commonwealth and State governments. Over the course of more than three decades, there have been strong trends from provider to funder control of care ‘systems’ and some movement towards a stronger focus on consumers. Notable achievements have included the development of high-quality aged care services, increased skills amongst aged care providers and workers, and a major shift from residential to community as the primary bases of care. A more detailed timeline of key developments in recent aged care reforms is presented in Fig. 14.1.

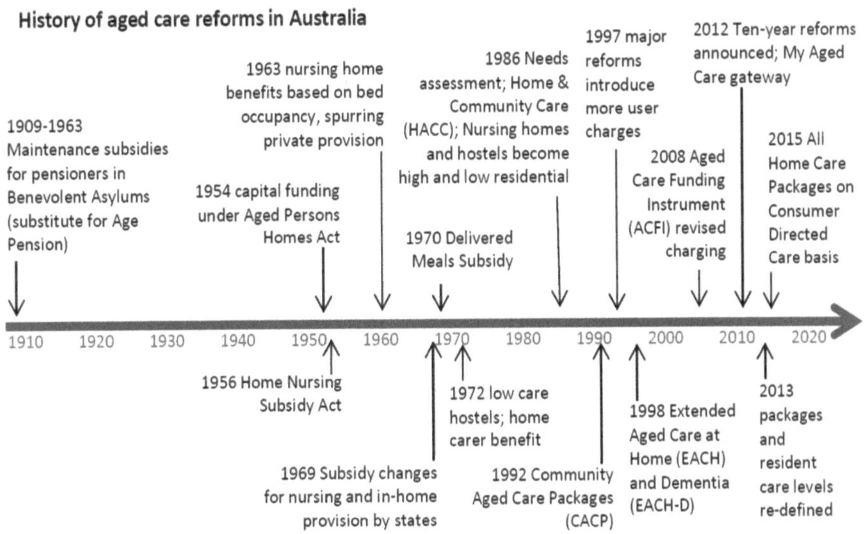

Fig. 14.1 History of aged care reforms in Australia (*Source*: Chomik and MacLennan 2014a)

The current era in aged care had its origins in the Productivity Commission's *Caring for Older Australians* Report (2011). The Report was strengthened by the independence and expertise of the Commission as a policy review body and by the Commission's extensive consultations with care services, peak bodies, governments, and individual experts. The Report's detailed recommendations had the overall aims of increasing choice, flexibility, and quality in accommodation and care services. Some key developments after the Commission's Report included the *Living Longer, Living Better* plan (Commonwealth of Australia 2012), legislated with bipartisan support in 2013, and programme reforms in the the Coalition Government's *Healthy Life, Better Ageing* (Coalition Government 2013).

The key directions in these aged care reforms include:

- Consumer Directed Care (CDC) initiatives that mark new directions in community and residential care.
- The new Commonwealth Home Support programme and home care packages which have fixed budgets set on the basis of level of need, while care providers are encouraged to work with clients and carers to develop and implement individualised care plans.
- Simplified planning and allocation controls on new care places with increasing provision for community care relative to residential care.
- A focus on the aged care workforce, with some effort to increase the supply of high-quality services.
- Stronger income and means testing through Centrelink for home care packages and residential aged care, with more attention to assets, including equity in owner-occupied housing.

- Increased accommodation payments for residential aged care and provision for accommodation bonds to fund capital expenditure on 'high care' residential care places.
- Development of the My Aged Care website which aims to inform consumer choice, while the policy framework has been strengthened through the Aged Care Quality Agency (former Aged Care Standards Accreditation Agency) and the Aged Care Financing Authority.

The Coalition Government's contentious 2014–2015 budget limited funding growth for aged care in the context of fiscal restraint, while continuing efforts to increase the equity and effectiveness of support programmes. The lead-up to the 2014–2015 budget included work on new fee-paying policies, and the Commonwealth announced that from 2015 all Home Care Packages would be funded as 'consumer directed' initiatives, as discussed later in the chapter.

14.2.1.1 Programmes and Services

The options for support and care services for older Australians are wide ranging. They include formal support, funded primarily by governments, with some user charges, and delivered by not-for-profit, for-profit, and some local and State governments, as outlined in Chomik and MacLennan (2014a). The HACC Program was, until 2015, the main avenue for funding and delivering services aimed at improving the quality of life of older people and their carers, and also for providing alternatives to inappropriate entry to residential care. The objectives of the new Commonwealth Home Support Programs (CHSP) and Home Care Packages, introduced in mid-2015, were to better assist people to remain living at home and to enable consumers to have choice and flexibility in the way that their aged care and support is provided at home.

The main directions for the new aged care programmes introduced in 2015 include:

- CHSP: incorporates the HACC Program, National Respite for Carers programme, Day Therapies programme, and the Assistance for Care and Housing for the Aged Program; interfaces with other related programmes (e.g. disability, transition care, residential respite).
- The Gateway (through the My Aged Care Website): takes responsibility for initial assessment, while decisions on eligibility for services will be separated from service providers.
- Regional Assessment Services: conduct assessments and determine resources on the basis of consumers' assessed needs, for example, levels of CHSP and Home Care Packages.
- National fees policy: user contributions to be determined consistently on a national basis, while consumers and families will be able to 'top-up' government resources in order to secure more support.[1]

[1] See the My Aged Care website for detailed information of currently offered aged care services http://www.myagedcare.gov.au/help-home; for the aged care reforms https://www.dss.gov.au/our-

Residential aged care provides permanent accommodation and short-term respite for older people with varying care needs. Australia's *Aged Care Act 1997* underscores the aged care system's role in providing seamless and safe care for older people by helping them to 'age in place'. This Act brought all aged care facilities under the same regulatory framework and facilitated residents being able to progress from low care to high care in the one facility, depending on the facility's resources to service higher needs. Since 2008, all permanent residential aged care clients have been required to undergo an Aged Care Funding Instrument (ACFI) appraisal process that determines their care needs. Government subsidies to residential aged care providers are determined by this ACFI process, which replaced the former Resident Classification Scale (Department of Health and Ageing 2011). The introduction of the ACFI has led to an increase in funding for residents with the highest care needs and some simplification of the classification procedures. However, the ACFI system has been criticised for failing to provide equitable subsidies, in particular for residents with special and complex care needs, and for those with low care needs that attract less financial benefit to the provider (Department of Health and Ageing 2011; Productivity Commission 2011). The distinction between high and low care allocations was removed from 1 July 2014 to ensure more flexible and transparent arrangements for the individual's ageing in place.

14.3 Who Uses and Who Provides 'Support and Care'?

14.3.1 Aged Care Consumers and Patterns of Service Use

Most older Australians enjoy reasonably healthy, active lifestyles, either with or without the support of others, and continue to make positive contributions to the cultural, social, and economic fabric of the country (Australian Institute of Health and Welfare, AIHW, 2014a). They overwhelmingly strive to maintain their sense of identity—striving to continue to 'be themselves'—notwithstanding health and other limitations that may define how they are seen by care providers (Minichiello et al. 2012). They typically evince a fierce will to remain as independent as possible while 'ageing in place' in the family home with its meanings, memories, habits, and social connections (Mackenzie et al. 2014). The ageing baby boomer cohort has a strong orientation towards independence and evidences increasing diversity in terms of lifestyles and housing choices through later life (see Chap. 10; Beer and Faulkner 2009).

At any one time, less than 20 % of the older population (i.e. those aged 65 and older) will be using care and support from formal health and aged care services, in addition to any informal network of support they may have around them. However, more than three out of four older people will make use of some form/s of aged care

---

responsibilities/ageing-and-aged-care/aged-care-reform; and for ageing and aged care programmes https://www.dss.gov.au/our-responsibilities/ageing-and-aged-care/programs-services

Box 14.1: A Profile of Aged Care Service Users

- About 77 % of residential and 65 % of community care clients are 80 years or older.
- Over 27 % of permanent residential aged care and 23 % of community care package clients require high-level care, across all care needs domains.
- Almost 55% of permanent residential aged care clients are reported as having mental health or behavioural conditions and about 52% as having dementia.
- Permanent residents stay in residential aged care for less than 3 years on average, with death being the main reason for separations (over 82 % of all separations). In contrast, the main reason for separation among community care clients is relocation to residential aged care (35 % of all separations), with about 17 % being due to death.
- Some 33 % of aged care clients (30 % of permanent residential aged care and 36 % of community care population) were born overseas although almost 88 % of clients nominate English as their preferred language. Compared to community care, a relatively lower proportion of non-English speaking clients use residential aged care than English speaking clients.

Sources: AIHW (2014b, 2016).

One of the most powerful dynamics underlying care provision is the balance between older people who wish to remain independent and in control of their own homes (and lives), and their well-intended family and paid caregivers who, in aiming to address their needs, may encourage or pressure them towards residential care (Brooke and Kendig 2004).

There is also great variability in the circumstances of the relatively small groups of older people who have multiple vulnerabilities which may result in them having a range of financial, accommodation, and care needs that cut across multiple support systems (Chomik and MacLennan 2014a, b). Less intense needs can be met by services in the open market (e.g. transport, home, and garden maintenance) as long as people have the cognitive capacities to manage for themselves, the finances to afford these services, and basic market protections.

The great variability among older people extends beyond their self-care capacities and carer availability to their financial resources, with the most vulnerable among them experiencing all of these limitations. Financial capacities for self-provision of support and care will continue to be affected by limited superannuation coverage of older people for another decade or more (Chomik and MacLennan 2014a), as well as by the continuing decline in home ownership and hardship for private tenants (Chap. 10). The decreasing availability of informal caregivers, a result of social and demographic change (Nepal et al. 2011), will further increase pressure and demand for support and care services. The latest figures of government spending on aged care ($15.8 billion) reflect this trend, with a 44% increase in their expenditure for various aged care support services and programs during 2014-15, compared to that of 2009-10 (AIHW 2016). This is a notable phenomenon given that this period mirrors the beginning of the aged care reforms introduced since the 2012 *Living Longer Living Better* plan.

14.3.1.1 Family Carers

Family members provide long-term support in the form of assistance with housing, personal care, and management of the person's illness and well-being. Without this informal unpaid care, many older individuals' chances of recovery or survival would be markedly reduced, and social issues such as homelessness would be far greater. The role of carers in assisting their family members, friends, or significant others to stay in their home environment, despite sometimes having severely limited daily activities, is particularly significant (AIHW 2013). With Australia's ageing population, escalating rates of chronic disease and disability, and continued emphasis on community-based care, the number of carers and the burden imposed on them is likely to continue to rise. Around 2.7 million Australians provide care for family and friends, of whom almost 30 % have primary caring roles[2]. Key findings from the Australian Bureau of Statistics (ABS) 2012 Survey of Disability, Ageing and Carers (SDAC) are presented in Box 14.2.

Box 14.2: A Profile of Informal Carers

- Females make up the majority of carers, representing 70 % of primary carers and 56 % of carers overall.
- Carers are most likely to be aged 55–64 years (21 % of carers).
- Just over one third of primary carers have a disability compared with 16 % of people who aren't carers.
- Around four in ten (39 %) carers report spending 40 hours or more per week providing care. This is more likely to be the case for female primary carers (42 %) than male carers (33 %).
- The labour force participation rate for primary carers (42 %) and carers (63 %) is lower than that for non-carers aged 15 years or more (69 %). Proportionally, carers are more likely than non-carers to have household incomes in the lowest 20 % for equivalised gross household income in 2012.

Source: ABS (2013)

[2] 'Primary carer' refers to an individual who is aged 15 and over and who meets the following criteria: has been providing help, or are likely to provide help, for at least six months; 2) provides help with one or more tasks associated with the core activities of mobility, self-care and communication, and 3) feels they provide the most care to the recipient for those activities (ABS 2013).

Challenges and negative consequences of caregiving roles on the carer's personal health and well-being, quality of life, and economic position have been documented in many studies in Australia and internationally (see Chap. 2). Almost six out of ten carers in Australia have been identified as being moderately depressed compared to 6 % of the general population, and carers were more likely than the general population to experience physical health problems such as chronic pain (Cummins et al. 2007). More recent statistics also report that just under one-third of primary carers frequently felt lethargic, worried, or depressed (AIHW 2013).

In recent years, the need to support informal carers has been recognised as central to policies encouraging care in the community rather than institutional care for older people. The introduction of the *Carer Recognition Act* 2010 formally acknowledged the critical role of carers, their rights, and need for support. The National Carer Strategy further articulated six priority areas for action: recognition and respect, information and access, economic security, services for carers, education and training, and health and well-being (Australian Government 2011). Honouring the *Carer Recognition Act* 2010 and the National Carer Strategy is critical if successful realisation of person-centred care, reablement, and consumer direction is to occur in all aged care contexts (Carers Australia 2013).

14.3.2 Aged Care Workforce

To meet current and future capacity demands with quality care services for aged care clients, it is essential that there is an adequate and well-equipped workforce. The 2012 national aged care workforce census (King et al. 2012) estimated over 240,000 aged care employees working in direct care roles (i.e. registered nurses, enrolled nurses, nurse practitioners, care assistants/workers, allied professionals/assistants). Box 14.3 describes the characteristics of the aged care workforce in Australia.

Box 14.3: Characteristics of the Aged Care Workforce

- Majority of direct care workers are: *female* (about 90 %); *personal care workers* (68 % in residential care, 81 % in community care); employed on a *permanent part-time* basis (72 % of residential care employees, 62 % in community care); have *post-secondary qualifications* (over 85 %). The proportion of allied health professionals employed is miniscule (1.7 %).
- The median age of this workforce is about 50 years, which is higher than the national workforce median age.
- The proportion of the direct care workforce born overseas is continuing to increase (35 % in 2012), with about a quarter of the workforce speaking a language other than English.

Source: The Aged Care Workforce 2012 (King et al. 2012).

The 2012 census data suggest that the Australian aged care workforce is reasonably stable: well over a quarter of the workforce had been in the sector for 15 years or more and only 5 % of the workforce indicated an intention to leave the sector in the next 12 months. However, the continued decline in the proportion of registered nurses, now less than 15 % of the direct care workforce, remains a major issue (King et al. 2012). Shortages in the health care workforce generally, especially nurses, are widespread and the issue of registered nurse shortages is most evident in the aged care sector. Key reasons why the aged care sector is being hardest hit by staff shortages include remuneration inequities between the aged and health care sectors, and between the staff working in aged care and non-health care industries with comparable skill-sets; limited opportunity for continued education and training; and poor working conditions with limited resources that impact on staff capacity to provide quality care (Chenoweth et al. 2010).

The issues around skilled workforce shortages are complex and have been a major public concern for more than two decades. The National Aged Care Workforce Strategy 2005 (Aged Care Workforce Committee 2005) provided a strategic approach with a focus on the need for long-term structural reform in the Australian aged care sector. Since then, progress, in terms of improved understanding of the workforce profile, has been achieved through the national aged care workforce census every 3 years. More opportunities for staff training and education through various scholarships and financial incentives, and the development of workplace practice models have also been realised. The findings of the 2012 aged care workforce census suggest some improvement in staff training and education: a substantial increase in the proportion of personal/community care workers with Certificate IV qualifications; four out of five direct care workers engaged in one or more training courses in the previous 12 months; and one in two workers involved in continuing and professional development (King et al. 2012).

However, a number of key actions under the 2005 Strategy have not been fully realised, for example, ensuring adequate skill mix and staffing, improving the status and image of aged care, and developing capacity of leadership and management. Creating a supportive work environment is one of the key strategies to improve staff retention in a sector marginalised by ageist social attitudes (see Chapter 3). Research evidence, albeit weak, points to the significant influence of leadership and management skills on improving job satisfaction, workforce retention, care quality, patient outcomes, and reducing the associated costs (Jeon et al. 2010). The Aged Care Leadership Development Project, a new initiative funded by the Commonwealth Department of Industry, has shown some promise in addressing the need to improve leadership capacity through the development of the Aged Care Leadership Development Strategy in 2013, the Aged Care Leadership Development Framework in 2014, and the Aged Care Leadership Development Portal in 2014 (Community Services & Health Industry Skills Council, 2015).

14.4 Advancing Aged Care

14.4.1 Principles and Philosophies of Care

Reflecting on recent developments in aged care provision, it is important to consider key issues to which aged care needs to be responsive. The term 'system' is used here because, although most services and other supports are delivered separately, they need to come together in an integrated way for individuals in their local communities.

Three key issues from the perspective of older people and their carers are as follows:

- The requirement for person-centred delivery, focusing on continuity, integration, transitions, and timeliness for older people with multiple needs.
- Having the information necessary to understand the possible care trajectories, taking into account individual diversity, alongside ageing changes, so that they can assess the responsiveness of service systems to their particular needs.
- Recognition of family/informal carers as active partners in care and support of older persons and their crucial role as advocates and in quality control 'surveillance'. Support of caregivers has improved substantially over time, but caregiving relationships continue to require close attention to the shared and, in some cases, divergent interests of both parties (Chomik and MacLennan 2014a,b).

From the perspective of service providers and the service system both overall and at local level, the key issues are as follows:

- Recognising the importance of 'upstream' action to preserve and strengthen capacities, notably health promotion to maintain health and independence and efforts to enable recovery, to complement the 'downstream' provisions of care and support as people's needs increase.
- The value of 'whole of government' approaches that aim to provide appropriate housing, health, income, and other forms of support extending well beyond the specific focus of aged care programmes.
- Active mechanisms that address and overcome the fragmentation of programme direction and delivery across departments and levels of government.

A logical starting point for developing responsive services is defining a 'good life' from the perspectives of older people and their loved ones, without necessarily focusing on their 'illness' or 'disability'. People develop their own frame of thought throughout their life course about what is a good life, and strive to live a good life no matter where they are on their health or illness trajectory. It is a fundamental human desire to have a good life, and the concept may help health and aged care professionals, service providers and administrators and policy makers in setting goals for better care and services overall, rather than focusing narrowly on a single medical condition or disability.

Six key elements of a good life, originally elicited from conversations with over 770 older Australians and their families, and subsequently confirmed through a research review (Barnett and Dean 2012) are as follows: uniqueness, being optimistic, being healthy, being in control, belonging, contribution, and engagement. These elements have a striking resemblance to the psychosocial theory of 'Personhood in Dementia': despite physical, psychological, neurological, and social impairments, people with dementia still retain a great deal of capacity to enjoy life, fulfilled by having their essential needs met (love, identity, comfort, attachment, occupation, and inclusion) (Kitwood 1997).

Given this fundamental human desire to have a good life, it is not surprising to see the rise in popularity of care philosophies such as person-centred care (or patient-centred care in hospital), reablement, or restorative care, and more recently a new service delivery model of consumer directed care, instead of provider directed or menu-driven care. In the following sections, these key philosophies of care and service models that underpin present and future aged care services are discussed. The underlying principles and the evidence supporting each model are outlined, as well as the implications for practice, and some of the conceptual limitations and challenges for implementation are also taken up.

14.4.1.1 Person-centred Care

The notion of putting patients/persons/clients first within the health system is receiving renewed emphasis in Australia and overseas, making the person a central and enabling agent in the system. Policy directions and clinical guidelines in place at all levels of government and non-government organisations, and within professional bodies, acknowledge the need for care to be person centred. The leading recommendation made by the Australian Commission on Safety and Quality in Health Care (ACSQHCA) is that: "Policy makers and regulators should include patient-centred care as a dimension of quality in its own right in strategic and other policy documentation" (ACSQHCA 2011, p. 8).

Person-centred care aims to maintain the dignity and respect of older people, facilitate information sharing, and encourage participation and collaboration in health (ACSQHCA 2011). Similarly, the principles of person-centred dementia care include valuing the uniqueness of each person, having respect for the past, considering the whole person, emphasising the positives and abilities, staying in communication, nourishing attachment, creating a sense of community, maximising freedom while minimising controls, valuing reciprocity, and maintaining an environment of trust (Kitwood 1997). Within a person-centred approach to care it is vital to recognise the person's perspective, based on the unique experiences, strengths, and limitations each brings to their own health and its management, and respecting their needs, wishes, and preferences in planning and implementing care (Edvardsson et al. 2008). Further, growing evidence indicates that when aged care clients experience care services and programmes that are person centred, their experiences with the health care system become more positive and are more likely to lead to better health outcomes (ACSQHCA 2011; Edvardsson et al. 2008).

14.4.1.2 Reablement

Reablement, also known as 'restorative care', is closely aligned to the notions of 'successful ageing', 'healthy ageing', 'active ageing', and 'wellness approach', all of which emphasise the importance of maximising health and well-being of older people through active and meaningful engagement in their physical, social, and community activities (see Chaps. 5, 10, 12). Reablement is defined as "(short-term) services for people with poor physical or mental health to help them accommodate their illness by learning or relearning the skills necessary for daily living" (Francis et al. 2011, p. 2). When care and support is underpinned by reablement, providers and practitioners collaborate and encourage the older person to learn, restore, and regain their functional and psychosocial capacity and independence as best as they are able (Cartwright et al. 2009; Francis et al. 2011).

Parallels can be drawn between person-centred care and reablement: their shared focus on encouraging the person's active participation in their own care and decision-making and enabling the person's innate positives and abilities, rather than negatives and disabilities. The emphasis in reablement on assisting people to regain functional capacity and improve independence is also found in rehabilitation, which is goal oriented, with an ultimate outcome of a full recovery if possible, or enabling people to live their lives to the fullest (Randall and Ford 2011).

Research evidence to date on reablement models of care suggest that reablement reduces the need for ongoing traditional home and social care; is cost effective in the long term despite its initial high costs; and improves outcomes for service users, although such evidence from the perspectives of service users and their family/carer is still limited (Francis et al. 2011; Lewin et al. 2014). More rigorous research is needed to support evidence relating to long-term effects or consequences, cost effectiveness, sustainability, acceptability among minority groups, and best models of delivery (Francis et al. 2011).

The concepts of person centredness and reablement may be seen as a 'logical principle' in modern society where consumer expectations are increasing as they become more knowledgeable and vocal about both their rights and their health. However, what is involved in the provision of person-centred care is, in fact, a radical move from the traditional understanding of the 'patient' or 'client' as a passive recipient of advice and care from the 'expert' health care practitioners. Similarly, reablement means that those who provide care need to understand the client's current level of capacity as well as their scope for future self-improvement, as it is the carer's role to facilitate the client's developing independence. Reablement is one of the key principles driving the new home care and support programmes. Good preparation is necessary for developing the aged care workforce who may be used to doing things *for* the person, rather than doing things *with* the person and finding optimal ways to engage family carers.

14.4.1.3 Consumer Direction

Ensuring that older consumers and carers have choice and voice as well as control over the type, delivery, and provider of care and services is essential to the core concepts of a good life and person centredness and underpins the Living Longer Living Better agenda (Commonwealth of Australia 2012). Under the new initiative of Consumer Directed Care (CDC), home care programme recipients are provided with a personalised budget, which allows them to see how much funding is available for the care and services they require based on the aged care assessment, and how the money is being spent (Department of Social Services, DSS, 2016). Through this process consumers are able to determine the level of involvement they would like to have in managing their own package. The Australian Government started offering CDC in all Home Care Packages from July 2015 and plans to introduce CDC into residential care in the future (DSS 2016).

Introduced as a major shift in the aged care service delivery model, CDC packages have received a great deal of interest and diverse reactions from a number of professional and consumer advocacy groups in aged care. Observations from other countries suggest that, despite it being a popular concept, evidence as to the effectiveness and efficacy of CDC is still inconclusive (Howe 2003; Low et al. 2012).

In their review of overseas evidence on CDC, Low et al. (2012) summarise key issues for CDC implementation as it proceeds in Australia, including:

- A CDC model is not appropriate for all clients as capacity to assume responsibility for complex finance and service matters varies between individuals, often requiring a family (carer) to be a surrogate decision maker.
- Administration processes can be costly and cause delays in the service provision.
- Risk of exploitation or misuse of the allowance necessitates an appropriate quality monitoring system.

The Australian CDC model, offered since 2013, appears to address some of the shortfalls found in other countries, mainly by giving service providers more control, for example, over clients' funds and brokering services, which could minimise the administrative burden for clients and ease monitoring of the fund use (Low et al. 2012). Pilot programme evaluations have yielded some positive preliminary findings such as positive impacts on client and carer satisfaction and increased choice and control. This was more notable among those who had higher levels of need, with more substantial packages, than those with low-level CDC package clients (KPMG 2012). The implementation of CDC is just beginning, and the outcomes are yet to be demonstrated.

Box 14.4 provides some key strategies and directions reported for successful CDC. These draw on lessons from pilot programmes such as People at Centre Stage (PACS) by the Uniting Care Community Options (Victoria) (Ottman et al. 2012) and ACH Group Consumer-directed Care (South Australia) (Williams and Fidock 2012), as well as the KPMG evaluation (KPMG 2012).

Box 14.4: Key Strategies/Directions for Successful CDC Implementation

- Both the workforce and consumers should be supported to understand, and redefine if necessary, the power relationships between care providers and care recipients and their role as partners in care and shared decision-making.
- 'CDC literacy' needs to be improved for greater consumer understanding.
- Better supply and quality of information is needed at a local level, as well as greater consideration for special needs groups who have low literacy in information and technology.
- Deep skill development and training is required across all levels of care staff, coordinators, and management; there should be a focus on skills for delivery of person-centred care and restorative care, with special attention for those clients and/or carers with limited capacity for making choices such as those with dementia, cognitive impairment, and low English language skills.
- Careful choice and use of language is important: 'customers', rather than clients/consumers, and 'advisors' instead of coordinators
- A 'safety net system' should be provided for those who have difficulty with exercising options.
- Integrated solutions are required across whole of government services—'Bundling' support in ways people can create 'integrated solutions'.

14.5 Quality of Aged Care

14.5.1 What Does 'Quality' Aged Care Mean?

Quality can be termed "the extent to which a health care service or product produces a desired outcome/s" (Runciman et al. 2007, p. 297). The concept of quality of care is complex and multi-dimensional, and often depends on the perspectives and preferences of individual assessors and the nature of the programmes or services being assessed. In an aged care context, care quality is most often referred to as the quality of the workforce, in terms of their technical and interpersonal skills and attitudes, as well as their availability and amount of individual contact time. Other dimensions of care quality include the amenity of the physical environment, standards of everyday services such as meals/dining, cleaning, socialising opportunities, spiritual activities, and organisational and administrative performance (Fethney et al. 2013; Jeon et al. 2012). As these dimensions have been identified from the residential care setting, which represents a total care environment, some qualifications may apply in community care where outcomes are also dependent on the home as a setting for care and the contribution of family care.

To attain the overarching goal of the aged care system—to promote the health and well-being of older people and their carers—the aged care system must ensure that services are affordable, accessible, high quality, person centred, and appropriate to the needs of recipients (Steering Committee for the Review of Government Service Provision SCRGSP 2015). That care quality is of continuing concern, is highlighted by the repeated discussion in government reports: a Senate Committee on Community Affairs report, *Quality and equity in aged care* (The Senate Community Affairs References Committee 2005) and the Productivity Commission's (2011) inquiry. These reports illuminate concerns relating to the complexity of the aged care system for consumers, limited services and consumer choice, inconsistent and inequitable services and subsidies, skilled workforce shortages, and excessive and unnecessary regulatory aspects. Skilled and competent managers play a critical role in ensuring good clinical care and leadership that is critical to care quality (Jeon et al. 2010). Ensuring a skilled workforce is by far the most important aspect of the quality assurance measures in aged care, but also one of the most contentious issues. Earlier in this chapter attention has been drawn to the increasing complexity of care needs and the importance of providing care that is person centred, enabling, and consumer directed. However, how these ideals can be best delivered, incorporated into everyday practice and sustained remains a major challenge given the workforce constraints.

14.5.1.1 Measuring Quality

To ensure quality, a well-articulated set of measures of 'quality' is critical. These measures must address the perspectives of the key stakeholders, including users and providers of care and services as well as policy makers. Assessing care quality is complex. For quality indicators to be effectively and reliably used for ongoing monitoring and quality improvement, measurement of care quality should be harmonised with what constitutes quality, holistic care. Following Donabedian (1988), they should be inclusive of multi-dimensional factors influencing care, such as: organisational and environmental features—structure/characteristics of each care setting (i.e. material and human resources, case-mix, organisational structure); process and delivery of care; outcomes of care to the person's health and well-being (Donabedian 1988).

As part of the aged care reforms, separate quality indicators for residential aged care and community care programmes were proposed with the results made publicly available through the My Aged Care website. This 'report card' process was proposed as a way to assist older people, families, and carers to make informed choices about care and support. The introduction of publicly available results of the quality indicators and the rating system aims to contribute to ensuring that government subsidised aged care programmes and services are more transparent and accountable for their spending and services (Productivity Commission 2011; SCRGSP 2015). However, debates continue about the potential for this system to bring about undesired influence, with concerns largely arising from the lack of evidence-based measures that capture 'quality', and possible (mis)interpretation of the results of the report card. The interpretation of the results of the quality indicator appraisal process is context specific and should be risk adjusted (Mor 2007).

National Aged Care Alliance (2014) has proposed six domains of residential aged care quality indicators including: person-centred interactions, health and well-being, engaging socially, daily services, physical environment, and organisational and governance matters. These domains, and the principles of the quality indicators proposed, extend beyond benchmarking and compliance to focus on promoting continuous quality improvement based on practical, valid, reliable, and person-centred measures of quality. Much has been debated as to how these domains will be captured in individual indicators and used to maintain and improve quality of care.

14.6 Conclusion

Issues of support and care for older Australians need to be appreciated in the context that the majority will enjoy reasonably healthy, active lifestyles, with or without support of others, through most of later life. Most will never enter residential care and fewer than one out of five will make use of any formal services. However, health and other limitations on older people are increasing with population ageing (Chap. 5) and there is a great diversity in terms of preferences, family and accommodation arrangements, and financial means among other factors influencing independence and well-being among older people.

There are major opportunities as well as challenges ahead as Australia moves forward in implementing consumer and carer directions in the 10 years of aged care reform under the Living Longer Living Better agenda (Commonwealth of Australia 2012). The Aged Care Roadmap produced by the Government's Aged Care Sector Committee (2016), which includes consumer, provider, and professional body representatives, outlined principles and priority actions towards achieving 'a consumer driven, market based, sustainable aged care system' (p. 3). Notable challenges include the capacity to build integrated 'systems' of support and care which are responsive to changing needs and capacities as people grow older. A comprehensive agenda for enhancing community care knowledge is included in the Community Care Research Agenda (Lewin et al. 2011). The impetus towards more community care and more consumer directed care is expected to continue, with resources being directed as per the choices of older people and carers themselves. Further, knowledge gaps still exist concerning the essential relationships between the mix of services and staff skills and expertise and the costs of care and, most importantly, the outcomes and satisfaction of consumers.

Directions for support and care for older people will be heavily influenced by societal goals for ageing well and community support as well as broader contextual issues that extend well beyond the fiscal constraints and programme adaptations that preoccupy the aged care sector at any particular point in time (Chap. 15). The balance of Commonwealth and State government responsibilities for the provision and funding of health and welfare services, and integration at local and regional levels, will be crucial influences. Major questions remain for the balance between individual and public responsibilities in paying for care services.

The period 2013–2015 signified the early stages of articulating transitions and inter-relationships between ageing and disability care and support. These and other essential issues continue to require the active and informed engagement of consumers, providers, and funders—as well as the broader community.

14.7 Case Study: Older People Who Are Ageing in Their Familiar Environment

14.7.1 John and Mary

John is a 79-year-old, retired teacher living in a two-story house with his wife Mary, who is 75 years old. They have been together for 55 years and have three children and eight grandchildren; Nora is one of their children and she and her family live close by. John and Mary have always enjoyed their active social lifestyle and have regular gatherings with their circle of friends. John often helps Mary around the house, spends a lot of time gardening, and likes to be in charge of the social club he and Mary have been attending for the last 30 years. His overall health is good, although he has some difficulty with vision (dry macular degeneration), while Mary has had Type 2 diabetes and arthritis in her knees for some time. John has recently been diagnosed with Alzheimer's disease and has started an anti-dementia drug to relieve symptoms of this condition. Mary is finding it increasingly difficult to rely on John's support but has not told any of her children about John's Alzheimer's disease.

Discussion Points

- What are the present and future needs of John and Mary in these circumstances?
- What care and support programmes can be offered to enable them to continue, as best they can, the lifestyle they have enjoyed? How can reablement approaches to care assist the couple?
- What is the role of consumer directed care in this case?

14.7.2 Sylvia

Sylvia is a 93-year-old French widow living in a 63-bed residential care facility in a rural town. She came to this facility 6 months ago after having a major fall resulting in a hip fracture, which required surgery at a hospital close to where she had been living for the last 30 years. She has no children, nor any relatives living nearby as she and her late husband migrated to Australia 50 years ago. During her hospitalisation she developed pneumonia and became extremely frail. The local Age Care Assessment Team determined that she was not fit to live at home alone as she would require support for her daily living, and therefore recommended admission to this residential care facility. Sylvia is sharing a room with another female resident who has dementia and has been in the facility for 3 years. Until she had the fall, Sylvia had always been a volunteer, helping other people in need: she used to help at local school canteens and churches. One of the care staff, Nell, has noticed Sylvia seemingly depressed and frequently forgetful and refusing to join any group activities offered. Nell is new to the facility and struggling to learn ways to care for residents with dementia. She finds it difficult to pay any special attention to Sylvia as the facility has had staff shortages for some time. Nell has not mentioned this to her care supervisor, Matthew, the only registered nurse in the facility, as he appears to be too busy to listen to her concerns.

Discussion Points

- What are the present and future needs of Sylvia in these circumstances?
- What does person-centred care mean for Sylvia and how does it help her situation?
- What are the challenges and opportunities the manager of this facility needs to consider?

Acknowledgement We thank Dr. Anna Howe for her constructive feedback on the draft chapter and Ms Emily Conway for her administrative support.

References

Aged Care Sector Committee. (2016). Aged Care Roadmap. Retrieved July 28, 2016, from https://agedcare.health.gov.au/aged-care-reform/aged-care-roadmap.

Aged Care Workforce. (2005). *National aged care workforce strategy*. Canberra, Australia: Department of Health and Ageing

Australian Bureau of Statistics. (2013). *Disability, ageing and carers, Australia: Summary of findings, 2012* (Cat. no. 4430.0). Canberra, Australia: ABS

Australian Commission on Safety and Quality in Health Care. (2011). *Patient centred care: Improving quality and safety through partnerships with patients and consumers*. Sydney, Australia: ACSQHC.

Australian Government. (2011). *National Carer Strategy*. Retrieved February 1, 2015, from http://www.comlaw.gov.au/Details/C2010A00123.

Australian Institute of Health and Welfare. (2013). *Australia's welfare 2013. Australia's welfare series no.11. (Cat. no. AUS 174)*. Canberra, Australia: AIHW.

Australian Institute of Health and Welfare. (2014a). *Australia's health 2014*. Canberra, Australia: AIHW.

Australian Institute of Health and Welfare. (2014b). *Residential aged care and aged care packages in the community 2012–13*. Canberra, Australia: AIHW. Retrieved February 1, 2015, from http://www.aihw.gov.au/aged-care/residential-and-community-2012-13/.

Australian Institute of Health and Welfare. (2015). *Use of aged care services before death. Data linkage series no. 19*. (Cat. no. CSI 21). Canberra, Australia: AIHW.

Australian Institute of Health and Welfare. (2016). Residential aged care and Home Care 2014–15. Canberra, Australia: AIHW. Retrieved July 29, 2016, from http://www.aihw.gov.au/aged-care/residential-and-home-care-2014-15/.

Barnett, K., & Dean, M. (2012). *Identifying the evidence base for the concept 'What is a Good Life'? Final Report: Matching research evidence to the Good Lives model*. Adelaide, Australia: Australian Workplace Innovation and Social Research Centre, The University of Adelaide.

Beer, A., & Faulkner, D. (2009). *21st Century housing careers and Australia's housing future. AHURI Final Report No. 128.*. Melbourne, Australia: Australian Housing and Urban Research Institute.

Brooke, L., & Kendig, H. (2004). Understanding community nursing for older individuals and carers. In R. Nay & S. Garrett (Eds.), *Nursing older people: Issues and innovations* (pp. 135–150). Melbourne, Australia: MacLennon and Petty.

Carers Australia. (2013). Carers Australia submission in response to questions contained in the key directions for the Commonwealth Home Support Program (CHSP) Discussion Paper (June 2013). Retrieved February 1, 2015, from http://carers-sa.asn.au/Assets/Files/Carers%20Australia%20Submission%20Commonwealth%20Home%20Support%20Program%20%20Key%20Directions.pdf.

Carer Recognition Act 2010 (Commonwealth of Australia) Act No. 123, 2010 (Australia).

Cartwright, C., Cosgrave, C., Gooden, J., & Carpenter, L. (2009). *Re-ablement of older people in north coast NSW*. Lismore, Australia: Aged Services Learning and Research Centre, Southern Cross University.

Chenoweth, L., Jeon, Y.-H., Merlyn, T., & Brodaty, H. (2010). A systematic review of what factors attract and retain nurses in aged and dementia care. *Journal of Clinical Nursing, 19*(1-2), 156–167.

Chomik, R. & MacLennan, M. (2014a, April). *Aged care in Australia: Part I—Policy, demand and funding*. CEPAR Research Briefing. Retrieved February 1, 2015 from http://cepar.edu.au/research-briefs.aspx.

Chomik, R. & MacLennan, M. (2014b, April). *Aged Care in Australia: Part II—Industry and Practice*. CEPAR Research Briefing. Retrieved February 1, 2015, from http://cepar.edu.au/research-briefs.aspx.

Coalition Government. (2013). *The coalition's policy for healthy life, better ageing*. Barton, ACT: Brian Loughnane. Retrieved March 30, 2015, from http://lpaweb-static.s3.amazonaws.com/Coalition%202013%20Election%20Policy%20%E2%80%93%20Better%20Aged%20Care%20-%20final.pdf.

Commonwealth of Australia. (2012). *Living longer living better*. Canberra, Australia: Department of Health and Ageing.

Community Services & Health Industry Skills Council. (2015). *Aged care leadership development project*. Retrieved February 1, 2015, from http://www.cshisc.com.au/develop/workforce-planning-and-development/workforce-development-projects/aged-care-leadership-development-project/.

Cummins, R., Hughes, J., Tomyn, A., Gibson, A., Woerner, J., & Lai, L. (2007). Australian Unity Well-Being Index: Report 17.1 *The well-being of Australians—Carer health and well-being*. Retrieved February 1, 2015, from http://www.acqol.com.au/reports/survey-reports/survey-017-1-report.pdf.

Department of Health and Ageing. (2011). *The review of the aged care funding instrument*. Canberra, Australia: Department of Health and Ageing.

Department of Social Services. Ageing and Aged Care: About the Home Care Packages. Retrieved July 29, 2016, from https://www.dss.gov.au/ageing-and-aged-care/programs-services/home-care/about-the-home-care-packages-programme.

Donabedian, A. (1988). The quality of care: How can it be assessed? *Journal of the American Medical Association, 260*(12), 1743–1748

Edvardsson, D., Winblad, B., & Sandman, P. O. (2008). Person-centred care of people with severe Alzheimer's disease: Current status and ways forward. *Lancet Neurology, 7*, 362–367.

Fethney, J., Jeon, Y.-H., Ludford, I., & Stewart, K. (2013). Validity and reliability of the client perception of value questionnaire. *Contemporary Nurse, 43*(2), 183–190.

Fine, M., & Pross, E. (2009). A human service revolution: Community Options and the Home and Community Care Program. In E. Moore (Ed.), *Case management for community practice* (pp. 298–332). Melbourne, Australia: Oxford University Press.

Francis, J., Fisher, M., & Rutter, D. (2011). *Reablement: A cost effective route to better outcomes*. London: Social Care Institute for Excellence.

Howe, A. (1997). The aged care reform strategy: A decade of changing momentum and margins for reform. In A. Borowski, S. Encel, & E. Ozanne (Eds.), *Ageing and social policy in Australia* (pp. 301–327). Melbourne, Australia: Cambridge University Press.
Howe, A. (2003). *Is consumer directed care a direction for Australia?* Retrieved February 1, 2015, from https://fightdementia.org.au/sites/default/files/ConsumerDirectedCare.pdf.
Jeon, Y.-H., Fethney, J., & Ludford, I. (2012). Measuring client satisfaction in residential aged care settings: A narrative review of instruments. *Internet Journal of Healthcare Administration, 8*(1).
Jeon, Y.-H., Glasgow, N., Merlyn, T., & Sansoni, E. (2010). Policy options to improve leadership of middle managers in the Australian residential aged care setting: A narrative synthesis. *BMC Health Services Research, 10*, 190.
Kendig, H. L., & Duckett, S. (2001). *Australian directions in aged care: The generation of policies for generations of older people*. Sydney, Australia: The Australian Health Policy Institute, University of Sydney.
King, D., Mavromaras, K., Wei, Z., He, B., Healy, J., Macaitis, K., et al. (2012). *The Aged Care Workforce, 2012*. Canberra, Australia: Department of Health and Ageing.
Kitwood, T. (1997). *Dementia reconsidered: The person comes first*. Buckingham: Open University Press.
KPMG. (2012). *Evaluation of the consumer-directed care initiative—Final Report*. Canberra, Australia: Department of Health and Ageing.
Lewin, G., Allan, J., Patterson, C., Knuiman, M., Boldy, D., & Hendrie, D. (2014). A comparison of the home-care and healthcare service use and costs of older Australians randomised to receive a restorative or a conventional home-care service. *Health and Social Care in the Community, 22*(3), 328–336.
Lewin, G., Burton, E., Sparrow, P., Carroll, M. & Kendig, H. (2011), Development of a community care research agenda for Australia. *Australasian Journal on Ageing*, 30, 37–40.
Low, L.-F., Chilko, N., Gresham, M., Barter, S., & Brodaty, H. (2012). Policy and Practices Updates: An update on the pilot trial of consumer-directed care for older persons in Australia. *Australasian Journal on Ageing, 31*(1), 47–51.
Mackenzie, L., Curryer, C., & Byles, J. (2014). Narratives of home and place: Findings from the Housing and Independent Living Study. *Ageing and Society, 35*(8), 1684–1712. doi:10.1017/S0144686X14000476.
Minichiello, V., Browne, J., & Kendig, H. (2012). Perceptions of ageism: Views of older people'. In J. Katz & S. Spurr (Eds.), *Adult lives: A life course perspective* (pp. 332–340). Bristol: The Policy Press.
Mor, V. (2007). Defining and Measuring Quality Outcomes in Long-Term Care. *Journal of the American Medical Directors Association, 8*(3), e129–e137.
National Aged Care Alliance. (2014). *Quality Indicators for Residential Care Advisory Paper.* . Retrieved February 1, 2015, from https://www.acs.asn.au/wcm/documents/ACS%20Website/Resources/Aged%20Care%20Reforms/NACA/NACA%20Advisory%20Paper%20-%20Quality%20Indicators%20in%20Residential%20Care%20-%20Jan%202014.pdf.
Nepal, B., Brown, L., Kelly, S., Percival, R., Anderson, P., Hancock, R. & Ranmuthugala, G. (2011). *Projecting the need for formal and informal aged care in Australia: A dynamic microsimulation approach*, NATSEM working paper. Retrieved February 1, 2015, from http://www.natsem.canberra.edu.au/publications/?publication=projecting-the-need-for-formal-and-informal-aged-care-in-australia-a-dynamic-microsimulation-approach.
Ottman, G., Laragy, C., Allen, J., & Feldman, P. (2012). *People at the centre stage: Summary report*. Retrieved February 23, 2015, from http://apo.org.au/node/31936.
Productivity Commission. (2011). *Caring for older Australians: Overview, Report No. 53, Final Inquiry Report*. Canberra, Australia: Productivity Commission.
Randall, S., & Ford, H. (2011). *Long-term conditions: A guide for nurses and healthcare professionals*. Oxford: Wiley-Blackwell.
Runciman, B., Merry, A., & Walton, M. (2007). *Safety and ethics in health care: A guide to getting it right*. Burlington, VT: Ashgate.

The Senate Community Affairs References Committee (2005). Quality and equity in aged care. Retrieved February 1, 2015, from http://www.aph.gov.au/Parliamentary_Business/Committees/Senate/Community_Affairs/Completed_inquiries/2004-07/aged_care04/report/index.

Steering Committee for the Review of Government Service Provision (SCRGSP). (2015). *Report on Government Services 2015, vol. F, Community services*. Canberra, Australia: Productivity Commission.

Williams, R., & Fidock, A. (2012). *Evaluation of service transformation to consumer directed care project*. Unley, SA: JFA Purple Orange.

Part IV
Actions and Directions

Chapter 15
Directions and Choices on Ageing for the Future

Hal Kendig

15.1 Introduction

This concluding chapter raises ideas and evidence that can guide constructive responses to societal ageing and older people in Australia. The fundamental argument is that positive actions can emerge purposefully as new thinking develops, gathers support, and proves to be influential over the middle and longer term. Good initiatives can eventually prove their worth and be taken up by broader coalitions of the wider public, influential interest groups, governments, and political leaders. Important past examples from which we are currently benefiting include community care, recognition of caregivers, 'rights' approaches, and positive ageing. Each of these areas developed initially with little support and considerable ignorance, yet over recent decades they have become central to progressive developments on ageing. Ideas, as tested and taken up by interest groups and the broader public, can contribute to societal changes setting new possibilities for an ageing Australia.

Contemporary circumstances and interests inevitably shape views on the future, and they will be further shaped by unknown events. This chapter was written at a time of deep public concern and mixed priorities after the first 2 years of the conservative Coalition Government elected in 2013. The optimism of the Labor government elected in 2007 had been set back by the Global Financial Crisis, accumulation of long-term public debt, and increasing political dissension in national government

H. Kendig (✉)
Centre for Research on Ageing, Health and Wellbeing (CRAHW), Research School of Population Health, Australian National University (ANU), Canberra, ACT 0200, Australia

ARC Centre of Excellence in Population Ageing Research (CEPAR), University of New South Wales, Kensington, NSW 2033, Australia
e-mail: hal.kendig@anu.edu.au

K. O'Loughlin et al. (eds.), *Ageing in Australia*, International Perspectives on Aging 16, DOI 10.1007/978-1-4939-6466-6_15

and Commonwealth-State funding tensions. The ongoing policy and political instability has continued and ageing, with various interpretations, is being invoked in policy contests at a time of deep political and economic uncertainty for the future.

This chapter returns to some of the overview issues raised in the book's introductory two chapters and it builds on the substantive chapters. Further research evidence and policy discussions are being reported in *Population Ageing and Australia's Future,* released by the Academy of Social Sciences in Australia (ASSA) and the Australian Research Council (ARC) Centre of Excellence in Population Ageing Research (CEPAR) (Kendig et al. 2016). Given the increasingly global significance of population ageing, we also consider some perspectives and comparative insights from the USA and Europe as well as international organisations. Individuals, social institutions, and governments are reconsidering societal aspirations and generational commitments in the context of population ageing and economic uncertainty.

15.2 Prospects for the Future and Budgetary Approaches

At this point in our history there is serious questioning as to Australia's social and economic directions now that the strong economic growth of the past several decades has slowed. The International Monetary Fund's World Economic Outlook 2015 (IMF 2015) forecast the continuation of a low 2% growth rate for Australia over the next 3 years. At this point the economy, while fragile and uncertain, appears to still be basically sound in terms of employment rates, at least relative to comparable countries. The key point is that Australia can no longer simply rely on being the 'lucky country' (if it ever could) for its economic and social well-being. Global ageing as well as national ageing needs to be incorporated into comprehensive strategies for the country's future. Traditionally in Australia, population ageing has been characterised as a cause for concern and a burden on the future prosperity of society. However we need to challenge this way of thinking and focus on the contributions that older people can make to society.

Australia's prospects as an ageing country remain favourable relative to our European and North American comparators. As outlined in Chap. 1, Australia is continuing to benefit from the 'longevity dividend' (Chomik and Piggott 2015) in which there are record proportions of people in the mid-age groups of peak productivity relative to the younger and older groups in which people are most likely to be dependent. Further, Australia has recently experienced increases in the birth rate as well as migration, and migration policy presents options to bring in people who could meet specific workforce shortages and generate demand in rural areas with declining populations (Chap. 4; McDonald 2016). While migration raises political and social challenges, it offers some prospects for targeted population growth with relatively fewer long-term investment costs to government. Beyond these population considerations, Australia also has the advantage, relative to Europe and the USA, of having less accumulated public debt and favourable (from a fiscal viewpoint) age-related income support provision.

15.2.1 The Intergenerational Reports

The Intergenerational Reports (IGRs), most recently IGR 2015 (Australian Treasury 2015), produced every 3–5 years, provide indications of the national Government's views on ageing and projections of the longer-term fiscal outlook. The IGRs need to be assessed critically, not only in terms of their longer-term projections but also their short-term political purposes, for example, as tactical tools in attempts at negotiating budgets in Parliament. The IGRs examine intergenerational equity defined as 'fairness in the distribution of public resources between generations of Australians' (Australian Treasury 2002, p. 14). A related Treasury definition of fiscal sustainability is that 'all obligations, current and future, can be met without changing current policy settings' (McKissack and Comley 2005: fn3, p. 2). The IGRs recognise the fundamental importance for economic growth of the '3Ps'—Population (growth), Participation (in the workforce), and Productivity.

The IGRs were initially developed by the Howard Coalition Government (1996–2007), and were invoked at that time as a rationale 'to reign in Labor debt', yet they continued under the Rudd/Gillard Labor Government (2007–2010/2010–2013) as well as the Abbott/Turnbull Coalition Government (2013–2016). Their basic approach is simple: first, identify government expenditure by age groups; second, project the age structure of the population into the future (to 2046 in the latest IGR); and third, project consequent dramatic rises in government expenditure as a rationale for moderating ageing-related expenditure. The underlying message of the IGRs is that government expenditure on older people raises problems of fiscal sustainability that would reduce the life chances of future generations. Overall, the IGRs have been important for raising appreciation in the wider public of the significance of population ageing, and for providing views on governments' political and fiscal priorities.

There are major concerns, however, with the IGRs (Woods and Kendig 2015; Kendig 2010). They attribute, to population ageing, projected increases in health expenditure that in reality also reflect rising utilisation rates and costs for all age groups. Older people are at risk of being 'scapegoated' for the consequences of demographic change, with governments attributing fiscal pressures to older people who are widely portrayed as a drain on public expenditure. IGR 2015 is in some ways the most political of these reports because it compared budgetary deficit prospects of the current Coalition and former Labor governments; but it also includes the Treasurer's positive comments on increasing mature age labour force participation.

15.2.2 Alternative Views

The Productivity Commission (PC) (2013) presented a valuable research paper *An Ageing Australia: Preparing for the Future,* aimed at informing the next IGR. The PC also released the *Productivity Update Report* in 2015. In contrast to the Treasury, which produces the IGRs on behalf of the Government, the PC has legislated independence in its statutory responsibilities to advise governments in the long-run interests of the community as a whole. The PC Report (2013) notes that 'slow but profound shifts in the nature of a society do not elicit the same scrutiny as immediate policy issues. The preferable time to contemplate the implications is while these near-inevitable trends are still in their infancy' (p. 2).

Some of the highlights of the 2013 Report are:

- Population ageing is largely a positive outcome. A female born in 2012, for example, will on average live for an estimated 94.4 years.
- Population growth and ageing will affect labour supply, economic output, infrastructure requirements, and government budgets.
- The population aged 75 or more years is expected to rise by 4 million from 2012 to 2060.
- Total private and public investment requirements over the 50-year period ahead are estimated to be more than five times the cumulative investment made over the last half century.
- Labour participation rates are expected to fall from around 65 %–60 % from 2012 to 2060.
- Labour force productivity is projected to grow at a relatively low 1.5 % per annum (much lower than earlier periods).
- Real disposable income per capita is expected to grow at 1.1 % per annum compared with the average 2.7 % annual growth over the last 20 years.

The PC (2013) report also notes possible new reform approaches that are not on the current policy horizon. These include:

- The design of the age pension and broader retirement income system might be linked to life expectancy after completion of the current transition in pension eligiblity to 67 years in 2023.
- Using some of the annual *growth* in the housing equity of older Australians could help ensure higher quality options for aged care services and lower fiscal costs.
- Wide-ranging healthcare reforms could improve productivity in the sector that is the largest contributor to fiscal pressures. Even modest improvements in this area would reduce fiscal pressures significantly.
- Projections that Australian governments (at all levels) will face additional pressures on their budgets equivalent to around 6 % of national GDP by 2060, principally reflecting the growth of expenditure on health, aged care, and the age pension.

The independent Grattan Institute has also provided views on the budgetary outlook and the fiscal options potentially available to a Commonwealth government in dealing with population ageing and revenue shortfalls. Daley and Wood (2015) have noted that the last 6 years of deficits are 'largely due to a rapid increase in net spending on older households. The costs of repaying this debt are likely to fall on younger households. The next ten years are likely to be even more difficult' (p. 1). They also note that savings to the Commonwealth are proposed to be achieved partly through reduced grants to the States, and hence would involve either service cuts, particularly in health, or a shifting of taxation responsibilities. They suggest four priority reforms for 'repairing' Commonwealth and State government revenues: reducing superannuation tax concessions, changing capital gains tax concessions, broadening the GST, and introducing a broad-based property levy. They argue that this would limit '...collateral damage to the economy and the most vulnerable in our society' (p. 1). These proposals have significant age-specific and generational impacts, especially for ageing people who have accumulated significant resources over their lives or who are in the process of doing so (see Chap. 2).

Additional approaches, of particular concern for older people, have been raised by the coalition of business, community, and worker representatives who formed the National Reform Summit (2015). The Summit recommended directions for reforms aimed at achieving sustained and equitable economic performance and quality of life. These included taxation of superannuation and other areas of 'tax expenditure', targeting expenditure on individuals most in need, and further efforts to increase workforce participation and productivity. A consistent implication of the Summit's work is that social equity and fairness issues extend beyond simple age and generational comparisons. These reforms also aim to take into account life span investments and inequalities across the life course. Advocates for a comprehensive Retirement Income Review, for example, argue for actions that aim for adequate support for all pensioners in need, balanced by reduced tax support for advantaged superannuants.

15.2.3 Towards Budgets and the Future

Government IGR projections and the (current) Coalition Government's first two budgets (2014, 2015) have aimed to address fiscal problems primarily through expenditure restraint. There has been less focus on government revenue or regressive distributional impacts. For these and a range of other reasons, many of the government's fiscal measures have, to date, been politically unacceptable as major budget measures could not pass the Senate. Government faces major challenges to turn around rapidly accumulating deficits in the context of what appears to be a tepid outlook for economic growth.

Prospects for real incomes and wealth are central in assessing generational equity between age groups and birth cohorts into the future. In his application of the rigorous and independent modelling capacities of the National Centre for Social and Economic Modelling (NATSEM), Phillips (2015) investigated trends in living standards, taking into account the effects of the 2014 and 2015 Coalition budgets. He found that over the previous 10 years the gap between the rich and poor had widened: the living standards of the top 20 % of households grew almost twice as fast as for the bottom 20 %. Further, older households had enjoyed stronger gains in living standards compared to younger households over the past 10 years. Over the next 10 years, however, it is projected that lower income older households will experience relatively slower growth in living standards as a result of recent budget decisions, including the tightening of asset and income means testing.

As was noted in Chap. 1 and developed in Chap. 9, housing wealth is central to mitigating the adverse effects of Australia's low rate of the old age pension (by international standards). A comprehensive review by Stebbing and Spies-Butcher (2015) noted that the high rate of home ownership among older Australians explains why their comparatively high rates of 'income poverty' contrast sharply with low rates of 'after housing costs'—the lowest in the OECD. They conclude that 'home-ownership can be seen as an important pillar of the Australian retirement income system' (p. 4), but also note the inequalities arising: home ownership is heavily subsidised through the tax system, as well as advantaged in pension means tests, but these benefits are available only to those who have had sufficient private means to buy. They also note the regressive effects on intergenerational equity for future groups of older people given projected declines in home ownership rates.

Taken as a whole, these findings indicate that Australia may be at a turning point in which many in future generations could have lower lifelong economic prospects than the baby boom birth cohort now entering later life. At the same time, inequalities within each of the generations remain strong and arguably are increasing over the life course. Policy options on pension means tests and related policies need to take account of widely held views recognising home ownership as central to the living standards of pensioners. Yet they also need to take account of the disadvantage experienced by those who have not had the advantages of home ownership late in life.

15.3 International Directions in Ageing

In this section we outline some key international contributions advancing progressive thinking and constructive debate on behalf of ageing people and global ageing. A generation ago the World Bank (1994) published its pessimistic, influential report, *Averting the Old Age Crisis,* that analysed population ageing as a major risk to economic growth and security for young and old. Over subsequent years important leadership on constructive approaches has been provided by a number of international organisations including the International Federation on Ageing (IFA), The International Association of Gerontology and Geriatrics (IAGG), the United Nations (UN), the World Bank, and the World Economic Forum. It is notable that the overall tone of this discourse over recent years has been increasingly positive, notwithstanding ongoing financial and economic concerns.

In turn, Australia has been an important contributor to international developments on ageing, notably Professor Gary Andrews' leadership as IAGG President (1997–2001) and his service as a driving force in UN and WHO international plans on ageing (Kendig et al. 2013). Australia also provided the Chair and Secretariat for the IAGG Asia/Oceania region (2011–2015) and, through the Australian Association of Gerontology, established the International Longevity Centre in 2014 (ILC-Australia 2014) as part of an international coalition. Australian researchers and service providers are contributing to a range of developments in Asia, for example, to building aged care services in China by the Royal District Nursing Service (RDNS) and policies for pensions and long-term care financing in China (CEPAR, 2014; Lu et al. 2014).

15.3.1 The United Nations

Since the turn of the century the UN's Vienna and Madrid International Plans of Action on Ageing (MIPAA), developed in collaboration with the IAGG, have been valuable in widening horizons for older people in developed, as well as developing, countries. For example, the MIPAA priority area of Older Persons and Development identified a range of older people's contributions important for their families, communities, and wider global economic and social development (UN 2002). Yet ageing, as a focus, receded to almost insignificance as poverty and children's and women's issues were foremost in the UN's widely disseminated Millennium goals established in 2000 (UN 2000). In September 2015, the UN, with the support of Age Concern and other international advocates, adopted a next generation of Sustainable Development goals as defined in *Transforming our World: the 2030 Agenda for Sustainable Development*; this included Goal 3 'Ensure healthy lives and promote well-being for all at all ages' (UN 2015).

UN efforts to establish a Convention on the Human Rights of Older People have been an important part of the global efforts to combat age and other forms of discrimination and to protect the rights of older people to access resources on the basis of their needs (Global Action on Aging 2009). The rights approach is valuable for moving beyond paternalistic or welfare approaches and recognises the dignity and social standing of older people. While Australia, the USA, and other countries have not supported this UN Convention—arguing that an age-specific approach was not necessary given generic UN human rights protection—the age rights approach has been influential in Australia, as reviewed in the Attitudes chapter of this book (Chap. 3). In 2015 ILC-Australia supported efforts to establish, in the UN World Assembly, a 'Human Rights Council Independent Expert on the enjoyment of all human rights by older persons' (http://www.ohchr.org/EN/Issues/OlderPersons/IE/Pages/IEOlderPersons.aspx).

15.3.2 The World Health Organisation

The World Health Organisation (WHO) initiated a constructive stance on ageing in its landmark Active Ageing Strategy launched at the Second United Nations World Assembly on Ageing held in April 2002. WHO defines 'active ageing' as 'the process of optimising opportunities for health, participation, and security in order to enhance quality of life as people age' (WHO 2002, p. 12). Importantly, this constructive approach conceptualises ageing across the life course, considers social determinants, and the value of supportive local environments on autonomy and dignity. WHO's former director of the Ageing and Life Course Programme, Doctor Alexandre Kalache, has recently updated an extension of the Active Ageing approach to encompass major contemporary trends including urbanisation, globalisation, growing inequities, age friendliness, feminisation of ageing, migration, technological innovation, and environmental and climate change (ILC-Brazil 2015).

In October 2015 WHO launched its comprehensive *World Report on Ageing and Health,* which focussed on healthy ageing and well-being, age-appropriate health systems, long-term care, age-friendly environments, and supporting information and research management (WHO 2015). The report is supported by references to extensive research evidence and articulates specific strategies for moving ahead in each of the priority areas.

15.3.3 The European Union

The European Union (EU) has an older population than Australia and has a number of ageing policies that relate closely to those of international organisations based in Europe. The European Innovation Partnership in Active and Healthy Ageing aims to increase average healthy life years by 2 years by 2020 (European Commission 2014). The 2015 Ageing Report, which parallels the Australian IGRs, concludes that the EU is projected to move from four to two working age people for every person aged 65 and older in 2060 (European Commission 2015a). In explaining this report, the European Commission stated that 'Strong reforms now introduced in a majority of Member States, especially in the field of pensions, show that the long term trends in public spending strongly influenced by demographic changes can be curbed through determined policy action' (European Commission 2015b, p. 3).

15.3.4 The United States of America

The United States of America (USA) is of course a highly diverse country and it is likely to age more slowly than Australia due to its higher birth rates and migration (Chap. 1). For the purposes of this book, a good summary of constructive developments is provided in 'Successful Ageing of Societies' by American and European experts (Rowe 2015). This edited collection presents progressive interpretations moving beyond the usual themes of ageing in the USA (e.g. the future costs of Medicare and Social Security). The collection of papers provides constructive analyses of the opportunities as well as the challenges to '... offer policy options to address... the substantial benefits and potential of an ageing society ...' and to '... enhance the transition to a cohesive, productive, secure, and equitable aging society. Such a society would not only function effectively at the societal level but will provide a context that facilitates the capacity of individuals to age successfully' (p. 5).

15.4 Ongoing Issues for the Future in Australia

We now turn to some 'big ideas' that are emerging as constructive directions for reconceptualising actions on ageing in Australia. They are advanced as key ways for addressing the challenges of the long-term 'IGR projections' for guiding coherent policy action, notwithstanding the incremental nature of annual budgets. These issues build on initiatives examined in earlier chapters and present opportunities that extend to greater productivity, more efficient and effective service delivery, and more equitable targeting of services and taxation support across the life span.

15.4.1 Human Capital Over the Life Span

Fundamental to advancing policy for an ageing Australia and for future generations is the value of developing and drawing on human capital over the life span, taking into account diversity and unequal life chances. Investing in education, health, and other social resources early in life—reinforced by additional opportunities (and where necessary recovery) during family and work lives in mid and later life—can yield considerable individual and social returns, including increased capacities and resources in later life. Life span investment at every age, transition, and life stage will strengthen the foundation for productive ageing and ageing well. It provides a powerful way of mitigating the adverse effects of demographic ageing and the real risk of declining economic resources for future cohorts.

A carefully developed life span approach has several major strengths. First, it cuts through simplistic and interest-ridden divisions in contests between young and old: one can invest in life resources among the young *and also* take different, more appropriate actions as people grow older. The life span approach can invest in potential, and ameliorate adversity, equitably and appropriately across all spheres of life. Investments in human capital over the coming decades may contribute crucially not only to improved workforce participation, but also to productivity gains that are essential yet elusive.

Increasing older people's health and social capacities and opportunities for productive contributions—be they paid or not (and whether counted in GDP or not)—provides win-win solutions. A particular point to emphasise is the complementarity between younger and older workers as the latter can generate and preserve opportunities for and along with the former. Retraining and workplace changes can facilitate adaptations that enable continued productivity. Further, satisfying and safe work is good for health and well-being at all ages.

15.4.2 Ageing Well

The concept of ageing well is emerging as a valuable way ahead in action towards what older (and younger) people want most: to feel well and to have the capacity to remain independent and able to continue to contribute and to participate in families and communities (Kendig and Browning 2016). Wellness is a holistic state that is related to (but extends beyond) health conceptualised in terms of usual definitions such as absence or amelioration of disease. The ageing well approach is based on evidence that health and well-being can be achieved, maintained, and where necessary regained when it is set back on a short-term basis (Chaps. 12 and 13). Along with living well is the value of eventual 'dying well' with personal control and dignity.

The ageing well approach recognises that health extends far beyond the treatment approaches, which are the primary focus of health professions and health industries. It encompasses lifelong healthy ways of life and recovery from adversity, as well as self-management of illness and partnerships between older people and health professionals. It extends to the ways in which social environments, health promotion, and age-friendly communities can provide appropriate supports that enable independence without the expense or dependency that can undermine older people's capacities and morale. An ageing society can recognise health as a resource and investment that can enhance well-being and yield sound economic returns.

'Enabling active ageing' provides a proactive, social approach to ageing well. A workshop of leading policymakers, ageing advocates, and researchers—convened by the ASSA—examined the concept and strategies to achieve it (Podger 2016). They challenged the idea of a fixed retirement age and considered the importance of facilitating continuing workforce participation, family and community contributions. They pointed to the importance of public health messages of 'making healthy choices the easy choices'. They underscored the importance of maintaining cognitive capacities (Anstey 2016) and the value of UK approaches to promote mental well-being: Connect; Be active; Take notice; Keep learning; Give (Foresight 2008, p. 24).

15.4.3 Community Support and Consumer-Directed Care

Enabling people to continue to live satisfying and independent lives notwithstanding age-related challenges is of self-evident importance yet it involves major dilemmas and obstacles. Building age-friendly communities and 'whole of life' adaptable dwellings can enable continuing independence for those with mobility impairments at all ages. Consumer-directed care aims to maximise older people's choices and their control of services when they require assistance in their households (see Chap. 14). The accommodation base for care as well as the management of support and care can rest primarily with older people and families when they have the decision-making capacities and the financial means available from public or private resources. A fully developed choice model of accommodation and care also requires the availability of well-developed local markets for accommodation and care, including quality information on available options and strong consumer protection.

The development of consumer-directed care is facing major challenges. The language of 'Client-directed care' makes clear that frail older people remain as the responsibility of organisations that retain powers to determine care in line with professional judgements. 'Duty of care' responsibilities underscore the fine and delicate balance required when older people's capacities and needs compromise their decision-making. In line with the strong disability rights movement for younger people, one can ask if older adults (along with other adults) should have the right to make their own choices, even when these choices are not deemed by 'experts' to be the best ones for their health and care. 'Entitlements' to care, as advocated by COTA (Australia), are not easily implemented during times of severe fiscal restraint.

The dilemmas of care are accentuated by the necessary accountability to funders of service providers and clients. Governments have responsibilities to ration public expenditure and ensure quality of care, but standardised packages of funding and routinised care procedures cannot take much account of older people's choices and the variability between them. The usual machinery of government risks major loss of efficacy and effectiveness, as well as choice, when 'clients' have no alternative but to accept fixed forms of assistance that may not be their priorities. More funding from clients themselves can increase flexibility and adequacy for many older people, but this approach does have risks for vulnerable people. Private payments (including top-up) may be desirable to better meet preferences, and can stretch resources farther through means testing, but they can increase inequalities experienced by those having fewer financial resources.

15.4.4 Age and Financial Need and Capacity

Ageing and retirement from paid work involve significant financial risks, but over time more people are bringing significant superannuation and investments to later life, while others remain on basic pensions only (Chap. 9). Superannuation concessions on contributions (and tax-free drawdowns) substantially benefit those on higher lifelong incomes, while mandated Superannuation Guarantees at present rates will not bring widespread benefits for at least a decade. Widening and increasing the Goods and Services Tax would impact relatively more on older people, collecting more tax on retired people who pay little or no income tax. This could increase equity between working and nonworking groups if compensatory adjustments were included for those on low incomes at all age groups.

Home ownership and other forms of wealth, which are a focal point for current public policy, are areas where many older people overall are advantaged relative to younger people (Chaps. 2 and 9). Increasing socio-economic inequality, along with financial pressures on governments, is bringing into focus the treatment of age in public policy on taxation as well as expenditure. An emergent issue concerns the accumulation of wealth over the life span, its uses in later life, and its transmission through gifts and inheritances to the next generation (Rosenman et al. 2015). A recent national survey concluded that the majority of people making wills are making provision for immediate family (Tilse et al. 2015) but '…these considerations are not reducing provisions for living comfortably in older age' (p. 333). The authors observe the increasing wealth of the baby boom cohort and conclude:

> Although the value of some estates will be diminished by a long period of post-retirement living intersecting with aged care and retirement income policy initiatives in user charges, self-provision and asset testing, a much larger proportion of the population has some assets of financial value to leave (p. 334).

New policy approaches to wealth would be contentious and redistributive actions would understandably be resisted by strong interests. Tax concessions on capital gains as well as superannuation—which benefit higher income earners over the course of their lives (not just in later life)—are under public scrutiny. Housing wealth is increasingly taken into account in pension means tests, but proposals to better enable older people to draw on the wealth in their homes have not advanced very far. More generally, more reliance on property taxes would impact heavily on older people (as well as others) having substantial wealth in their homes. The impact of inheritance taxes of various forms would fall on the middle generations in advantaged family lineages, but these taxes were removed from the political agenda 30 years ago with the abolition of both Commonwealth and State death duties.

The complexity and importance of tax and financial means extend the scope of traditional conceptions of ageing policy beyond health and welfare matters, but are central to a deeper recognition of ageing, age relationships, and generational and social change. Trends towards market approaches to aged care and financial resources for retirement underscore the value of financial literacy and consumer protection. The choices and power of those older people who do have substantial financial resources to match their life aspirations can be expected to have a transforming impact on the image and facts of ageing.

15.4.5 A 'Rights' Approach

New conceptions of the rights of ageing are working to advance positive attitudes and opportunities for older people (Chap. 3). The rights approach provides a means of confronting the negative images of older people that can be demoralising and, in their own way, disabling for ageing people across spheres of life. The strengths of the rights approach are its importance for dignity and respect, and also to facilitate participation and contributions in public and private life. People are to be 'treated' equally at all ages while taking into account what is appropriate for different age groups and indeed different individuals.

The rights approach typically emerges in preamble statements and occasionally in hard legal action on discrimination in Australia. The symbolic and social leadership value of the rights approach has been applied in Australia in terms of public images of ageing, treatment in residential care, and most effectively in terms of the workplace (Chap. 8). In the next section we will review factors that have led to the rights approach emerging as one of the most effective strategies for advancing positive action on ageing in Australia.

15.5 Australian Leadership on Change (or Making it Happen)

National leadership on ageing needs to take into account the Australian context.

In contrast to our Asian neighbours, or even the European Union, Australia like the USA does not have much of a focus on national development planning on ageing. Indeed, the National Strategy on Ageing, released by the Prime Minister in 2001 (Department of Health and Ageing 2001), had lofty ideals and some sensible directions, but it never had a chance for 'implementation' in the annual budget contests of the powerful departments that form the Australian government. Further complexities arise in Australia with many of the responsibilities on ageing divided (and often contested) between Commonwealth, State, and (to a limited extent) local government. Reviews under way of federalism and of taxation are central to integrated and appropriate treatment of older people in policy development.

Several recent Australian developments have demonstrated ways in which comprehensive advances can be achieved. The *Living Longer, Living Better* reforms found a way through a gridlock of interest groups, earlier failed efforts, financial restraint, and a change of government during implementation (Chap. 14). How was this success achieved? First, the expert and independent Productivity Commission review consulted widely and then tested draft recommendations with a range of constituencies. Second, both the national and State Councils on the Ageing (COTA Australia) brought a strong consumer voice to join with government in conducting advance consultations and then leading the National Aged Care Coalition (NACA) representing consumer, provider, professional, and other interests in guiding implementation. Third, there was the leadership of the Minister on Mental Health and Ageing at the time, as well as bipartisan support in the transition to implementation led by the Coalition's Minister on Ageing and Disability.

Another instance of national progress has been work by the Australian Human Rights Commission (AHRC) in improving attitudes and confronting ageism, particularly in the workforce (Chaps. 3 and 8). A number of factors contributed to this progress. First, the times had been right given the national priority to raise labour force participation, employer needs to retain good workers, and older people's wishes to be able to stay in the workforce. Second, there was a carefully planned process of lodging discrimination complaints on the AHRC website (and efforts made to resolve them), as well as supportive research projects, consultations, and demonstrations that pointed the way towards the value and feasibility of win-win actions. Third, there was the high profile leadership of the Age Discrimination Commissioner, particularly in challenging age discrimination in the workplace (Chap. 3).

As for future directions, the Department of Social Services states: 'We aspire to be Australia's pre-eminent social policy agency. Our mission is to improve the lifetime wellbeing of people and families in Australia' (Department of Social Services, 2015). The portfolio has central responsibilities for benefits and services across the life span including families and children, housing, seniors, communities and vulnerable people, disability and carers and, until recently, aged care which has moved to the Department of Health. Ageing is also of major importance for central departments (Prime Minister and Cabinet, Treasury, and Finance) and information agencies (notably the Australian Institute of Health and Welfare and Australian Bureau of Statistics). Working together these agencies have potential for cross portfolio development and long-term planning that could enhance ongoing policy coherence through the Departments and across budget and electoral cycles.

The Department of Human Services (DHS) (with Centrelink) aims to include older Australians in the mainstream of delivering Commonwealth payments and services (DHS 2015). Flexible options are provided for in-centre, telephone, and web communication concerning pensions, Medicare, carer benefits, and aged care services supplemented by 'News for Seniors'. These systems potentially can tailor integrated information in an appropriate and timely way, for individuals in varied circumstances, such as those with limited mobility, those making transitions (e.g. retirement and widowhood), and those who rely on carers. An Older Persons Advisory Group of national consumer representatives and service delivery organisations, along with ongoing service quality monitoring, contribute to DHS efforts working with relevant departments in 'co-planning' developments that aim to improve access and efficiency. These directions recognise the community's increasing (and varied) organisational literacy, information technology capacities, and expectations for a next generation of quality and choice in government engagement with older people.

State government leadership has primarily consisted of Ministers and Offices on Ageing who continue to articulate the value of older people in the community and provide leadership in a range of pilot programmes, consultations, and advocacy or coordination within mainstream health and care delivery. An ambitious *NSW Ageing Strategy*, and related developments in age-friendly policies in Victoria and the ACT, aimed to bring ageing and older people as central concerns into the mainstream of state and local government (Kendig et al. 2014).

15.6 Conclusion

Far-sighted actions on ageing will depend on recognising the importance of life span development, social inequalities, and social change as central to Australia's future. The scale of population ageing and increasing life expectancy, along with anticipated slowing of economic growth and continuing fiscal pressures on government, will have increasing impacts as the diverse baby boom cohort moves fully into later life over the coming decades to mid-century. While the broad sweep of economic and social developments sets the context of change, it is important to recognise that advances in attitudes to, and policies on, ageing will be led by older people themselves, employers, advocates, and other social institutions as much as by governments. For example, while efforts to build health promotion into government policy have not been very successful to date, the fact remains that ageing well has been advancing over recent decades largely through the actions of older people themselves, enabled by increasing personal resources and improvements in attitudes and expectations for ageing.

Over the coming decades, later life will continue to be shaped by further developments in Australian society. We can expect, and indeed work towards creating, opportunities for more choice in learning, working, family life, leisure, and other activities flexibly over the life course. We can expect the status and social power of older people to rise along with the social and economic resources of ascendant cohorts in later life. Inequalities over the life course and between families and social classes would appear to be increasing, as Australia experiences highly variable economic development, but more equitable outcomes can be achieved through purposeful actions. We have suggested some 'big ideas' for advancing ways of thinking and guiding equitable and effective actions on ageing and development over the life course.

We can envisage major societal tensions and contradictions as well as opportunities and challenges ahead for ageing. Continuing resource constraints in the public sector and increasing economic disparities among older (and younger) people may be in conflict with the new aspirations emerging for the diverse baby boom cohort. We can expect ongoing advocacy by consumer representatives, notably COTA (Australia) and other peak bodies. We can also expect lobbying by industries concerning public subsidies and regulation in the expanding markets for ageing-related products and services. *There is the reasonable prospect that older people themselves can be an important part of the solutions to the challenges of population ageing.* The push towards constructive change also needs to harness the momentum of employer and other economic interest groups that are themselves engaged in political action and image-making on ageing. This complex of institutions and interests forms the ongoing grist of public affairs and social change.

Public policy on ageing requires an in-depth and ongoing understanding of cohort life chances, disadvantaged groups, and the changing income and consumption of age groups over the life span. Caution is essential to avoid misleading characterisations of groups as a whole, be they '*the* baby boomers' or '*the* elderly'. Diversity of resources, opportunities, and vulnerabilities are influenced by many social and individual factors including gender, social class positions along with other social divides through earlier as well as later life. A strong, engaged research base and education on ageing are vital to the building of knowledge and its translation into action. Informed action by governments, employers, and other social institutions—as well as by individuals and families—is essential for effectively and equitably building capacities and meeting needs over the life span in a climate of social and economic uncertainty for the future.

As Australia faces the challenges ahead, it is important to appreciate that, in comparison to relatively wealthy European countries, Australia is facing a gentle pace of population ageing and our economic and fiscal outlook is more encouraging. There will be challenges and adjustments, but arguably more opportunities than crises ahead. Australia has several decades or more before it reaches the proportions of older people that are already being managed in Europe and Japan. There is a strong case for investing in human capital at *all* points across the life span—in education, good health, workforce skills and opportunities, and age-friendly built environments. Actions in these areas arguably are the most effective, efficient, and equitable way to protect against demographic pressure and economic shocks in Australia's future. There is every prospect that constructive, positive attitudes and purposeful, whole-of-community actions can transform Australia into an age-friendly society that realises the benefits of an older society.

References

Anstey, K. (2016). Enhancing cognitive capacities over the lifespan. In H. Kendig, P. McDonald, & J. Piggott (Eds.), *Population ageing and Australia's future*. Canberra: ANU Press for the Academy of Social Sciences in Australia.

Australian Treasury. (2002). *Intergenerational Report 2002-03*. Canberra: Australian Government.

Australian Treasury. (2015). *2015 Intergenerational Report, Australia in 2055*. Canberra: Australian Government.

Centre of Excellence in Population Ageing Research (2014). *Annual report, 2014*. Australian Research Council CEPAR. UNSW; USYD; ANU: Australian Government

Chomik, R., & Piggott, J. (2015). Population Aging and Social Security in Asia. *Asian Economic Policy Review, 10*(2), 199–222.

Daley, J. & Wood, D. (2015). *Fiscal Challenges for Australia*. GRATTAN Institute. Working Paper No. 2015-4, July 2015. Retrieved from http://grattan.edu.au/wp-content/uploads/2015/07/824-Fiscal-challenges-for-Australia2.pdf.

Department of Health and Ageing. (2001). *National Strategy for an Ageing Australia: An older Australia, Challenges and opportunities for all*. Canberra: Australian Government.

Department of Human Services (DHS) (2015). *Older Australians*, Australian Government. Retrieved from: http://www.humanservices.gov.au/customer/themes/older-baustralians?from=theme-bar

Department of Social Services (2015). *About the department*, Australian Government. Retrieved from https://www.dss.gov.au/about-the-department.

European Commission (2014). *Investments in health: Policy guide for the European structural and investment funds (ESIF) 2014-2020*. March 2014. Retrieved from http://ec.europa.eu/health/health_structural_funds/docs/esif_guide_en.pdf.

European Commission. (2015a). *The 2015 Ageing Report: Economic and Budgetary Projections for the 28 EU Member States (2013-2060)*. European Economy, March, 2015. Retrieved from http://ec.europa.eu/economy_finance/publications/european_economy/2015/pdf/ee3_en.pdf.

European Commission (2015b). *2015 Ageing Report: Frequently Asked Questions*. European Economy March, 2015. Retrieved from http://ec.europa.eu/economy_finance/structural_reforms/ageing/pdf/memo_2015_ageing_report_en.pdf

Foresight (2008). *Foresight Mental Capital and Wellbeing Project: Final Project Report*. London: The Government Office for Science.

Global Action on Aging (2009). *United Nations activities towards a convention*. Retrieved from http://globalaging.org/agingwatch/convention/index.htm

International Longevity Centre Australia (ILC-Australia) (2014). *Members: ILC-Australia*. Retrieved from http://www.ilc-alliance.org/index.php/members/details/ilc-australia.

International Longevity Centre Brazil (ILC-Brazil) (2015). *Active ageing: A policy framework in response to the longevity revolution*, ILC-Brazil; ILC. Retrieved from http://www.ilcbrazil.org/wp-content/uploads/2015/07/ActiveAgeingPolicyFramework_2015.pdf

International Monetary Fund. (2015). *World economic outlook: Adjusting to lower commodity prices (October)*. Washington, DC: IMF.

Kendig, H. (2010). The intergenerational report 2010: A double-edged sword. *Australasian Journal of Ageing, 29*(4), 145–146.

Kendig, H., & Browning, C. (2016). Social research and actions in ageing well. In H. Kendig, P. McDonald, & J. Piggott (Eds.), *Population ageing and australia's future*. Canberra: ANU Press for the Academy of Social Sciences in Australia.

Kendig, H., Lucas, N., & Anstey, K. (2013). Thirty years of the United Nations and global ageing: an Australian perspective. *Australasian Journal of Ageing, 32*(Suppl. 2), 28–34.

Kendig, H., Elias, A., Matwijiw, P., & Anstey, K. (2014). Developing age-friendly cities and communities in Australia. *Journal of Aging and Health, 26*(8), 1490–1414.

Kendig, H., McDonald, P., & Piggott, J. (Eds.). (2016). *Population ageing and Australia's Future*. Canberra: ANU Press for the Academy of Social Sciences in Australia.

Lu, B., He, W., & Piggott, J. (2014). Should China introduce a social pension? *Journal of the Economics of Aging, 4*, 76–87.
McDonald, P. (2016). Ageing in Australia: Population changes and responses. In H. Kendig, P. McDonald, & J. Piggott (Eds.), *Population ageing and Australia's future*. Canberra: ANU Press for the Academy of Social Sciences in Australia.
McKissack, A. & Comley, B. (2005). Fiscal sustainability and pre-funding strategies in OECD countries. *Economic Roundup* Winter 2005, Commonwealth Treasury. Retrieved from http://www.treasury.gov.au/documents/1009/PDF/Winter_2005_ Economic_Roundup.pdf.
Ministerial Advisory Committee on Ageing. (2015). About us Family and Community Services: Ageing, Disability and Home Case, N.S.W. Government. Retrieved from http://www.maca.nsw.gov.an/about_us.
National Reform Summit (2015). *Statement from the National Reform Summit*. The Australian Financial Review, 26 August 2015. Retrieved from http://www.kpmg.com/AU/en/IssuesAndInsights/ArticlesPublications/national-reform-summit/Documents/national-reform-summit-statement-26-august-2015.pdf.
New South Wales Government. (2012). *NSW ageing strategy*. Retrieved from http://www.adhc.nsw.gov.au/about_us/strategies/nsw_ageing_strategy&gt.
Phillips, B. (2015). *Living standard trends in Australia: Report for Anglicare Australia*. NATSEM, University of Canberra, September 2015. Retrieved from http://www.natsem.canberra.edu.au/storage/Living%20Standard%20Trends%20Final.pdf.
Podger, A. (2016). Conclusions an historic: triumph that present big challenger for public policy. In H. Kendig, P. McDonald, & J. Piggott (Eds.), *Population ageing and Australia's future*. Canberra: ANU Press for the Academy of Social Sciences in Australia.
Productivity Commission. (2013). *An ageing Australia: Preparing for the future. Research Report*. Canberra: PC.
Productivity Commission. (2015). *PC productivity update report*. Canberra: PC.
Rosenman, L., Tilse, C., Wilson, J., Feeney, R. & White, B. (2015). *Wills and the intergenerational transfer of wealth*, Paper presented at APRU Symposium, 28-29 September, 2015. University of NSW.
Rowe, J. W. (2015). Successful Aging of Societies. *Journal of the American Academy of Arts and Sciences*, Spring 2015, 5–12.
Stebbing, A., & Spies-Butcher, B. (2015). The decline of a homeowning society? Asset-based welfare, retirement and intergenerational equity in Australia. *Housing Studies, 31*(2), 190–207. doi:10.1080/02673037.2015.1070797.
Tilse, C., White, B., Rosenman, L., & Feeney, R. (2015). Will-making prevalence and patterns in Australia: keeping it in the family. *Australian Journal of Social Issues, 50*(3), 319–338.
United Nations. (2000). *United Nations Millennium Declaration, Resolution Adopted by the General Assembly*, 18 September 2000, A/RES/55/2. Retrieved from http://www.refworld.org/docid/3b00f4ea3.html.
United Nations. (2002). *Political declaration and Madrid international plan of action on ageing*. The Second World Assembly on Ageing, Madrid, Spain, 8-12 April 2002, New York: UN.
United Nations. (2015). *Transforming our world: the 2030 Agenda for sustainable development*. UN Sustainable development summit 2015, 25-27 September. New York City, USA. Retrieved from https://sustainabledevelopment.un.org/post2015/transformingourworld.
Woods, M., & Kendig, H. (2015). The Intergenerational Report 2015: A limited and political view of our future. *Australasian Journal on Ageing, 34*(4), 217–219.
World Bank. (1994). *Averting the old age crisis: Policies to protect the old and promote growth*. Washington, DC: Oxford University Press.
World Health Organization. (2002). *Active Ageing: A policy framework*. WHO/ NMH/NPH/02.8. Geneva: WHO.
World Health Organization. (2015). *World report on ageing and health*. Geneva: WHO.. Retrieved from http://www.who.int/ageing/publications/world-report-2015/en/

Appendix: Key Websites Related to Chapter Content for Students to Access

Australian

Academy of the Social Sciences in Australia (ASSA): http://www.assa.edu.au/

Aged and Community Services Australia (ACSA): http://www.agedcare.org.au/

Anti-Discrimination Board of NSW: http://www.antidiscrimination.justice.nsw.gov.au/

Australian Association of Gerontology (AAG): https://www.aag.asn.au/

Australian Bureau of Statistics (ABS): http://www.abs.gov.au/

Australian Human Rights Commission (AHRC): https://www.humanrights.gov.au/

Australian Institute of Health and Welfare (AIHW): http://www.aihw.gov.au/

Centre of Excellence in Population Ageing Research (CEPAR): http://www.cepar.edu.au/

Council on the Ageing, Australia (COTA): http://www.cota.org.au/australia/

Department of Health: http://www.health.gov.au/

Department of Social Services: https://www.dss.gov.au/

Diversity Council of Australia: http://www.dca.org.au/

Health Services Research Association of Australia and New Zealand (HSRAANZ): http://www.hsraanz.org/Home.aspx

Independent Living Centres, Australia (ILC): http://ilcaustralia.org.au/

Leading Age Services (LASA): http://www.lasa.asn.au/

My Aged Care: http://www.myagedcare.gov.au

National Aged Care Alliance (NACA): http://www.naca.asn.au/index.html

National Seniors Australia (NSA): http://www.nationalseniors.com.au/

The Royal Australian College of General Practitioners (RACGP): http://www.racgp.org.au/

K. O'Loughlin et al. (eds.), *Ageing in Australia*, International Perspectives on Aging 16, DOI 10.1007/978-1-4939-6466-6

International

Global Aging Times (GAT): http://www.globalagingtimes.com/aging/

Help Age International: http://www.helpage.org/global-agewatch/

International Association of Gerontology and Geriatrics (IAGG): http://www.iagg.info/

International Federation on Ageing (IFA): http://www.ifa-fiv.org/

United Nations (UN): http://www.un.org/en/index.html

World Health Organization (WHO): http://www.who.int/en/

Index

K. O'Loughlin et al. (eds.), *Ageing in Australia*, International Perspectives on Aging 16, DOI 10.1007/978-1-4939-6466-6

B

Kolibri 360 GmbH
Ferdinand-Jühlke-Straße 7
99095 Erfurt, Deutschland
produktsicherheit@kolibri360.de